# COMPUTATIONAL PSYCHIATRY

# COMPUTATIONAL PSYCHIATRY

## MATHEMATICAL MODELING OF MENTAL ILLNESS

*Edited by*

ALAN ANTICEVIC
*Yale University School of Medicine*
*New Haven, CT, United States*

JOHN D. MURRAY
*Yale University School of Medicine*
*New Haven, CT, United States*

ACADEMIC PRESS
An imprint of Elsevier

Academic Press is an imprint of Elsevier
125 London Wall, London EC2Y 5AS, United Kingdom
525 B Street, Suite 1800, San Diego, CA 92101-4495, United States
50 Hampshire Street, 5th Floor, Cambridge, MA 02139, United States
The Boulevard, Langford Lane, Kidlington, Oxford OX5 1GB, United Kingdom

**Notices**
Knowledge and best practice in this field are constantly changing. As new research and experience broaden our understanding, changes in research methods, professional practices, or medical treatment may become necessary.

Practitioners and researchers must always rely on their own experience and knowledge in evaluating and using any information, methods, compounds, or experiments described herein. In using such information or methods they should be mindful of their own safety and the safety of others, including parties for whom they have a professional responsibility.

To the fullest extent of the law, neither the Publisher nor the authors, contributors, or editors, assume any liability for any injury and/or damage to persons or property as a matter of products liability, negligence or otherwise, or from any use or operation of any methods, products, instructions, or ideas contained in the material herein.

**Library of Congress Cataloging-in-Publication Data**
A catalog record for this book is available from the Library of Congress

**British Library Cataloguing-in-Publication Data**
A catalogue record for this book is available from the British Library

ISBN: 978-0-12-809825-7

For information on all Academic Press publications visit our
website at https://www.elsevier.com/books-and-journals

Working together
to grow libraries in
developing countries

www.elsevier.com • www.bookaid.org

*Publisher:* Nikki Levy
*Acquisition Editor:* Melanie Tucker
*Developmental Editor:* Kristi Anderson
*Production Project Manager:* Sujatha Thirugnana Sambandam
*Cover Designer:* Greg Harris

Typeset by TNQ Books and Journals

# Contents

## I

## APPLYING CIRCUIT MODELING TO UNDERSTAND PSYCHIATRIC SYMPTOMS

# II

# MODELING NEURAL SYSTEM DISRUPTIONS IN PSYCHIATRIC ILLNESS

# III

# CHARACTERIZING COMPLEX PSYCHIATRIC SYMPTOMS VIA MATHEMATICAL MODELS

## 10. Bayesian Approaches to Learning and Decision-Making
QUENTIN J.M. HUYS

## 11. Computational Phenotypes Revealed by Interactive Economic Games
P. READ MONTAGUE

# Contributors

**Rick A. Adams**   University College London, London, United Kingdom

**Matthew A. Albrecht**   University of Maryland School of Medicine, Baltimore, MD, United States; Curtin University, Perth, WA, Australia

**Alan Anticevic**   Yale University School of Medicine, New Haven, CT, United States

**Deanna M. Barch**   Washington University in St. Louis, St. Louis, MO, United States

**Albert Compte**   Institut d'Investigacions Biomèdiques August Pi i Sunyer (IDIBAPS), Barcelona, Spain

**Adam Culbreth**   Washington University in St. Louis, St. Louis, MO, United States

**Gustavo Deco**   Universitat Pompeu Fabra, Barcelona, Spain; Institució Catalana de Recerca i Estudis Avançats (ICREA), Barcelona, Spain

**Murat Demirtaş**   Yale University, New Haven, CT, United States; Universitat Pompeu Fabra, Barcelona, Spain

**Gregory Dumont**   Ecole Normale Superieure PSL Research University, Paris, France

**Michael J. Frank**   Brown University, Providence, RI, United States

**James M. Gold**   University of Maryland School of Medicine, Baltimore, MD, United States

**Boris Gutkin**   Ecole Normale Superieure PSL Research University, Paris, France; National Research University Higher School of Economics, Moscow, Russia

**Jakob Heinzle**   University of Zurich and Swiss Federal Institute of Technology (ETH), Zurich, Switzerland

**Quentin J.M. Huys**   University Hospital of Psychiatry, Zürich, Switzerland; University of Zürich and Swiss Federal Institute of Technology (ETH), Zürich, Switzerland

**John H. Krystal**   Yale University School of Medicine, New Haven, CT, United States

**Reinoud Maex**   Ecole Normale Superieure PSL Research University, Paris, France

**John D. Murray**   Yale University School of Medicine, New Haven, CT, United States

**Juan P. Ramirez-Mahaluf**   P. Universidad Católica de Chile, Santiago, Chile

**P. Read Montague**   Virginia Tech, Roanoke, VA, United States; University College London, London, United Kingdom

**A.D. Redish**   University of Minnesota, Minneapolis, MN, United States

**Julia Sheffield**   Washington University in St. Louis, St. Louis, MO, United States

**Klaas E. Stephan**   University of Zurich and Swiss Federal Institute of Technology (ETH), Zurich, Switzerland; University College London, London, United Kingdom

**Cody J. Walters**   University of Minnesota, Minneapolis, MN, United States

**James A. Waltz**   University of Maryland School of Medicine, Baltimore, MD, United States

**Xiao-Jing Wang**   New York University, New York, NY, United States; NYU Shanghai, Shanghai, China

# Preface

Psychiatry aims to understand the neural bases and mechanisms underlying mental disorders and ultimately aims to treat individuals suffering from debilitating behavioral symptoms. Mechanistic understanding of brain-based mechanisms that generate such symptoms is integral to the development of new diagnostic and therapeutic approaches. A great challenge faced by the field of psychiatry is bridging the vast explanatory gaps across levels of analysis: from neurons to circuits and to complex human behavior. This challenge opens up a massive experimental search space, and the field currently lacks an effective "road map" for developing the empirical knowledge base in the face of such a massive search space. This profoundly limits the field's capacity to develop novel and rationally guided therapies that are grounded in well-established neural mechanisms. This knowledge gap drove the development of the emerging cross-disciplinary framework of "computational psychiatry"—a new branch of psychiatry that uses mathematical principles and formalism to generate consistent, rigorous, and testable hypotheses that can lead to better understanding of mechanism across levels of analyses, spanning theoretical neuroscience, and formal modeling of behavior. The emerging range of such computational psychiatry approaches and methods has evolved to the point where this volume is warranted to provide a synthesis of what the field has done, where it is going, and what it needs to achieve its goals. Thus, the overarching objective of this volume is to provide a compressive survey of the current "state-of-the-art" in computational psychiatry with the vision toward the future of the field. It will be of service to researches, clinicians, and students entering the field of psychiatry looking for a synthesis of leading work in the field of computational psychiatry. The volume is broadly organized into three sections, with several chapters covering each topic area:

1. Studies that leverage biophysically based models, which are constrained by realistic physiological properties of synapses, neurons, and circuits. An important feature of these models is that they also enable one to formally test theories about synaptic signaling and microcircuit mechanisms that are not accessible to investigators in human experimental research
2. A branch of modeling that can be broadly referred to as connectionist models, which are designed to capture the function of

large-scale neural networks and generate behavioral predictions in conjunction with human cognitive neuroscience experiments. Connectionist models help to build bridges between understanding of low-level properties of neural systems and their participation in higher-level (systems) behavior

3. Finally, mathematical models of behavior formalize the algorithms underlying cognitive computations, which can in turn formalize behavioral symptoms, such as deficits in various aspects of cognition, decision making, or motivation to name a few. These models are well suited to quantitatively fit complex behavior and symptoms, which are at times out of reach of the first two approaches.

In spanning these computational domains, this cross-disciplinary volume is designed to provide a nexus for the interface with preclinical neuroscience, cognitive neuroscience, and behavioral experiments in humans, respectively. Therefore, it will be of broad interest to basic re-searchers and students in the complementary experimental and theoret-ical neuroscience looking to learn how computational and experimental approaches in their areas are being leveraged to understand psychiatric disease. In summary, the simple fact is that formal mathematic models are now badly needed to synthesize the massive available information in psychiatry and help generate dissociable cross-level mechanistic strong inference predictions for the next generation of experiments. This framework, if executed successfully, can help us accelerate the mapping of the vast multilevel search space between neural mechanism and behavioral symptom. Collectively, this volume summarizes successes, opportunities and challenges ahead for the field of computational psychiatry.

*Alan Anticevic*
*John D. Murray*

# Meeting Emerging Challenges and Opportunities in Psychiatry Through Computational Neuroscience

*Alan Anticevic, John H. Krystal, John D. Murray*
Yale University School of Medicine, New Haven, CT, United States

## PATHS TOWARD MECHANISTIC DISCOVERY IN PSYCHIATRY

Psychiatry, and clinical neuroscience more generally, has faced major challenges in mechanistically mapping the gaps between underlying neural computations, complex behavioral symptoms characteristic of mental illness, and therapeutics. In essence, this "levels of analysis" mapping problem presents the central roadblock toward improving neurobiologically informed diagnosis and rational discovery of novel therapeutics. Historically, psychiatry has generally tackled this challenge by reverse-engineering mechanisms behind serendipitous observations of therapeutic effects. A cardinal example is the eventual discovery that antipsychotics alleviate positive symptoms of patients diagnosed with schizophrenia via antagonism of the dopamine (DA) receptor. The original observation was made in 1949 for an antipsychotic drug chlorpromazine by Henri Laborit—a French surgeon who reported "sedation without narcosis" in an attempt to alleviate shock-related symptoms in injured soldiers (Tost et al., 2010). This application was eventually extended to treat patients suffering from psychosis, which in turn revolutionized psychiatry and opened an era of neuropharmacology (Nestler, 2013). Fast-forward 60 years to the present and the DA model of schizophrenia features prominently in our understanding of the etiology and treatment of psychosis symptoms in particular (Abi-Dargham, 2014). This example highlights that, in essence, serendipity has not been without its successes (Keshavan et al., 2017). In fact, the DA model of schizophrenia has been highly informative, particularly in the realm of medication target engagement. For instance, increased amphetamine stimulated DA release appears to be a risk marker for psychosis and occupancy of type 2 DA receptors could be used to optimize the dosing of antipsychotic medications. In fact, there are a growing number of medications where positron emission tomography (PET) receptor imaging can be used to evaluate target engagement. These examples notwithstanding, the field of clinical neuroscience still does not posses reliable neural markers that can comprehensively map onto disease risk, predict diagnosis, inform treatment response, or offer viable treatment targets for the majority of psychiatric disease (Woo et al., 2017).

These massive knowledge gaps across the psychiatric spectrum illustrate how the serendipity-based approach toward discovery in clinical neuroscience is not adequate to support hypothesis-driven research across levels of analysis. This challenge is magnified by the size of the search space that needs to be mapped to comprehensively close the knowledge gaps in the effort to link molecules, receptors, circuits, brain regions, neural systems, and ultimately symptoms (Krystal et al., 2017). Let's consider for a moment the massive range of possible "malfunctions" that could alter neural computations and ultimately produce abnormal behavior: these could involve architectural deficits in cell structure, deformation in arborization of dendrites, localized synaptic deficits on specific neuronal types, genetic alterations in expression of proteins that form receptors on the cell surface, dysfunctional synthesis of specific neurotransmitters, or malfunction in long-range feedback and feedforward interactions between large-scale systems (Anticevic and Lisman, 2017). These focused examples, which illustrate only a few of the plethora of potential alterations, highlight the emerging challenge faced by clinical neuroscience: While our understanding of basic neurobiology is pushing the field closer to mechanism, the number of causal upstream pathogenic alterations that could modify neural computations and consequently lead to a downstream harmful behavioral dysfunction is daunting. Put differently, the field faces many-to-one and many-to-many mapping problems when attempting to link mechanisms governing malfunctioning neural computations to behavior. To express this more formally, neural computations could malfunction in $N$ ways due to $P$ mechanisms (Anticevic and Murray, 2017), which could in turn lead to $R$ distinct neural abnormalities. These abnormal features, which may exhibit across a set of convergent neural pathways. To make matters even more complex, the $N \times P$ mechanisms may vary over time due to the time-dependent nature of neural development and gene expression. In other words, it is quite likely that temporal neural dynamics matter, making the mapping of neural mechanisms in psychiatric illness a 4-dimensional problem (Krystal and Anticevic, 2015).

Given this seemingly intractable challenge, clinical neuroscience needs a theoretically guided framework that can help generate formal hypotheses that constrain the search space. There are promising examples. For instance, theoretical models now contextualize DA contributions to the neurobiology of schizophrenia within more complex networks that also include glutamate and gamma-aminobutyric acid signaling (Krystal et al., 2003; Coyle, 2006; Lisman et al., 2008). This rationally guided refinement of schizophrenia neurobiology has largely emerged through experimental pharmacology (Krystal et al., 1994), postmortem studies (Lewis et al., 2012), animal translational work (Homayoun and Moghaddam, 2007), and human neuroimaging (Marsman et al., 2013), complemented by cognitive neuroscience observations (Horga et al., 2016; Cassidy et al., 2016; Slifstein et al., 2015).

Parallel successes in novel therapeutics are emerging in the field of major depression research, supported via noninvasive neuroimaging research, building on our understanding of the complex glutamate pharmacology cutting across the psychiatric spectrum (Abdallah et al., 2017; Lener et al., 2017a,b; Sanacora et al., 2008; Zarate et al., 2006; Krystal et al., 2013). Nevertheless, translational progress toward therapeutics has been frequently unsuccessful and success has generally been too slow. Consequently, while important genetic insights that hold the promise a causal mapping to symptom continue to emerge, the field still lacks an effective rationally guided treatment for specific neural targets for even a subset of schizophrenia patients—a situation that again generalizes across the psychiatric spectrum. A promising emerging way to meet these challenges involves leveraging advances from theoretical and computational neuroscience—a rapidly growing enterprise to formalize neurobiology via mathematic principles.

## TACKLING COMPLEXITY OF MECHANISM VIA COMPUTATIONAL NEUROSCIENCE

This effort is timely because it is no longer controversial to conceptualize severe mental illness as largely a brain-based disease, which arises from malfunctioning computations in one or more of the aforementioned possible ways. The rapid proliferation of basic research in neuroscience has established that the brain is essentially a massively complex, nonlinear computational device that governs expression of an organism's behavior. This perspective is also captured by the ongoing paradigm shift within the field of psychiatry, from categorical, behavioral psychiatric diagnoses toward understanding of dimensional disruptions in complex behaviors that map onto well-defined neural substrates. Critically, this formulation of mental illness is not designed to argue exclusively for reductionist bottom-up models of complex psychiatric symptoms based entirely on the impact of genetic variation expressed as perturbations in neural network properties. In fact, this "computational" view allows for the environmental perturbations to play a causal role in network disturbances underlying the expression of symptoms. The goal is to identify "final common pathways" for genetic and environmental mechanisms to produce symptoms that govern the neural networks or, more precisely, the computations performed by these networks. Given the aforementioned complexity of the brain and the vast number of possible disruptions, across distinct levels of analysis, the field of clinical neuroscience is best served by an attempt to harness all available tools to mechanistically understand abnormal behavior, regardless of whether the upstream cause lies in genes or environmental forces. Therefore, integrating theoretical and computational neuroscience into the emerging paradigm of "Computational Psychiatry" can help to accelerate the path to mechanistic translational insight (Krystal et al., 2017; Anticevic et al., 2015; Redish and Gordon, 2016; Wang and Krystal, 2014; Stephan and Mathys, 2014; Montague et al., 2012; Rolls et al., 2008).

In line with this goal, this volume is a collection of contributions that synthesize progress in this emerging field of "Computational Psychiatry." The key logic behind the emergence of the field of computational psychiatry is the use of mathematical principles and formalism to generate consistent, rigorous, and testable hypotheses that can lead to better understanding of mechanism across levels of analyses, spanning theoretical neuroscience, and formal modeling of behavior. The volume combines a variety of contributions that collectively illustrate how mathematical formalism can help constrain and guide the experimental search space, whether it is applied for instance at the level of synaptic deficits in glutamatergic signaling, neuroimaging biomarker development, or complex behavioral decision making deficits across psychiatric conditions. In turn, the experimental insights can map right back onto computational models, iteratively refining them based on new evidence across levels of analyses.

In that sense, computational psychiatry offers the opportunity to interface with experimental and treatment studies at the level of neurons, neural systems, and ultimately abnormal behavior (Fig. 1). Disrupted mechanisms that contribute to a particular psychiatric disease can occur at the level of neurons and synapses, whereas psychiatric symptoms express at the level of behavior, which involves large-scale brain networks. Formally and mathematically linking these disparate levels is vital for gaining mechanistic insight into mental illness, and for rational development of therapeutics, which act at the synaptic level. In the same vein, it is critical for experiments conducted at each of these levels to guide computational model refinement. Thus, the overarching mission of

## Experimental & Computational Modeling Approaches

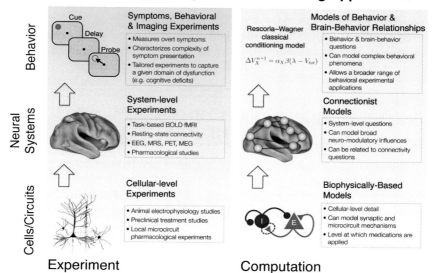

**FIGURE 1**    **Conceptual illustration of the computational modeling and experimental interplay across levels of analysis.** The utility of computational modeling, particularly in the study of schizophrenia, is its ability to inform a given level of experimental study. Because we study abnormalities in schizophrenia from the cellular level at the neural system level, and ultimately at the level of behavior, we have to utilize our modeling approaches to best fit the experimental framework. For instance, cellular-level experiments use techniques and produce measurements that are best captured using models that contain the necessary level of biophysical realism (bottom panels). Such models can, for instance, inform synaptic processes that may govern the microcircuit phenomena under study such as neural oscillations. In turn, a number of neuroimaging studies have focused on characterizing system-level disturbances in schizophrenia both using task-based paradigms and resting-state functional connectivity approaches. To best inform such system-level cognitive neuroscience experiments, models should capture the relevant detail and complexity of larger-scale neural systems (middle panel). Such models can perhaps better inform the role of systemic pharmacological manipulations on blood oxygen level dependent (BOLD) functional magnetic resonance imaging (fMRI) or can be used to predict results of functional connectivity studies in schizophrenia. Finally, schizophrenia produces complex and devastating behavioral symptoms, which can be measured via increasingly sophisticated behavioral paradigms. Here the use of models that formalize complex behavior can provide a powerful tool to quantitatively examine a given behavioral process in patients (e.g., reinforcement learning). EEG, electroencephalography; MRS, magnetic resonance spectroscopy; PET, positron emission tomography; MEG, magnetoencephalography. *Figure adapted with permission from Anticevic, A., Murray, J.D., Barch, D.M., 2015. Bridging levels of understanding in schizophrenia through computational modeling. Clin. Psychol. Sci. 3 (3), 433–459.*

computational psychiatry is to bridge the mechanistic gaps and to help formalize how computations across levels of analysis can lead to abnormal behavioral expression. The emerging range of such computational psychiatry approaches and methods has evolved to the point where this volume is warranted to provide a roadmap of what the field has done, where it is going, and what it needs to achieve its objectives.

# THE CURRENT STATE OF COMPUTATIONAL PSYCHIATRY

Mathematical models of neural systems contribute to our understanding of how neural dynamics and behavior arise from underlying mechanisms. In particular, researchers can evaluate which models or model parameters better describe empirical findings in a disease state compared to healthy individuals. By examining the disruptions within models, we can describe key differences in underlying processes in a mathematically precise way. This approach can also test hypotheses about lower-level phenomena (e.g., specific synaptic disruptions) with experimentally accessible noninvasive measures at higher levels in humans (e.g., neuroimaging or precise psychophysical behavioral measures that capture variation in a symptom). Indeed, in the past several decades the field of psychiatry has profited tremendously from the ongoing development of noninvasive neuroimaging technology, more sophisticated human cognitive neuroscience, and advances in basic preclinical neuroscience. The combination of these approaches has directly impacted the broad evidence base in psychiatry, providing support for neural alterations across psychiatric conditions (Nestler, 2013). Such evidence for neural alterations in psychiatric disease has emerged from a wide range of experimental modalities. Analysis of postmortem brain tissue reveals differences in neuronal and synaptic structure across psychiatric disorders (Lewis and Hashimoto, 2007). Genetic studies implicate specific genetic pathways, with promising links to neuronal and synaptic alterations (Sekar et al., 2016; Schizophrenia Working Group of the Psychiatric Genomics, 2014). Neurochemical alterations can be studied in vivo using noninvasive methods such as PET (Abi-Dargham et al., 2000) and magnetic resonance spectroscopy (Marsman et al., 2013; Bustillo et al., 2014; Napolitano et al., 2014). The spatiotemporal properties of brain activity during various tasks and at rest can be measured with noninvasive neuroimaging such as functional magnetic resonance imaging (Barch et al., 2013), electroencephalography (Hirano et al., 2015), and magnetoencephalography (Rivolta et al., 2014; Uhlhaas, 2013). Neuroimaging can also reveal structural differences in gray-matter distribution and long-range, white matter pathways (Sotiropoulos et al., 2013), which can be applied to study the "macro-structure" alterations in disease. As noted, these neural measurements are complemented by rapid and parallel advances in cognitive neuro-science that allow us to quantitatively characterize, at the behavioral level, cognitive, affective, and perceptual impairments in patients.

Nonetheless, the field of psychiatry has yet to produce systematic and mechanistic disease models that are amenable to rational development of therapies. To achieve such a multilevel understanding of psychiatric diseases, the field of psychiatry must ultimately link these levels of analysis from synapses, to cells, to neural circuits, to large-scale systems, and to behavioral disturbances. The field of computational psychiatry is in a unique position to help provide the theoretical foundation needed to close these gaps. Computational psychiatry benefits from the diverse range of complementary modeling approaches within the broader field of computational neuroscience and behavioral modeling. Different types of models are best suited to address different questions and different types of experimental data. In that sense, the level of computational modeling is matched to the appropriate levels of experimental inquiry, enabling the interplay between experiment and theory. This is a theme that is revisited throughout the chapters. Consequently, multilevel understanding of complex abnormalities in behavior (i.e., psychiatric symptoms) directly benefits from these complementary modeling approaches (Fig. 1).

This volume generally outlines three broadly defined levels of computational psychiatry:

1. Studies that leverage biophysically based models, which are constrained by realistic physiological properties of synapses, neurons, and circuits (Krystal et al., 2017; Wang, 2010, 2001; Compte et al., 2000; Durstewitz et al., 1999; Durstewitz and Seamans, 2002, 2008; Brunel and Wang, 2001). These models are well suited for testing synapse-level hypotheses and to allow comparison to physiological and pharmacological studies in psychiatry (Wang and Krystal, 2014). This type of modeling is also well suited to study neurophysiological biomarkers of disease states, such as modeling oscillations. Critically, this branch of computational modeling is distinguished by building on the physical properties of real neurons and synapses, producing realistic cellular-level dynamics. This level of detail is ultimately vital to generate mechanistic predictions at the level of treatments or to elucidate synaptic mechanisms governing complex behaviors in psychiatric patients. An important feature of these models is that they also enable one to formally test theories about synaptic signaling and microcircuit mechanisms that are not accessible to investigators in human experimental research.

For example, one can estimate the computational impact of a cell-specific manipulation in a modeled human neural circuit even though this could not be studied in humans. In this way, computational neuroscience provides a unique bridge between human experimental translational neuroscience and the explosion in cell-specific manipulations and biochemical assessments ongoing within molecular and cellular neuroscience. Chapters 1–4 discuss application of such biophysical models to psychiatric disease.

2. A parallel line of computational psychiatry research has emerged in lockstep with the cognitive neuroscience revolution, building on the rapid and productive advances in noninvasive neuroimaging technology in humans. In fact, the past two decades have witnessed an explosive growth of knowledge regarding the neural correlates of various cognitive and affective processes in healthy individuals. This field of cognitive neuroscience has generated an increasingly robust platform for interpreting clinical neuropsychiatric phenomena (Barch and Ceaser, 2012). This is primarily accomplished by garnering an increased understanding of neural systems known to be involved in various cognitive operations in healthy individuals and translating these to clinical studies. This understanding in turn constrains our search space for what aspects of brain circuitry may be abnormal in clinical populations exhibiting deficits in these same cognitive operations. Therefore, using a cognitive neuroscience framework with the specific application toward understanding computational mechanisms offers a promising tool for elucidating and ultimately treating psychiatric symptoms by delineating abnormalities in neural circuits whose functions are increasingly understood in healthy populations. This cognitive neuroscience framework has emerged in parallel with the development of computational modeling in psychiatry. Historically, the types of computational psychiatry models that first appeared were most directly aligned with the neural systems and behavioral levels of analyses. Therefore, there exists a productive and ongoing interplay between cognitive neuroscience in psychiatry and neural system-level modeling approaches (Anticevic et al., 2015). Specifically, this branch of modeling can be broadly referred to as *connectionist models* that are designed to capture the function of large-scale neural networks and generate behavioral predictions in conjunction with human cognitive neuroscience experiments. Connectionist models help to build bridges between understanding of low-level properties of neural systems and their participation in higher-level (systems) behavior. They are able to capture a wide range of complex behaviors and neural systems interactions, which cannot at present be modeled as effectively by biophysically based

models. Computational psychiatry studies that build on connectionist models use more abstract neural elements but incorporate a systems-level neural architecture. These models can be applied to more complex behaviors and are also well suited to interface with task-based neuroimaging and behavior—namely cognitive neuroscience approaches in psychiatry. Chapters 5–7 discuss application of such system-level connectionist models to psychiatric disease.

3. Finally, mathematical models of behavior formalize the algorithms underlying cognitive computations, which can in turn formalize behavioral symptoms and cognitive deficits. These models are well suited to quantitatively fit complex behavior and symptoms, which are at times out of reach of the first two approaches. The computational psychiatry approaches discussed above, biophysically based circuit modeling and connectionist modeling, aim to develop models with elemental components that instantiate how neural systems transform their inputs into outputs. These approaches are often limited in the range of cognitive behaviors to which they can be applied, due to our limited knowledge of the relevant neurophysiological mechanisms. An alternative approach to modeling cognition is to develop mathematical models that provide an account of the psychological computations underlying a given cognitive function. These models are often grounded in so-called normative accounts that aim to define the algorithms that optimize performance at a task and to characterize constraints that limit such task performance. Although developed to explain behavior, these models can potentially make important links to neural circuits. For example, experiments can reveal that activity in specific brain areas represents the internal variables of the model, providing support for this computational account and providing insight into the division of labor among brain systems. In turn, such neuroscience findings can inform the algorithms and modules that compose the computational architecture of the model. Chapters 8–11 discuss application of such models to psychiatric disease.

Collectively, this volume provides a synthesis of the current state-of-the-art in computational psychiatry considering these levels, spanning from local neuronal circuits to complex behavior such as reinforcement learning and decision-making, which feature prominently across psychiatric diagnoses. The advances outlined in the volume offer the promise to achieve advances in a number of areas that can yield major impact in clinical neuroscience, which are briefly considered next.

## The Potential Impact of Computational Psychiatry and What Is Needed to Get There

### *Integration Across Levels of Analysis*

This introductory chapter outlines how computational psychiatry has offered a formal framework to bridge levels of analysis, from local neuronal circuits to large-scale neural systems and ultimately psychiatric symptoms. The challenge that still faces this rapidly growing field is how to provide a formal bridge between these levels of inquiry for most psychiatric symptoms. This volume will highlight some cardinal example cases where such progress has occurred—such as in the context of working memory and schizophrenia (see Chapters 1 and 2). In such cases, computational psychiatry has begun to establish formalisms for how to link local circuits to large-scale neuronal dynamics and expression of behavior. However, the current models in this area of computational psychiatry are still simplistic in their repertoire of behavior and are designed to capture very limited range of experimental contexts. Therefore, to accurately capture more complex cognitive and affective behaviors, models must incorporate multiple interacting brain areas possessing various distributed and modular computations. In this area, biophysically based modeling will be informed by connectionist-style models that are more abstract and removed from biophysical detail, but that can be readily applied to modeling psychological processes and interactions among distributed neural systems. Similarly, a highly related challenge facing computational psychiatry is to provide a formal extension of biophysically based modeling to large-scale distributed brain networks, beyond the microcircuit level, as this level of analysis offers the potential for development of disease-related biomarkers. For instance, a key extension of such models would be to capture, in a biophysically realistic way, the large-scale thalamocortical functional loops that may play a critical role across a range of psychiatric disorders. The next generation of computational psychiatry studies will be vital to formalize these bridges across levels of analysis.

### *Computational Phenotyping and Biomarker Refinement*

It is widely appreciated by both researchers and clinicians that psychiatric symptoms are remarkably heterogeneous, a reality that cuts both within and across currently defined diagnostic categories. This conceptual problem is captured in the case of existing tension between category-based DSM-style diagnostic rubrics and the dimensional Research Domain Criteria framework (Casey et al., 2013; Insel et al., 2010). Instead of making an a priori decision regarding which framework may be better for a given psychiatric symptom or syndrome, we posit that perhaps there should be a parallel movement toward "precision psychiatry." That is, it may be possible to harness advances in computational psychiatry to generate individual-level inferences regarding the type of underlying computational mechanisms that are governing the observed patterns of neuroimaging effects or behavioral symptom expression. The current state-of-the-art in this area of research focuses on group effects. That is, the majority of work that has emerged in computational psychiatry has focused on explaining group-level findings across a given diagnosis. We argue that one highly promising future application of computational psychiatry may be to provide individual-level computational phenotyping based on mechanistic models. Such level of "precision medicine" may possibly provide a method to manage the massive clinical heterogeneity that exists both within and across our current clinical rubrics in psychiatry. Ultimately, mapping such clinical heterogeneity onto underlying computational mechanisms will be vital to develop better mechanistically informed treatments for individual patients. In computational phenotyping, the parameters of a model are fit to best capture the empirical data for a single subject. Computational phenotyping could take advantage of a range of empirical data and modeling approaches. For instance, a reinforcement learning model could be fit to a subject's decision making behavioral data. Similarly, a biophysically based network model could be fit to a subject's resting-state neuroimaging data and in turn a measured symptom expression (see Chapter 4). The promise of this approach is that these model parameters may have diagnostic value, potentially leading to more predictive biomarkers. These model parameters may allow effective classification according to diagnostic criteria, and may reveal clinically relevant clusters or dimensions within a diagnostic category, or be predictive of a patient's response to treatment. Longitudinal studies will be critical to validate the diagnostic potential of computational phenotyping for mapping risk prediction and treatment response.

### *Modeling Treatments*

Of utmost relevance to psychiatry in general is that computational modeling has the potential to provide a method for simulating possible effects of treatments (or compensations) that can occur across the described levels of analysis. For instance, some modeling efforts have already begun to generate predictions for how compensations onto N-methyl-D-aspartate receptor or GABA receptor (GABAR) synapses in local circuits can impact cortical disinhibition that adversely affects working memory computations. However, this area is at present underdeveloped and has future potential for guiding intuition in cases where complex neural dynamics are involved. Here further development and refinement of biophysically based models, in particular, has particular advantages (see Chapter 1). Given the relevant cellular detail, such models can actually generate both biomarker and behavioral predictions of disruptions on specific synaptic sites. In turn, such models can simulate "compensations" to specific synaptic perturbations and ultimately generate sweeps of vast ranges of parameters, which is experimentally not tractable (Murray et al., 2014). Such level of detail may be out of reach in cases that are not informed by animal models. Nevertheless, this is precisely the point of cutting-edge interplay between future research in computational psychiatry and clinical neuroscience. We posit that future studies in this area hold the exciting promise of rationally guiding treatment development in psychiatry, grounded in basic neuroscience.

# WHAT COMPUTATIONAL PSYCHIATRY NEEDS TO SUCCEED?

While the various chapters in the volume present progress in the field of computational psychiatry to date, we also want to take the opportunity to outline key challenges and areas of emerging need that have to be overcome for this promising framework to render its full impact:

## Common Language

At present there is a language disconnect among theoretical neuroscientists, preclinical researchers, translational researchers, and ultimately clinicians. For the proposed framework to succeed, we will need an active emphasis on developing a unified language and nomenclature that facilitates the communication across these areas of expertise, as they need to seamlessly inform one another. Here it will be important to establish knowledge base of "fundamental computational constructs" that can be studied across levels of analyses.

## Generating Multilevel Data

Consequently, there is an urgent need for additional multilevel data across species (Nestler and Hyman, 2010), in the same organism, and ultimately in same individual patient. The problem here is that we cannot map circuit function in humans with appreciable spatiotemporal precision. While promising, we are still far from being able to utilize the full potential of high-field magnetic resonance-based imaging to achieve subcolumnar cortical resolutions in humans (Ugurbil, 2016). Nevertheless, there is still much to be gained by generating multimodal and multilevel data on a more consistent basis with an eye for application in clinical studies. Here it will be key to consider areas of translational potential where such data are already being generated (e.g., studies of working memory across the primate and human that can link circuit observations with translational behavior that can be deployed in clinical studies) (Wang et al., 2013).

## Mapping Categorical Versus Continuous Alterations: RDOC Versus DSM

The topic of whether human psychopathology fits better along "categorical" rubrics, as mandated by the current diagnostic classification systems, or always maps onto continuous alterations is simply to broad to cover here and we refer the reader to original and recent pieces on the topic (Casey et al., 2013; Insel et al., 2010). This is an important and evolving issue that continues to motivate proposed solutions (e.g., the recently proposed Hierarchical Taxonomy of Psychopathology framework Kotov et al., 2017). However, this "tension" may represent a possibly important conceptual roadblock for computational psychiatry or could be turned into an opportunity. Instead of debating on whether we should study categories or continua (often at the level of committees), it will be vital to leverage computational psychiatry to inform this tension. Here computational psychiatry approaches need to combine with "big data" science to empirically establish and ultimately map which alterations (neural or behavioral) exhibit truly a "binary" classification patterns versus a "continuously" distributed patterns with no clear bimodality that can distinguish health from disease. Here a productive and ongoing interplay between theory and "big data" analytics will be the key.

## Conceptually Separate "Big Data Analytics" From "Theory"

Building on the argument presented above, it is vital that we draw the distinction between development and application of theoretical models (derived from neuroscience or cognitive science) used to generate predictions (Wang and Krystal, 2014) from efforts of deploying analyses on big data via large-scale computation (Cohen et al., 2017). While the later is certainly a "computational" effort, it is important to avoid conflating these efforts as they are designed to achieve separate objectives. For instance, theoretical neuroscience that studies the function of local microcircuits can generate "strong inference" predictions for how specific interneuron population alterations may produce distinct dynamics. This can be experimentally tested. In turn, computational efforts from big data analytics are not conceptually positioned to do this. However, such big data analyses can help map the search space following theoretical model predictions. Put simply, these efforts need to function in concert as opposed separately to allow the overall "computational psychiatry" initiative to thrive.

## Sharing, Standardization, and Reproducibility

For theoretical advances to succeed, they need to rely on access to shared data that are standardized and reproducible. In other words, there needs to be an effort on generating publically shared and accessible datasets that, ideally, would strive for standardization. A great example here is the Human Connectome Project—a 5-year effort to comprehensively map the human brain via noninvasive multimodal neuroimaging methods and behavior (Barch et al., 2013; Smith et al., 2013; Marcus et al., 2013; Glasser et al., 2013; Van Essen et al., 2012). This effort has been vital for the neuroscience community both from a big data analysis standpoint and as a data repository to provide a test bed for computational model refinements. To ensure fluid interface between the experimental and theoretical arms of computational psychiatry, the field needs rapid and seamless access to data that is organized in a unified and standardized way across studies, modalities, species and experiments. We are still far from this goal. Nevertheless, the bioinformatics advances coupled with the proliferation of cloud-based, high-performance computing stand to make a major impact in the coming years (Hodge et al., 2016).

## Ground-Truth Datasets for Benchmarking of Models

The Human Connectome Project dataset is a strong starting point, but the field needs many more "ground-truth" datasets across species and levels of analyses to validate models. This goal connects back to the argument above concerning data sharing. Such sharing efforts need to generalize to studies of clinical populations as well, which can help provide benchmarking for model performance.

## Infrastructure

Admittedly, this is a broad goal that can be conceptualized in several ways. Here, we specifically refer to data storage, growth of public databases, and access to high performance computational resources. One way to solve this challenge is to actively seek out opportunity for partnership between academia, federal funding agencies, and industry. For instance, the expansive growth of "cloud computing" enterprises such as Amazon Web Services may ultimately render the "brick and mortar" University-level high performance computing platforms not only obsolete due to performance and scalability but also intractable in terms of access, sharing, and generalizability. Therefore, the field needs active efforts to "grow" an infrastructure that can facilitate more rapid discovery.

## Not Losing Track of Time

Above, we note that psychiatric clinical phenomena are ultimately a 4-dimensional problem that takes place over time, a reality that is particularly important to consider for development of therapeutics (Krystal and Anticevic, 2015). Therefore, the field needs to generate more careful longitudinal data across development and across distinct illness stages of various neuropsychiatric syndromes. This information will be critical to accurately model neural adaptations and compensations versus time-dependent disease mechanism.

## Linking Theoretical Approaches

More concerted efforts are needed to establish "links" between different modeling methods discussed in this volume. Put simply, there needs to be active cross talk between modelers themselves who work across distinct levels of analyses. To achieve such synthesis of modeling approaches we need better information on how complex psychiatric symptoms (e.g., delusions) map onto specific quantifiable mathematical behavioral measure (e.g., prediction error). In turn, such mapping needs to be then linked to a specific circuit that implements a computation (e.g., striatal-prefrontal circuits involving DA signaling). Such a mapping has yet to be achieved across various complex psychiatric symptoms.

## Training and Development of the Scientific Workforce

The field needs an active focus on training the next generation of "computational clinical neuroscientists," an issue that is becoming pervasive in basic sciences (Hyman, 2017). This effort is critical to produce a skilled and growing research workforce that is capable of traversing the interface of computational techniques, neurobiology, and clinical phenomena. For this to occur, both graduate programs and residency programs will likely need to move beyond the "silos" of psychiatry and clinical psychology training models.

## Facilitating Dialogue Between Computational and Experimental Neuroscience

The computational psychiatry community needs to actively communicate a "priority list" for specific experimental evidence that is needed for active model refinement. For instance, we need more systematic pharmacological manipulations of behavior. We need to know how positive and negative prediction errors are encoded. We need detailed information on neocortical lamina and feedforward and feedback connections for development of Bayesian models. We need information on large-scale and local structural connectivity as well as morphology of specific microcircuits. The fields need continuous refinement of evidence detailing neuronal and cellular diversity both in terms of structure and function. We need to understand differences in physiology in local and long range—neural inputs. In turn, we need evidence for brain gene expression patterns across development in humans. Again, this list is by no means exhaustive and not ordered by priority. Instead, it should serve to highlight how computational psychiatry and experimental/clinical studies should be considered in a constant interplay of iterative refinement.

## CONCLUDING REMARKS

The overarching objective of this volume is to provide a compressive survey of the current state-of-the-art in computational psychiatry with the vision toward the future of the field. We have presented the argument that the sheer complexity of the evidence base in clinical neuroscience is such that a formal neurobiologically grounded framework for its integration is needed. The simple fact is that theoretical models are now needed to synthesize the massive available information and help generate dissociable cross-level mechanistic, strong inference predictions for the next set of experiments. This framework, if executed successfully, can help us accelerate the mapping of the vast multilevel search space between neural mechanism and behavioral symptom. Computational psychiatry is uniquely positioned to help inform this challenge.

## References

Abdallah, C.G., et al., 2017. Ketamine treatment and global brain connectivity in major depression. Neuropsychopharmacology 42 (6), 1210–1219.

Abi-Dargham, A., 2014. Schizophrenia: overview and dopamine dysfunction. J. Clin. Psychiatry 75 (11), e31.

Abi-Dargham, A., et al., 2000. Increased baseline occupancy of D2 receptors by dopamine in schizophrenia. Proc. Natl. Acad. Sci. U.S.A. 97 (14), 8104–8109.

Anticevic, A., Lisman, J., 15 May, 2017. How can global alteration of excitation/inhibition balance lead to the local dysfunctions that underlie schizophrenia? Biol Psychiatry 81 (10), 818–820. http://dx.doi.org/10.1016/j.biopsych.2016.12.006. Epub Jan 4, 2017. PubMed PMID: 28063469.

Anticevic, A., Murray, J.D., 2017. Rebalancing altered computations: considering the role of neural excitation and inhibition balance across the psychiatric spectrum. Biol. Psychiatry 81 (10), 816–817.

Anticevic, A., Murray, J.D., Barch, D.M., 2015. Bridging levels of understanding in schizophrenia through computational modeling. Clin. Psychol. Sci. 3 (3), 433–459.

Barch, D.M., Ceaser, A., 2012. Cognition in schizophrenia: core psychological and neural mechanisms. Trends Cogn. Sci. 16 (1), 27—34.

Barch, D.M., et al., 2013. Function in the human connectome: task-fMRI and individual differences in behavior. Neuroimage 80, 169—189.

Brunel, N., Wang, X.J., 2001. Effects of neuromodulation in a cortical network model of object working memory dominated by recurrent inhibition. J. Comput. Neurosci. 11 (1), 63—85.

Bustillo, J.R., Chen, H., Jones, T., Lemke, N., Abbott, C., Qualls, C., Canive, J., Gasparovic, C., Mar, 2014. Increased glutamine in patients undergoing long-term treatment for schizophrenia: a proton magnetic resonance spectroscopy study at 3 T. JAMA Psychiatry 71 (3), 265—272. http://dx.doi.org/10.1001/jamapsychiatry.2013.3939. PubMed PMID: 24402128.

Casey, B.J., et al., 2013. DSM-5 and RDoC: progress in psychiatry research? Nat. Rev. Neurosci. 14 (11), 810—814.

Cassidy, C.M., et al., 2016. Dynamic connectivity between brain networks supports working memory: relationships to dopamine release and schizophrenia. J. Neurosci. 36 (15), 4377—4388.

Cohen, J.D., et al., 2017. Computational approaches to fMRI analysis. Nat. Neurosci. 20 (3), 304—313.

Compte, A., et al., 2000. Synaptic mechanisms and network dynamics underlying spatial working memory in a cortical network model. Cereb. Cortex 10 (9), 910—923.

Coyle, J.T., 2006. Glutamate and schizophrenia: beyond the dopamine hypothesis. Cell Mol. Neurobiol. 26 (4—6), 365—384.

Durstewitz, D., Seamans, J.K., 2002. The computational role of dopamine D1 receptors in working memory. Neural Netw. 15 (4—6), 561—572.

Durstewitz, D., Seamans, J.K., 2008. The dual-state theory of prefrontal cortex dopamine function with relevance to catechol-o-methyltransferase genotypes and schizophrenia. Biol. Psychiatry 64 (9), 739—749.

Durstewitz, D., Kelc, M., Gunturkun, O., 1999. A neurocomputational theory of the dopaminergic modulation of working memory functions. J. Neurosci. 19 (7), 2807—2822.

Glasser, M.F., et al., 2013. The minimal preprocessing pipelines for the human connectome project. Neuroimage 80, 105—124.

Hirano, Y., et al., 2015. Spontaneous gamma activity in schizophrenia. JAMA Psychiatry 72 (8), 813—821.

Hodge, M.R., et al., 2016. ConnectomeDB—sharing human brain connectivity data. Neuroimage 124 (Pt B), 1102—1107.

Homayoun, H., Moghaddam, B., 2007. NMDA receptor hypofunction produces opposite effects on prefrontal cortex interneurons and pyramidal neurons. J. Neurosci. 27 (43), 11496—11500.

Horga, G., et al., 2016. Dopamine-related disruption of functional topography of striatal connections in unmedicated patients with schizophrenia. JAMA Psychiatry 73 (8), 862—870.

Hyman, S., 2017. Biology needs more staff scientists. Nature 545 (7654), 283—284.

Insel, T., et al., 2010. Research domain criteria (RDoC): toward a new classification framework for research on mental disorders. Am. J. Psychiatry 167 (7), 748—751.

Keshavan, M.S., et al., 2017. New drug developments in psychosis: challenges, opportunities and strategies. Prog. Neurobiol. 152, 3—20.

Kotov, R., et al., 2017. The hierarchical taxonomy of psychopathology (HiTOP): a dimensional alternative to traditional nosologies. J. Abnorm. Psychol. 126 (4), 454—477.

Krystal, J.H., Anticevic, A., 2015. Toward illness phase-specific pharmacotherapy for schizophrenia. Biol. Psychiatry 78 (11), 738—740.

Krystal, J.H., et al., 1994. Subanesthetic effects of the noncompetitive NMDA antagonist, ketamine, in humans. Psychotomimetic, perceptual, cognitive, and neuroendocrine responses. Arch. Gen. Psychiatry 51 (3), 199—214.

Krystal, J.H., et al., 2003. NMDA receptor antagonist effects, cortical glutamatergic function, and schizophrenia: toward a paradigm shift in medication development. Psychopharmacology (Berl.) 169 (3–4), 215–233.

Krystal, J.H., et al., 2017a. Computational psychiatry and the challenge of schizophrenia. Schizophr. Bull. 43 (3), 473–475.

Krystal, J.H., et al., 2017b. Impaired tuning of neural ensembles and the pathophysiology of schizophrenia: a translational and computational neuroscience perspective. Biol. Psychiatry 81 (10), 874–885.

Krystal, J.H., Sanacora, G., Duman, R.S., Jun 15, 2013. Rapid-acting glutamatergic antidepressants: the path to ketamine and beyond. Biol Psychiatry 73 (12), 1133–1141. http://dx.doi.org/10.1016/j.biopsych.2013.03.026. Review. PubMed PMID: 23726151; PubMed Central PMCID: PMC3671489.

Lener, M.S., Niciu, M.J., Ballard, E.D., Park, M., Park, L.T., Nugent, A.C., Zarate Jr., C.A., May 15, 2017a. Glutamate and gamma-aminobutyric acid systems in the pathophysiology of major depression and antidepressant response to ketamine. Biol Psychiatry 81 (10), 886–897. http://dx.doi.org/10.1016/j.biopsych.2016.05.005. Epub May 12, 2016. Review. PubMed PMID: 27449797; PubMed Central PMCID: PMC5107161.

Lener, M.S., Kadriu, B., Zarate Jr., C.A., 2017b. Ketamine and beyond: investigations into the potential of glutamatergic agents to treat depression. Drugs 77 (4), 381–401.

Lewis, D.A., et al., 2012. Cortical parvalbumin interneurons and cognitive dysfunction in schizophrenia. Trends Neurosci. 35 (1), 57–67.

Lewis, D.A., Hashimoto, T., 2007. Deciphering the disease process of schizophrenia: the contribution of cortical GABA neurons. Int. Rev. Neurobiol. 78, 109–131.

Lisman, J.E., et al., 2008. Circuit-based framework for understanding neurotransmitter and risk gene interactions in schizophrenia. Trends Neurosci. 31 (5), 234–242.

Marcus, D.S., Harms, M.P., Snyder, A.Z., Jenkinson, M., Wilson, J.A., Glasser, M.F., Barch, D.M., Archie, K.A., Burgess, G.C., Ramaratnam, M., Hodge, M., Horton, W., Herrick, R., Olsen, T., McKay, M., House, M., Hileman, M., Reid, E., Harwell, J., Coalson, T., Schindler, J., Elam, J.S., Curtiss, S.W., Van Essen, D.C., WU-Minn HCP Consortium., Oct 15, 2013. Human Connectome Project informatics: quality control, database services, and data visualization. Neuroimage 80, 202–219. http://dx.doi.org/10.1016/j.neuroimage.2013.05.077. Epub May 24, 2013. PubMed PMID: 23707591; PubMed Central PMCID: PMC3845379.

Marsman, A., et al., 2013. Glutamate in schizophrenia: a focused review and meta-analysis of 1H-MRS studies. Schizophr. Bull. 39 (1), 120–129.

Montague, P.R., et al., 2012. Computational psychiatry. Trends Cogn. Sci. 16 (1), 72–80.

Murray, J.D., et al., 2014. Linking microcircuit dysfunction to cognitive impairment: effects of disinhibition associated with schizophrenia in a cortical working memory model. Cereb. Cortex 24 (4), 859–872.

Napolitano, A., Shah, K., Schubert, M.I., Porkess, V., Fone, K.C., Auer, D.P., May, 2014. In vivo neurometabolic profiling to characterize the effects of social isolation and ketamine-induced NMDA antagonism: a rodent study at 7.0 T. Schizophr Bull. 40 (3), 566–574. http://dx.doi.org/10.1093/schbul/sbt067. Epub May 13, 2013. PubMed PMID: 23671195; PubMed Central PMCID: PMC3984514.

Nestler, E.J., 2013. In: Charney, D.S., et al. (Eds.), Neurobiology of Mental Illness, fourth ed. Oxford University Press, New York, NY, p. 1248.

Nestler, E.J., Hyman, S.E., 2010. Animal models of neuropsychiatric disorders. Nat. Neurosci. 13 (10), 1161–1169.

Redish, A.D., Gordon, J.A. (Eds.), 2016. Computational Psychiatry: New Perspectives on Mental Illness. MIT Press, Cambridge, Mass.

Rivolta, D., et al., 2014. Source-reconstruction of event-related fields reveals hyperfunction and hypofunction of cortical circuits in antipsychotic-naive, first-episode schizophrenia patients during Mooney face processing. J. Neurosci. 34 (17), 5909–5917.

Rolls, E.T., et al., 2008. Computational models of schizophrenia and dopamine modulation in the prefrontal cortex. Nat. Rev. Neurosci. 9 (9), 696–709.

Sanacora, G., et al., 2008. Targeting the glutamatergic system to develop novel, improved therapeutics for mood disorders. Nat. Rev. Drug Discov. 7 (5), 426–437.

Schizophrenia Working Group of the Psychiatric Genomics, C., 2014. Biological insights from 108 schizophrenia-associated genetic loci. Nature 511 (7510), 421–427.

Sekar, A., et al., 2016. Schizophrenia risk from complex variation of complement component 4. Nature 530 (7589), 177–183.

Slifstein, M., et al., 2015. Deficits in prefrontal cortical and extrastriatal dopamine release in schizophrenia: a positron emission tomographic functional magnetic resonance imaging study. JAMA Psychiatry 72 (4), 316–324.

Smith, S.M., et al., 2013. Resting-state fMRI in the human connectome project. Neuroimage 80, 144–168.

Sotiropoulos, S.N., et al., 2013. Advances in diffusion MRI acquisition and processing in the human connectome project. Neuroimage 80, 125–143.

Stephan, K.E., Mathys, C., Apr, 2014. Computational approaches to psychiatry. Curr Opin Neurobiol 25, 85–92. http://dx.doi.org/10.1016/j.conb.2013.12.007. Epub Dec 29, 2013. Review. PubMed PMID: 24709605.

Tost, H., Alam, T., Meyer-Lindenberg, A., 2010. Dopamine and psychosis: theory, pathomechanisms and intermediate phenotypes. Neurosci. Biobehav. Rev. 34 (5), 689–700.

Ugurbil, K., 2016. What is feasible with imaging human brain function and connectivity using functional magnetic resonance imaging. Philos. Trans. R. Soc. Lond. B Biol. Sci. 371 (1705).

Uhlhaas, P.J., 2013. Dysconnectivity, large-scale networks and neuronal dynamics in schizophrenia. Curr. Opin. Neurobiol. 23 (2), 283–290.

Van Essen, D.C., et al., 2012. The human connectome project: a data acquisition perspective. Neuroimage 62 (4), 2222–2231.

Wang, X.J., 2001. Synaptic reverberation underlying mnemonic persistent activity. Trends Neurosci. 24 (8), 455–463.

Wang, X.-J., 2010. Neurophysiological and computational principles of cortical rhythms in cognition. Physiol. Rev. 90 (3), 1195–1268.

Wang, M., et al., 2013. NMDA receptors subserve persistent neuronal firing during working memory in dorsolateral prefrontal cortex. Neuron 77 (4), 736–749.

Wang, X.-J., Krystal, J.H., 2014. Computational psychiatry. Neuron 84 (3), 638–654.

Woo, C.W., et al., 2017. Building better biomarkers: brain models in translational neuroimaging. Nat. Neurosci. 20 (3), 365–377.

Zarate, C.A., Singh, J., Manji, H.K., 2006. Cellular plasticity cascades: targets for the development of novel therapeutics for bipolar disorder. Biol. Psychiatry 59 (11), 1006–1020.

# APPLYING CIRCUIT MODELING TO UNDERSTAND PSYCHIATRIC SYMPTOMS

# 1

# Cortical Circuit Models in Psychiatry: Linking Disrupted Excitation–Inhibition Balance to Cognitive Deficits Associated With Schizophrenia

*John D. Murray[1], Xiao-Jing Wang[2,3]*

[1] Yale University School of Medicine, New Haven, CT, United States;
[2] New York University, New York, NY, United States; [3] NYU Shanghai, Shanghai, China

## OUTLINE

*Computational Psychiatry*
http://dx.doi.org/10.1016/B978-0-12-809825-7.00001-8

3

## 1.1 INTRODUCTION

Ultimately, a central goal of neuropsychiatric research is to explain how symptoms and cognitive deficits arise from neurobiological pathologies. At present, we do not yet adequately understand how a neural system generates complex symptoms of any psychiatric disorder. This is in large part due to the stark explanatory gaps between levels of analysis: mechanisms underlying a psychiatric disease occur at the level of neurons and synapses, whereas symptoms are manifested and diagnosed at the level of cognition and behavior, which involve collective computations in brain circuits. Linking these levels is vital for gaining mechanistic insight into mental illness, and for the rational development of pharmacological treatments, which have physiological impact at the biophysical level. An emerging interdisciplinary approach to this challenge, "computational psychiatry," uses mathematical models of neural systems, in close interplay with experimentation, to study how disruption at lower levels propagate upward to produce dysfunction at higher levels of behavior and function (Wang and Krystal, 2014; Anticevic et al., 2015; Huys et al., 2016).

Biophysically based neural circuit modeling is a framework particularly well suited to link synaptic-level disruptions to emergent brain dysfunction. Circuit models can simulate neural population activity and computations, incorporating key properties of neurons, synapses, and circuit connectivity. Dynamic neural activity can be simulated through systems of differential equations governing the biophysical properties of neurons and synapses. Emergent patterns of activity in the model can be informed by—and tested with—empirical measures of neural activity. In certain circuit models, neural activity can be mapped onto a behavioral response, thereby generating model predictions that can be tested with behavioral data from corresponding task paradigms. Neural circuit models can play a key role in translational neuroscience, because by virtue of their biophysical basis such models provide opportunities to mechanistically understand how synapse-level disruptions produce aberrant neural activity and deficits in cognition and behavior.

In this chapter, we focus our review on a set of studies that leverage biophysically based neural circuit models to understand how synaptic disruptions may induce cognitive deficits, with particular relevance to schizophrenia (Murray et al., 2014; Starc et al., 2017; Lam et al., 2017). These studies thereby provide a test bed for a computational psychiatry framework utilizing biophysically based neural circuit modeling. Specifically, we utilized spiking circuit models of microcircuits in association cortical areas (such as the prefrontal cortex and posterior parietal cortex), which can perform two core cognitive functions, working memory, and decision making (Compte et al., 2000; Wang, 2002). These models were developed to capture key neurophysiological correlates of cognitive function in association cortex and have been validated through their predictions for cognitive behavior, neural activity, and synaptic mechanisms. We applied these models to study the impact of alterations in the balance between synaptic excitation and inhibition (E/I balance) (Hensch and Fagiolini, 2004). Altered E/I balance is hypothesized to contribute to pathophysiological states across a number of neuropsychiatric disorders such as schizophrenia. Cognitive deficits, including in working memory and decision making, lie at the core of schizophrenia (Elvevåg and Goldberg, 2000). Yet it remains poorly understood how they may relate to the hypothesized neuropathological alterations such as disrupted E/I balance.

For both working memory and decision making, we found that relatively small perturbations of the E/I ratio in the models can profoundly impact cognitive function. Importantly, the models make dissociable predictions—testable at the level of cognitive behavior—for elevated versus lowered E/I ratio. These modeling studies illustrate general principles for computations in recurrent cortical circuits and elucidate distinct modes of cognitive dysfunction in disease-related states. This test bed for computational psychiatry can be applied to other disorders and other cognitive functions. Furthermore, we argue that to be most useful in computational psychiatry, neural circuit modeling should go hand in hand with basic neuroscience research by integrating findings into a formal model and making testable predictions, which can inform the design and analysis of experiments in animals and humans. Neural circuit models can thereby play a key translational bridge across various levels of analysis, spanning biophysics, systems neurophysiology, and cognitive neuroscience applied to clinical populations.

## 1.2 ROLES FOR BIOPHYSICALLY BASED NEURAL CIRCUIT MODELING IN COMPUTATIONAL PSYCHIATRY

From a computational psychiatry perspective, biophysically based neural circuit modeling can address key questions that are inaccessible to other levels of computational modeling, such as connectionist models or normative mathematical models of behavior (Wang and Krystal, 2014; Anticevic et al., 2015; Huys et al., 2016). For instance, circuit models are well suited to mechanistically study the impacts of synapse-level alterations. One can perturb specific synaptic parameters in the model that are physiologically interpretable, thereby allowing direct implementation of hypothesized perturbations related to disease mechanisms and pharmacological manipulations. One can then characterize the impact of these perturbations on emergent neural activity and behavior, and relate them to experimental findings from healthy and clinical populations as well as in animal models of disease states. Biophysically based models may also be able to inform the rational design of pharmacological therapies, by mechanistically linking molecular, cellular, circuit, and ultimately behavioral levels of analysis.

The specific scientific questions under study play a critical role in determining the level of biophysical detail included in a particular model. Circuit models typically incorporate certain biophysical details but not many others. For instance, some questions related to how dopaminergic dysregulation in schizophrenia impacts synaptic transmission could be addressed in a biophysically based model of an individual synapse that includes subcellular signaling pathways (Qi et al., 2010). In contrast, emergent circuit-level dynamics, such as oscillations or persistent activity, can be simulated in thousands of recurrently connected spiking neurons whose individual dynamics are simplified to include only certain channels and receptors (Wang, 2010, 2008). Modeling systems-level disturbances, such as large-scale connectivity alterations in schizophrenia, may entail coarse-grained mean-field models of local nodes organized in large-scale networks that still contain neurophysiologically interpretable parameters and enable study of questions related to E/I balance (Yang et al., 2014, 2016a).

An important area of research in clinical neuroscience is the discovery and characterization of predictive neurophysiological biomarkers for psychiatric disorders. Models can inform the circuit mechanisms underlying these biomarkers and their relations to putative synaptic disruptions. These biomarkers can include electrophysiological measurements of aberrant neural oscillatory dynamics and neuroimaging measurements of large-scale dysconnectivity in resting-state activity. Insofar as the circuit model is well constrained and validated through experiments, dissociable model predictions allow interpretation of biomarkers to generate mechanistic hypotheses for future experimental studies in animal models and human subjects.

One area of modeling progress with relevance to biomarkers explores the mechanisms underlying oscillatory neural activity that emerge at the network level in recurrent cortical circuits (Wang, 2010). Cortical oscillatory activity is found to be abnormal in a number of neuropsychiatric disorders. In particular, schizophrenia is associated with alterations in oscillatory activity in the gamma (30—80 Hz) range (Gonzalez-Burgos and Lewis, 2012; Uhlhaas, 2013). Computational models, in conjunction with physiological findings, support the idea that neocortical gamma oscillations arise from a feedback loop in a microcircuit of pyramidal cells reciprocally connected to perisomatic-targeting, parvalbumin-expressing interneurons (Buzsáki and Wang, 2012). Circuit models of gamma oscillations have been applied to explore the dynamical effects of putative synaptic perturbations associated with schizophrenia, including reduced production of $\gamma$-aminobutyric acid (GABA) and parvalbumin in inhibitory interneurons (Vierling-Claassen et al., 2008; Spencer, 2009; Volman et al., 2011; Rotaru et al., 2011). In each case, the models have provided specific hypotheses for how systems-level dynamics, which can be measured in humans through techniques such as electroencephalography (EEG) or magnetoencephalography (MEG), may be altered as a result of synaptic- or cellular-level changes.

Below, we focus on how circuit models of cognitive functions can be applied to understand cognitive deficits resulting from synaptic disruptions associated with schizophrenia. For certain core cognitive computations, we have knowledge of the neural circuit basis underlying these processes, which typically involve contributions from animal studies. For these cases, detailed circuit models can be developed rigorously to provide the link from synaptic disruptions to behavior (e.g., cognitive deficits discussed below). In other cases, psychiatric symptoms relate to complex cognitive functions for which we lack understanding of the underlying neuronal representations or circuit mechanisms. At present, these circuit models are limited and cannot be applied to complex behavioral tasks, for which we lack understanding of neural circuit correlates. We now turn to the conditions in which circuit models may be best suited to study cognitive deficits in psychiatric disorders.

## 1.3 LINKING PROPOSITIONS FOR COGNITIVE PROCESSES

A major goal in computational psychiatry research is for biophysically based neural circuit models to explain mechanistically how synaptic-level disruptions induce cognitive-level deficits. We argue that for this approach to be most effective, the circuit model should be grounded in a well-supported relationship between neuronal activity and a given cognitive process. Such relationships have been formalized by the concept of a *linking proposition* that states the nature of a statistical correspondence between a given neural state and a cognitive state. Related to the concept of the linking proposition is that of a *bridge locus*, which is the set of neurons for which this linking proposition holds (Teller, 1984). Convergent evidence supporting a linking proportion comes from a number of experimental methodologies applied to animal models, especially to behaving nonhuman primates, given the strong homologies of areas in the human and nonhuman primate brains (Schall, 2004). Single-neuron recordings can relate neuronal activity to computations posited in psychological processes. Further evidence can come from perturbative techniques such as microstimulation or inactivation.

As an exemplary application of this perspective to a nonsensory function, Schall (2004) considered the neural underpinnings of the preparation of saccadic eye movements. In the case of saccade preparation, a well-supported candidate for the bridge locus is a distributed network of cortical and subcortical areas, including the frontal eye field and superior colliculus. During saccade preparation, so-called "movement" neurons in these areas exhibit a location-selective ramping of their firing rates, and a saccade is initiated when their firing rates reach a threshold level. At the level of mental processes, a leading psychological model for response preparation is accumulation of a signal until reaching a fixed threshold level, which triggers the response. In such accumulator models, sequential sampling of a stochastic signal generates variability in the rate of rise to the fixed threshold, which can explain the observed variability in saccade reaction times. The linking proposition between a neural state (movement cell firing rates) and a psychological state (level of an accumulator) provides a framework for detailed hypothesis generation and experimental examination of psychological models.

What linking propositions do we have for core cognitive functions, and specifically for working memory and decision making? The neural correlates of working memory have been studied extensively through single-neuron recordings from monkeys performing tasks in which the identity of a transient sensory stimulus must be maintained internally across a seconds-long mnemonic delay to guide a future response. These studies revealed that a key neural correlate of working memory is stimulus-selective persistent activity, i.e., stable elevated firing rates in a subset of neurons that spans the mnemonic delay (Goldman-Rakic, 1995; Wang, 2001). These neuronal activity patterns are observed across a distributed network of interconnected brain areas with prefrontal cortex as a key locus. For instance, in one well-studied experimental paradigm, the oculomotor delayed response task, the subject must maintain in working memory the spatial location of a visual cue across a delay period to guide a saccadic eye movement toward that location

(Funahashi et al., 1989). During the mnemonic delay, a subset of prefrontal neurons exhibit stimulus-tuned persistent activity patterns, with single neurons firing at elevated rates for a preferred spatial location. These neurophysiological findings have grounded the leading hypothesis that working memory is supported by stable persistent activity patterns in prefrontal cortex that bridge the temporal gap between stimulus and response epochs.

The neural computations underlying decision making have been most studied in task paradigms in which a categorical choice is based on the accumulation of perceptual evidence over time. In one highly influential task paradigm, the subject must decide the net direction of random-dot motion stimuli, which encourages decision making based on the temporal integration of momentary perceptual evidence (Roitman and Shadlen, 2002). Behavior can be well captured by psychological process models of evidence accumulation to a threshold. Single-neuron recordings have found that in association cortex, such as the lateral intraparietal area, choice-selective ramping of neuronal firing rates reflects accumulated perceptual evidence, with activity crossing a threshold level reflecting the decision commitment (Gold and Shadlen, 2007). These neural correlates reflect two key computations needed for perceptual decision making: accumulation of evidence and formation of categorical choice.

Conceptually, a neural circuit model can instantiate a linking proposition for a cognitive process and propose circuit mechanisms underlying the computations. If associated with a hypothesized bridge locus, model predictions for these circuit mechanisms can be experimentally tested, such as through single-neuron recordings. For instance, in the case of working memory, experiments have tested how focal antagonism of specific synaptic receptors affects persistent activity, thereby informing the neuronal and synaptic mechanisms supporting the computations (Wang et al., 2013; Rao et al., 2000). The stronger these links are among (1) the synaptic and neuronal processes in circuit mechanisms, (2) neural activity, and (3) the cognitive function, the greater the potential for translational computational psychiatry. Once established, the model can then make rigorous predictions for the consequences of alterations in those circuit mechanisms. In this way, circuit models can iteratively contribute to our understanding of these links across levels of analysis and leverage them to study dysfunction in neuropsychiatric disorders.

## 1.4 ATTRACTOR NETWORK MODELS FOR CORE COGNITIVE COMPUTATIONS IN RECURRENT CORTICAL CIRCUITS

Biophysically based neural circuit modeling has provided mechanistic hypotheses for how working memory and decision making computations can be performed in recurrent cortical circuits (Wang, 2001, 2008).

As noted, a key neurophysiological correlate of working memory is stimulus-selective, persistent neuronal activity across the mnemonic delay in association cortical areas. Delays in working memory tasks (a few seconds) are longer than the typical timescales of neuronal or synaptic responses (10–100 ms). Similarly, perceptual decision making demands categorical selection and benefits from temporal integration of evidence over long timescales (hundreds of milliseconds). Both of these computations therefore implicate circuit mechanisms.

Motivated by experimental observations of stable persistent activity in single neurons, a leading theoretical framework proposes that working memory-related persistent activity states are dynamical attractors, i.e., stable states in network activity. In the mathematical formalism of dynamical systems, an attractor state is an activity pattern that is stable in time, so that following a small transient perturbation away from this state the network will converge back to the attractor state. A class of neural circuit models called attractor networks have been applied to explain the mechanisms that allow a recurrent network of spiking neurons to maintain persistent activity during working memory (Amit, 1995; Wang, 2001). An attractor network typically possesses multiple attractor states: a low-firing baseline state and multiple memory states in which a stimulus-selective subset of neurons are persistently active. Because the memory state is an attractor state, it is self-reinforcing and resistant to noise or perturbation by distractors, allowing the stimulus-selective memory to be stably maintained over time (Brunel and Wang, 2001; Compte et al., 2000).

In a typical attractor network, subpopulations of excitatory neurons are selective to different stimuli. Recurrent excitatory synaptic connectivity exhibit a "Hebbian" pattern such that neurons of similar selectivity have stronger connections between them (Fig. 1.1A). When the strength of recurrent excitatory connections is strong enough, the circuit can support stimulus-selective attractor states that can subserve working memory (Fig. 1.1B). Strong recurrent excitation thereby provides the positive feedback that sustains persistent activity. Wang (1999) found that incorporating physiologically realistic synaptic dynamics pose constraints on the synaptic mechanisms supporting this positive feedback. Strong positive feedback is prone to generate large-amplitude oscillations that can destabilize persistent states and can drive firing rates beyond physiologically plausible ranges. It was found that both of these problems can be solved if recurrent excitation is primarily mediated by slow N-methyl-D-aspartate (NMDA) receptors.

Critically, recurrent excitation must be balanced by strong feedback inhibition mediated by GABAergic interneurons. Feedback inhibition stabilizes the low-activity baseline state (Amit and Brunel, 1997; Wang, 1999). In a persistent activity memory state, lateral inhibition enforces selectivity of the working memory representation, preventing the spread

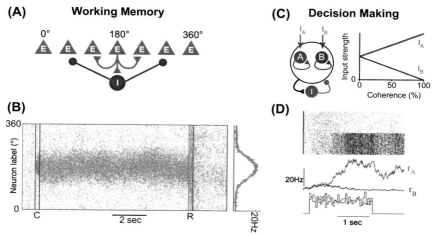

FIGURE 1.1 Biophysically based cortical circuit models of working memory and decision making computations. (A) Schematic of the network architecture for a model of spatial working memory. The model consists of recurrently connected excitatory pyramidal cells (E) and inhibitory interneurons (I). Pyramidal cells are labeled by the angular location they encode (0−360 degrees). Excitatory-to-excitatory connections are structured, such that neurons with similar preferred angles are more strongly connected. Connections between pyramidal cells and interneurons are unstructured and mediate feedback inhibition. (B) Spatiotemporal raster plot showing a bump attractor state in an example trial. A stimulus is presented at 180 degrees during the brief cue epoch (denoted C) and during the subsequent delay; the stimulus location is encoded by persistent activity throughout the working memory delay until the response epoch (denoted R). On right is shown the firing rate profile of the working memory bump attractor state. (C) Schematic of the network architecture for a model of perceptual decision making. The circuit contains two populations of pyramidal neurons which are each selective to one of the two stimuli (A and B). Within each pyramidal-neuron population there is strong recurrent excitation, and the two populations compete via feedback inhibition mediated by interneurons (I). Right: The selective populations receive sensory-related inputs determined by the stimulus coherence. (D) Example neuronal activity in a single trial for a zero-coherence stimulus. Top: Spatiotemporal raster plot for the two selective populations. Middle: Population firing rates $r_A$ and $r_B$. Bottom: Stochastic sensory-related inputs. During decision making, the circuit exhibits an initial slow ramping, related to temporal integration of evidence, which leads to categorical choice (for A in this trial). *Panels (B and D) adapted from Compte, A., Brunel, N., Goldman-Rakic, P.S., Wang, X.-J., 2000. Synaptic mechanisms and network dynamics underlying spatial working memory in a cortical network model. Cereb. Cortex 10, 910−923; Wang, X.-J., 2002. Probabilistic decision making by slow reverberation in cortical circuits. Neuron 36, 955−968, respectively.*

of excitation to the entire neuronal population (Murray et al., 2014). Attractor dynamics supporting working memory are thereby supported by recurrent E/I that are strong and balanced. These circuit models make predictions for the relationship between synaptic mechanisms and working memory activity, which are confirmed through experiments combining single-neuron recording and pharmacological manipulation in prefrontal cortex. Locally blocking excitation mediated by NR2B NMDA receptors attenuates persistent activity for the preferred stimulus (Wang et al., 2013). Locally blocking inhibition mediated by $GABA_A$ receptors reduces stimulus selectivity of delay activity by elevating responses to nonpreferred stimuli (Rao et al., 2000).

In addition to working memory computations, strong recurrent excitatory and inhibitory connections in cortical attractor networks provide a circuit mechanism for decision making, supporting temporal integration of evidence, and categorical choice (Wang, 2002, 2008; Wong and Wang, 2006). In this model, choice-selective neuronal populations receive external inputs corresponding to sensory information (Fig. 1.1C). Reverberating excitation enables temporal accumulation of evidence through slow ramping of neural activity over time (Fig. 1.1D). This property highlights that attractor networks not only support multiple stable states (representing categorical choices), but also support slow transient dynamics that can instantiate computations such as temporal integration. In these models, temporal integration via recurrent excitation benefits from the slow biophysical timescale of NMDA receptors (Wang, 2002). Feedback and lateral inhibition mediated by GABAergic interneurons mediates competition among neuronal populations underlying the formation of a categorical choice. Irregular neuronal firing, a ubiquitous feature of cortex, contributes to stochastic choice behavior across trials, even when presented with identical stimulus inputs.

These computational modeling studies demonstrate that an association cortical microcircuit model can support working memory and decision making computations through attractor dynamics. This therefore suggests a shared "cognitive-type" circuit mechanism for these functions, which may provide components on which more complex cognitive processes may be built (Wang, 2013). Because these functions rely on strong recurrent E/I, they are particularly well suited to study how cognitive deficits may arise from alterations in synaptic function, which are implicated in neuropsychiatric disorders.

## 1.5 CIRCUIT MODELS OF COGNITIVE DEFICITS FROM ALTERED EXCITATION—INHIBITION BALANCE

In a series of studies, we have applied cortical attractor network models of working memory and decision making function to characterize the impact of E/I disruptions in association cortex (Murray et al., 2014; Starc et al., 2017; Lam et al., 2017). Alteration of cortical E/I balance is implicated in multiple neuropsychiatric disorders, including schizophrenia, autism spectrum disorder, and major depression. A key strength of these circuit models is that they make explicit predictions not just for neural activity but also for behavior, which can be tested experimentally in clinical populations or after causal perturbation.

In schizophrenia, cortical microcircuit alterations are complex, with observed dysfunction in both glutamatergic excitation and GABAergic inhibition. Postmortem investigations of prefrontal cortex in schizophrenia find reductions in spines on layer-3 pyramidal cells, which potentially reflect reduced recurrent excitation. Such studies also have revealed multiple impairments in inhibitory interneurons, which potentially reflect reduced feedback inhibition. Pharmacological manipulations provide complementary evidence. One such approach is to use NMDA receptor antagonists (e.g., ketamine), which transiently, safely, and reversibly induce cardinal symptoms of schizophrenia in healthy subjects (Krystal et al., 2003). A leading hypothesis regarding ketamine's effects on neural function proposes a state of cortical disinhibition potentially via preferential blockade of NMDA receptors on GABAergic interneurons (Greene, 2001; Homayoun and Moghaddam, 2007; Kotermanski and Johnson, 2009). However, many questions remain regarding the neural effects of ketamine, such as which NMDA receptor subunits and neuronal cell types may be the preferential sites of action (Khlestova et al., 2016; Zorumski et al., 2016).

Mechanistic links between altered E/I ratio and cognitive impairment remain tenuous. A primary motivation for these modeling studies was to formulate dissociable behavioral predictions for distinct sites of synaptic perturbation. In these studies, E/I ratio was perturbed bidirectionally via hypofunction of NMDA receptors at two recurrent synaptic sites: on inhibitory interneurons that elevates E/I ratio via disinhibition; or on excitatory pyramidal neurons that lowers E/I ratio (Fig. 1.2A).

## 1.5.1 Working Memory

Working memory function is a promising candidate in clinical neuroscience as an endophenotype, a quantitatively measurable core trait that is intermediate between genetic risk factors and a psychiatric disorder (Insel and Cuthbert, 2009). It is important to consider the component processes and features involved in overall working memory function: encoding, maintenance, robustness to distraction, precision, and capacity. Ongoing work in clinical cognitive neuroscience aims at resolving how these processes are impaired in schizophrenia (Barch and Ceaser, 2012). Many studies have found a deficit in working memory encoding (Lee and Park, 2005). For visuospatial working memory, patients with schizophrenia exhibit deficits in encoding and in maintenance, which results in a graded loss of precision (Badcock et al., 2008; Starc et al., 2017). Other visual paradigms find reduced capacity but not necessarily precision (Gold et al., 2010).

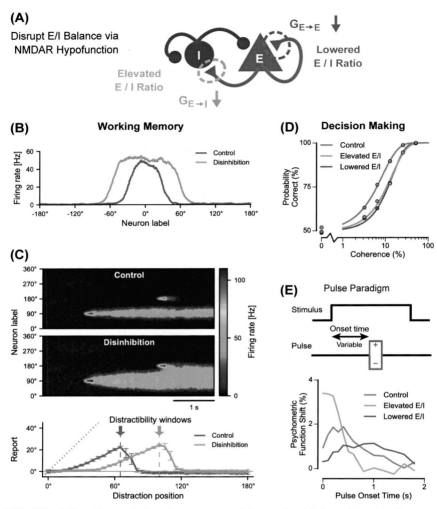

FIGURE 1.2 Effects of altered excitation–inhibition (E/I) balance in cortical circuit models of working memory and decision making. (A) E/I ratio was perturbed bidirectionally via hypofunction of NMDA receptors at two recurrent synaptic sites: on inhibitory interneurons that elevates E/I ratio via disinhibition; or on excitatory pyramidal neurons that lowers E/I ratio. (B) For the working memory circuit, the firing rate profile of the "bump" attractor activity pattern during working memory maintenance. Elevated E/I ratio via disinhibition results in a broadened working memory representation. (C) Disinhibition impairs the network's ability to filter out intervening distractors. Top: Spatiotemporal plot of network activity in response to a distractor presented during the delay at a distance of 90 degrees from the target. Bottom: Deviation of the read-out report as a function of the angular distance between the distractor and the target. The "distractibility window" is widened by disinhibition. (D) In the decision making circuit, performance as quantified by the psychometric function, i.e., the proportion of correct choices as a function of stimulus coherence. Both perturbations, elevated and lowered E/I ratio, can comparably degrade performance relative

Murray et al. (2014) examined the effects of altered E/I balance in a cortical circuit model of visuospatial working memory (Compte et al., 2000; Carter and Wang, 2007; Wei et al., 2012). Disinhibition, with results in an elevated E/I ratio, was implemented through antagonism of NMDA receptors preferentially onto interneurons. In this model, disinhibition leads to a broadening in the neural-activity patterns in the mnemonic attractor states (Fig. 1.2B). This neural change induced specific cognitive deficits. During maintenance, the mnemonic activity pattern undergoes random drift that leads to decreased precision of responses. Disinhibition increased the rate of this drift, thereby inducing a specific deficit in mnemonic precision during working memory maintenance.

Additionally, Murray et al. (2014) found that broadened neural representations make working memory more vulnerable to intervening distractors (Fig. 1.2C). In the model, a distractor is more likely to "attract" the memorandum toward it if the two representations overlap. Distractibility therefore depends on the similarity between the representations of the mnemonic target and the intervening distractor. Consistent with this model behavior, it has been found empirically that in visuospatial working memory, a distractor can attract the memory report toward its location, but only if the distractor appears within a "distractibility window" around the target location (Herwig et al., 2010). Because disinhibition broadens the mnemonic activity patterns, there is an increased range of distractors that can disrupt working memory.

To test the model prediction of broadened working memory representations under disinhibition, Murray et al. (2014) analyzed behavior from healthy humans administered ketamine during a spatial delayed match-to-sample task (Anticevic et al., 2012). The model predicted a pattern of errors depending on whether the probe was similar to a target held in working memory. Analysis of the behavioral data guided by the model revealed a similar specific pattern of errors under ketamine versus control conditions to that predicted by the computational model. Consistent with model predictions, ketamine increased the rate of errors specifically for distractors that would overlap with a broadened mnemonic representation. A similar pattern of errors has been observed in schizophrenia, with a selective increase in false alarms for near nontarget

---

to the control circuit. (E) A perceptual decision making task paradigms that characterizes the time course of evidence accumulation can test dissociable behavioral predictions from elevated versus lowered E/I ratio. Top: The pulse paradigm uses a brief pulse of additional perceptual evidence at different onset times. This pulse induces a shift the psychometric function, which quantifies the sensitivity of the choice on evidence presented at that time point. Bottom: Shift in the psychometric function as a function of pulse onset time, for the three E/I regimes. Relative to control, in the elevated E/I circuit the pulse has a stronger impact at early onset times, but less impact at later onset times. The lowered E/I circuit shows a flattened profile of the shift, with greater impact at late onset times.

probes but not for far nontarget probes (Mayer and Park, 2012). In contrast to the model predictions arising from disinhibition, insufficient recurrent excitation in the model leads to a collapse of persistent activity that would induce an error pattern of misses and spatially random errors.

To apply this model to schizophrenia and more directly to test model predictions, Starc et al. (2017) designed a working memory task to be explicitly aligned with the model and with the primate electrophysiology task paradigms for which the model was developed. Such an alignment, in which the clinical study is linked to basic neurophysiology findings through a computational model, allows stronger inferences and testing of hypotheses. In the working memory task of Starc et al. (2017), the memorandum is a single visuospatial location, and the response is a direct report of the remembered location that provides a continuous measure of mnemonic coding. To test the model prediction of increased drift during working memory maintenance, the duration of the mnemonic delay is varied, to characterize how response variability increases with the duration of maintenance. To test the model prediction of increased distractibility dependent on target–distractor similarity, a set of trials included a distractor during the delay with a variable distance from the target. Starc et al. (2017) found that results largely followed model predictions, whereby patients exhibited increased variance and less working memory precision as the delay period increased relative to healthy controls. Schizophrenia patients also exhibited increased working memory distractibility, with reports biased toward distractors at specific spatial locations. This study illustrates a productive computational psychiatry approach in which predictions from biophysically based neural circuit models of cognition can be translated into experiments in clinical populations.

## 1.5.2 Decision Making

Broadly, decision making function is impaired in multiple psychiatric disorders (Lee, 2013). To study dysfunction in neural circuit models, we focus on perceptual decision making in task paradigms similar to those studied via electrophysiology in nonhuman primates. As reviewed above, cortical attractor network models have been developed to capture behavior and neuronal activity from association cortex during random-dot motion paradigms (Wang, 2002; Furman and Wang, 2008). In these two-alternative forced choice tasks, a random-dot motion stimulus is presented, and the subject must report the net direction of motion (e.g., left vs. right). The coherence of the random-dot pattern can be parametrically varied to control the strength of perceptual evidence and thereby task difficulty. The psychometric function, giving the percent correct as a function of coherence, defines the discrimination threshold as the coherence eliciting a certain level of accuracy.

Random-dot motion paradigms have been applied to clinical populations, and have revealed impaired perceptual discrimination in schizophrenia, as measured by a higher discrimination threshold (Chen et al., 2003, 2004, 2005). Similar impairments in the discrimination threshold have also been observed in patients with autism spectrum disorder (Milne et al., 2002; Koldewyn et al., 2010). These impairments are typically interpreted as evidence of neural dysfunction in sensory representations (Butler et al., 2008). However, it is possible that such impairments may have contributions from dysfunction in evidence accumulation downstream from early sensory areas, within association cortical circuits.

To explore this issue, Lam et al. (2017) studied the effects of altered E/I balance in the association cortical circuit model of decision making developed by Wang (2002). E/I ratio was perturbed bidirectionally to compare the impact of elevated versus lowered E/I ratio via NMDA receptor hypofunction on inhibitory versus excitatory neurons, respectively. Interestingly, Lam et al. (2017) found that disruption of E/I balance in either direction can similarly impair decision making as assessed by psychometric performance, following an inverted-U dependence on E/I ratio (Fig. 1.2D). Therefore, the standard psychophysical measurements from clinical populations cannot dissociate among distinct circuit-level alterations: elevated E/I ratio, lowered E/I ratio, or an upstream sensory coding deficit.

Nonetheless, Lam et al. (2017) found that these regimes make dissociable predictions for the time course of evidence accumulation. The random-dot motion task paradigm promotes a cognitive strategy of evidence accumulation across the stimulus presentation. Both in the circuit model and in empirical psychophysical behavior, the choice is not uniformly sensitive to the stimulus value at all time points due to bounded accumulation (Kiani et al., 2008). Multiple more complex task paradigms have been developed to characterize the time course of evidence accumulation. For instance, in the "pulse" task paradigm (Huk and Shadlen, 2005; Wong et al., 2007), a brief pulse of additional coherence is inserted at a variable onset time during the otherwise constant-coherence stimulus (Fig. 1.2E). This pulse induces a shift of the psychometric function according to pulse coherence. The dependence of this shift on pulse onset time reflects the weight of that time point on choice.

The pulse paradigm, as well as other paradigms, was able to dissociate distinct decision making impairments under altered E/I ratio (Fig. 1.2E). Under elevated E/I ratio, decision is impulsive: perceptual evidence presented early in time is weighted much more than late evidence. In contrast, under lowered E/I ratio, decision making is indecisive: evidence integration and winner-take-all competition between options are weakened. These effects are qualitatively captured by modifying a widely used abstract model for decision making from mathematical psychology, the drift diffusion model (Ratcliff, 1978; Gold and Shadlen, 2007). The standard drift diffusion model assumes perfect integration with an infinite time constant for memory. Lowered E/I ratio in the circuit model can be captured by "leaky" integration with finite time constant for memory. In contrast, elevated E/I ratio can be captured by "unstable" integration, which has an intrinsic tendency to diverge toward the decision threshold. This study demonstrates the potential to link synaptic-level perturbations in neural circuit models to measurable cognitive behavior and to more abstract models from mathematical psychology.

## 1.6 CRITICAL ROLE OF EXCITATION–INHIBITION BALANCE IN COGNITIVE FUNCTION

As described in the above section, neural circuit models of cognitive functions can generate dissociable predictions for how distinct synaptic perturbations impact behavior under various task paradigms. Biophysically based models can also suggest what aspects of neural activity or behavior may be differentially sensitive or robust to particular manipulations by pathology, compensation, or treatment. Changes in certain network parameters, or the combinations of parameters, may have much stronger impact on model behavior than do changes in other parameter combinations. A "sloppy" axis in parameter space is one among which the model response is relatively insensitive to perturbations in that parameter combination, whereas a "stiff" axis is one in which the model response is highly sensitive to perturbations (Gutenkunst et al., 2007).

Murray et al. (2014) and Lam et al. (2017) characterized function in theses neural circuit models under parametric variation in E/I ratio. Specifically, they explored the parameter space of reductions of NMDA receptor conductance onto both inhibitory interneurons (elevating E/I ratio) and onto excitatory pyramidal neurons (reducing E/I ratio) (Fig. 1.3). For the working memory model, circuit function is determined by the width of the mnemonic persistent activity pattern. For the decision making model, circuit function can be measured through discrimination sensitivity (inverse of the discrimination threshold). Similarly for both circuit models for working memory and decision making, E/I ratio was found to be a key parameter for optimal network function. Following relatively small perturbations, circuit function is robust as long as E/I balance is preserved. Preserved E/I ratio therefore corresponds to a "sloppy" axis in this parameter space. In contrast, even subtle changes to E/I ratio (along a "stiff" axis) have a strong impact on model function.

If the imbalance is substantial, either elevated or lowered, the circuit can lose multistability. If disinhibition is too strong (via elevated E/I ratio), then the spontaneous state is no longer stable. Conversely, if recurrent excitation is too weak (via lowered E/I ratio), then the circuit cannot support persistent activity. Collectively, these analyses reveal that E/I balance is vital for optimal cognitive performance in these cortical circuit models. This suggests that despite the complexity of synaptic alterations in disorders such as schizophrenia, the impact on cognitive function in neural circuits may be understandable in terms of their "net

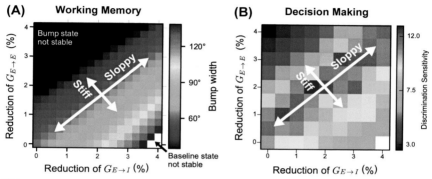

FIGURE 1.3   Dependence of circuit function on synaptic parameters: a critical role of excitation−inhibition (E/I) balance in both working memory and decision making. The plots illustrate a parameter space of reductions of two recurrent NMDAR conductance strengths from excitatory pyramidal neurons: onto inhibitory interneurons ($G_{E \to I}$) or onto excitatory pyramidal neurons ($G_{E \to E}$). This analysis characterizes the sensitivity of model function to joint perturbations of these two parameters. (A) For the working memory circuit, we measured the width of the working memory bump attractor state. Bump width affects mnemonic precision and distractibility during working memory maintenance. (B) For the decision making circuit, we measured the discrimination sensitivity, which is defined as the inverse of the discrimination threshold (i.e., coherence which yields 81.6% correct). A higher sensitivity corresponds to better performance. For both working memory and decision making circuits, within this range of perturbation, if $G_{E \to I}$ and $G_{E \to E}$ are reduced together in a certain proportion, circuit performance is essentially unaltered, because E/I balance is maintained. E/I balance defines a "sloppy" axis in parameter space along which the function is insensitive. In contrast, the function is highly sensitive to small orthogonal perturbations, along a "stiff" axis (Gutenkunst et al., 2007). Reduction of $G_{E \to I}$ in greater proportion elevates E/I ratio and can degrade performance: for working memory, due to broadened mnemonic representations; for decision making, due to highly unstable integration leading to impulsive selection. In contrast, reduction of $G_{E \to E}$ in greater proportion lowers E/I ratio and can degrades performance: for working memory, due to loss of the bump attractor state; for decision making, due to indecisive selection. These findings indicate that E/I ratio is a crucial effective parameter for cognitive function in these circuits, with an "inverted-U" dependence of function on E/I ratio. *Panels (A and B) adapted from Murray, J.D., Anticevic, A., Gancsos, M., Ichinose, M., Corlett, P.R., Krystal, J.H., Wang, X.-J., 2014. Linking microcircuit dysfunction to cognitive impairment: effects of disinhibition associated with schizophrenia in a cortical working memory model. Cereb. Cortex 24, 859−872; Lam, N.H., Borduqui, T., Hallak, J., Roque, A.C., Anticevic, A., Krystal, J.H., Wang, X.-J., Murray, J.D., 2017. Effects of altered excitation-inhibition balance on decision making in a cortical circuit model. bioRxiv. http://dx.doi.org/10.1101/100347, respectively.*

effect" on effective parameters, such as E/I ratio, to which the circuit is preferentially sensitive.

## 1.7 FUTURE DIRECTIONS IN NEURAL CIRCUIT MODELING OF COGNITIVE FUNCTION

In this chapter, we have primarily reviewed studies leveraging biophysically based neural circuit models to explore the effects of altered E/I balance on the core cognitive functions of working memory and decision making. These studies revealed that E/I ratio is a critical property for proper cognitive function in cortical circuits. Furthermore, they provide a test bed for computational psychiatry demonstrating that neural circuit models can play a translational role between basic neurophysiology and clinical applications. Here we turn to some critical areas for future modeling to address.

### 1.7.1 Integrating Cognitive Function With Neurophysiological Biomarkers

As noted above, biophysically based circuit models are well positioned to explore the mechanisms through which synaptic-level perturbations may be associated with neurodynamical biomarkers. In the context of schizophrenia, circuit models have been applied to studying mechanisms of disrupted gamma-band oscillations (Vierling-Claassen et al., 2008; Spencer, 2009; Volman et al., 2011; Rotaru et al., 2011), which can be related to EEG/MEG data from patients (Uhlhaas and Singer, 2010). At very different spatiotemporal scales, circuit models of large-scale "dysconnectivity" can be related to resting-state BOLD data (Yang et al., 2014, 2016a). Such biomarker-related models are mostly nonfunctional, in the sense that they do not directly relate to cognitive function or behavior. Future modeling work is needed in the integration of cognitive function with neurophysiological biomarkers across multiple scales of analysis.

### 1.7.2 Incorporating Further Neurobiological Detail

To address increasingly complex and detailed questions about neural circuit dysfunction, future models will need to incorporate further elements of known neurobiologically which can be constrained and tested with experiments. One notable limitation is that the cortical circuit models described above only contain a single type of inhibitory interneuron, and therefore are not able to speak to important questions regarding preferential dysfunction in specific interneuron cell types. There are key differences between parvalbumin-expressing and somatostatin-expressing interneurons, which differ in their synaptic connectivity and functional responses (Gonzalez-Burgos and Lewis, 2008). Microcircuit models that propose a division of labor among interneuron classes (Wang et al., 2004; Yang et al., 2016b) have the potential to make dissociable predictions for dysfunction in distinct cell types. Another aspect of microcircuitry for model extension is laminar structure in cortex (Mejias et al., 2016), which may address to mechanistic hypotheses of impaired predictive coding (Bastos et al., 2012). Beyond the level of local microcircuitry, further modeling work is needed on distributed cognitive computations across brain areas (Chaudhuri et al., 2015; Murray et al., 2017), and their application to alterations in large-scale network dynamics in psychiatric disorders (Yang et al., 2014, 2016a).

## 1.7.3 Informing Task Designs

These modeling studies suggest important considerations for the design of cognitive tasks applied to computational psychiatry. In each model, multiple modes of cortical dysfunction (e.g., elevated vs. lowered E/I ratio vs. upstream sensory coding deficit) can impair performance. Standard performance analyses in common task paradigms (e.g., match–nonmatch error rate in working memory, or psychometric threshold in decision making) may be insufficient to resolve dissociable predictions. More fine-grained analyses of task behavior should distinguish different types of errors or deficits, rather than simply measuring overall performance, which could be impaired due to deficits in distinct cognitive subprocesses (e.g., encoding vs. maintenance for working memory) or opposing deficits in a single subprocess (e.g., leaky vs. unstable integration in decision making). Circuit modeling can provide insight into the variety of potential "failure modes" in a cognitive function, and into which task designs can reveal them. In turn, alignment of a task design with a circuit model allows for generation of mechanistic neurophysiological hypotheses from behavioral measurements.

## 1.7.4 Studying Compensations and Treatments

Finally, of utmost relevance to psychiatry, biophysically based circuit modeling has the potential to provide a method for simulating possible effects of treatments (or compensations), which act at level of ion channels and receptors. As a proof-of-principle example of this, Murray et al. (2014) examined in the working memory circuit model how E/I balance can be restored through compensations acting on multiple parameters; for instance, elevated E/I ratio due to disinhibition can be compensated for by a treatment that strengthens inhibition or by one that attenuates excitation. In turn, restoration of E/I balance ameliorated the associated deficits in working memory behavior. However, further development and refinement of biophysically based models is needed to go beyond proof of principle. Future development in this area will benefit from the other directions noted above. Incorporation of more detailed microcircuitry and receptors will be needed to better capture pharmacological effects. Integration of biomarkers and behavior in the models will allow refinement through more direct testing with empirical data from pharmacological manipulations in animal models and humans. Future studies in this area hold exciting promise of contributing to the rational development of treatments in psychiatry, grounded in basic neuroscience.

## Acknowledgments

JDM was partly supported by the NIH grant TL1 TR000141, and X-JW was partly supported by the NIH grant R01 MH062349. JDM consults for BlackThorn Therapeutics.

## References

Amit, D.J., 1995. The Hebbian paradigm reintegrated: local reverberations as internal representations. Behav. Brain Sci. 18, 617−626.

Amit, D.J., Brunel, N., 1997. Model of global spontaneous activity and local structured activity during delay periods in the cerebral cortex. Cereb. Cortex 7, 237−252.

Anticevic, A., Gancsos, M., Murray, J.D., Repovs, G., Driesen, N.R., Ennis, D.J., Niciu, M.J., Morgan, P.T., Surti, T.S., Bloch, M.H., Ramani, R., Smith, M.A., Wang, X.-J., Krystal, J.H., Corlett, P.R., 2012. NMDA receptor function in large-scale anticorrelated neural systems with implications for cognition and schizophrenia. Proc. Natl. Acad. Sci. U.S.A. 109, 16720−16725.

Anticevic, A., Murray, J.D., Barch, D.M., 2015. Bridging levels of understanding in schizophrenia through computational modeling. Clin. Psychol. Sci. 3, 433−459.

Badcock, J.C., Badcock, D.R., Read, C., Jablensky, A., 2008. Examining encoding imprecision in spatial working memory in schizophrenia. Schizophr. Res. 100, 144−152.

Barch, D.M., Ceaser, A., 2012. Cognition in schizophrenia: core psychological and neural mechanisms. Trends Cogn. Sci. 16, 27−34.

Bastos, A.M., Usrey, W.M., Adams, R.A., Mangun, G.R., Fries, P., Friston, K.J., 2012. Canonical microcircuits for predictive coding. Neuron 76, 695−711.

Brunel, N., Wang, X.-J., 2001. Effects of neuromodulation in a cortical network model of object working memory dominated by recurrent inhibition. J. Comput. Neurosci. 11, 63−85.

Butler, P.D., Silverstein, S.M., Dakin, S.C., 2008. Visual perception and its impairment in schizophrenia. Biol. Psychiatry 64, 40−47.

Buzsáki, G., Wang, X.-J., 2012. Mechanisms of gamma oscillations. Annu. Rev. Neurosci. 35, 203−225.

Carter, E., Wang, X.-J., 2007. Cannabinoid-mediated disinhibition and working memory: dynamical interplay of multiple feedback mechanisms in a continuous attractor model of prefrontal cortex. Cereb. Cortex 17 (Suppl. 1), i16−i26.

Chaudhuri, R., Knoblauch, K., Gariel, M.A., Kennedy, H., Wang, X.-J., 2015. A large-scale circuit mechanism for hierarchical dynamical processing in the primate cortex. Neuron 88, 419−431.

Chen, Y., Bidwell, L.C., Holzman, P.S., 2005. Visual motion integration in schizophrenia patients, their first-degree relatives, and patients with bipolar disorder. Schizophr. Res. 74, 271–281.

Chen, Y., Levy, D.L., Sheremata, S., Holzman, P.S., 2004. Compromised late-stage motion processing in schizophrenia. Biol. Psychiatry 55, 834–841.

Chen, Y., Nakayama, K., Levy, D., Matthysse, S., Holzman, P., 2003. Processing of global, but not local, motion direction is deficient in schizophrenia. Schizophr. Res. 61, 215–227.

Compte, A., Brunel, N., Goldman-Rakic, P.S., Wang, X.-J., 2000. Synaptic mechanisms and network dynamics underlying spatial working memory in a cortical network model. Cereb. Cortex 10, 910–923.

Elvevåg, B., Goldberg, T., 2000. Cognitive impairment in schizophrenia is the core of the disorder. Crit. Rev. Neurobiol. 14, 1–21.

Funahashi, S., Bruce, C.J., Goldman-Rakic, P.S., 1989. Mnemonic coding of visual space in the monkey's dorsolateral prefrontal cortex. J. Neurophysiol. 61, 331–349.

Furman, M., Wang, X.-J., 2008. Similarity effect and optimal control of multiple-choice decision making. Neuron 60, 1153–1168.

Gold, J.I., Shadlen, M.N., 2007. The neural basis of decision making. Annu. Rev. Neurosci. 30, 535–574.

Gold, J.M., Hahn, B., Zhang, W.W., Robinson, B.M., Kappenman, E.S., Beck, V.M., Luck, S.J., 2010. Reduced capacity but spared precision and maintenance of working memory representations in schizophrenia. Arch. Gen. Psychiatry 67, 570–577.

Goldman-Rakic, P.S., 1995. Cellular basis of working memory. Neuron 14, 477–485.

Gonzalez-Burgos, G., Lewis, D.A., 2008. GABA neurons and the mechanisms of network oscillations: implications for understanding cortical dysfunction in schizophrenia. Schizophr. Bull. 34, 944–961.

Gonzalez-Burgos, G., Lewis, D.A., 2012. NMDA receptor hypofunction, parvalbumin-positive neurons, and cortical gamma oscillations in schizophrenia. Schizophr. Bull. 38, 950–957.

Greene, R., 2001. Circuit analysis of NMDAR hypofunction in the hippocampus, in vitro, and psychosis of schizophrenia. Hippocampus 11, 569–577.

Gutenkunst, R.N., Waterfall, J.J., Casey, F.P., Brown, K.S., Myers, C.R., Sethna, J.P., 2007. Universally sloppy parameter sensitivities in systems biology models. PLoS Comput. Biol. 3, 1871–1878.

Hensch, T., Fagiolini, M., 2004. Excitatory-Inhibitory Balance. Kluwer Academic/Plenum Publishers, New York.

Herwig, A., Beisert, M., Schneider, W.X., 2010. On the spatial interaction of visual working memory and attention: evidence for a global effect from memory-guided saccades. J. Vis. 10, 8.

Homayoun, H., Moghaddam, B., 2007. NMDA receptor hypofunction produces opposite effects on prefrontal cortex interneurons and pyramidal neurons. J. Neurosci. 27, 11496–11500.

Huk, A.C., Shadlen, M.N., 2005. Neural activity in macaque parietal cortex reflects temporal integration of visual motion signals during perceptual decision making. J. Neurosci. 25, 10420–10436.

Huys, Q.J.M., Maia, T.V., Frank, M.J., 2016. Computational psychiatry as a bridge from neuroscience to clinical applications. Nat. Neurosci. 19, 404–413.

Insel, T.R., Cuthbert, B.N., 2009. Endophenotypes: bridging genomic complexity and disorder heterogeneity. Biol. Psychiatry 66, 988–989.

Khlestova, E., Johnson, J.W., Krystal, J.H., Lisman, J., 2016. The role of GluN2C-containing NMDA receptors in ketamine's psychotogenic action and in schizophrenia models. J. Neurosci. 36, 11151–11157.

Kiani, R., Hanks, T.D., Shadlen, M.N., 2008. Bounded integration in parietal cortex underlies decisions even when viewing duration is dictated by the environment. J. Neurosci. 28, 3017–3029.

Koldewyn, K., Whitney, D., Rivera, S.M., 2010. The psychophysics of visual motion and global form processing in autism. Brain 133, 599–610.

Kotermanski, S.E., Johnson, J.W., 2009. $Mg^{2+}$ imparts NMDA receptor subtype selectivity to the Alzheimer's drug memantine. J. Neurosci. 29, 2774–2779.

Krystal, J.H., D'Souza, D.C., Mathalon, D., Perry, E., Belger, A., Hoffman, R., 2003. NMDA receptor antagonist effects, cortical glutamatergic function, and schizophrenia: toward a paradigm shift in medication development. Psychopharmacology (Berl.) 169, 215–233.

Lam, N.H., Borduqui, T., Hallak, J., Roque, A.C., Anticevic, A., Krystal, J.H., Wang, X.-J., Murray, J.D., 2017. Effects of altered excitation-inhibition balance on decision making in a cortical circuit model. bioRxiv. http://dx.doi.org/10.1101/100347.

Lee, D., 2013. Decision making: from neuroscience to psychiatry. Neuron 78, 233–248.

Lee, J., Park, S., 2005. Working memory impairments in schizophrenia: a meta-analysis. J. Abnorm. Psychol. 114, 599–611.

Mayer, J.S., Park, S., 2012. Working memory encoding and false memory in schizophrenia and bipolar disorder in a spatial delayed response task. J. Abnorm. Psychol. 121, 784–794.

Mejias, J.F., Murray, J.D., Kennedy, H., Wang, X.-J., 2016. Feedforward and feedback frequency-dependent interactions in a large-scale laminar network of the primate cortex. Sci. Adv. 2, e1601335.

Milne, E., Swettenham, J., Hansen, P., Campbell, R., Jeffries, H., Plaisted, K., 2002. High motion coherence thresholds in children with autism. J. Child Psychol. Psychiatry 43, 255–263.

Murray, J.D., Anticevic, A., Gancsos, M., Ichinose, M., Corlett, P.R., Krystal, J.H., Wang, X.-J., 2014. Linking microcircuit dysfunction to cognitive impairment: effects of disinhibition associated with schizophrenia in a cortical working memory model. Cereb. Cortex 24, 859–872.

Murray, J.D., Jaramillo, J.H., Wang, X.-J., 2017. Working Memory and Decision Making in a Fronto-Parietal Circuit Model.

Qi, Z., Miller, G.W., Voit, E.O., 2010. Computational modeling of synaptic neurotransmission as a tool for assessing dopamine hypotheses of schizophrenia. Pharmacopsychiatry 43 (Suppl. 1), S50–S60.

Rao, S.G., Williams, G.V., Goldman-Rakic, P.S., 2000. Destruction and creation of spatial tuning by disinhibition: GABA(A) blockade of prefrontal cortical neurons engaged by working memory. J. Neurosci. 20, 485–494.

Ratcliff, R., 1978. A theory of memory retrieval. Psychol. Rev. 85, 59–108.

Roitman, J.D., Shadlen, M.N., 2002. Response of neurons in the lateral intraparietal area during a combined visual discrimination reaction time task. J. Neurosci. 22, 9475–9489.

Rotaru, D.C., Yoshino, H., Lewis, D.A., Ermentrout, G.B., Gonzalez-Burgos, G., 2011. Glutamate receptor subtypes mediating synaptic activation of prefrontal cortex neurons: relevance for schizophrenia. J. Neurosci. 31, 142–156.

Schall, J.D., 2004. On building a bridge between brain and behavior. Annu. Rev. Psychol. 55, 23–50.

Spencer, K.M., 2009. The functional consequences of cortical circuit abnormalities on gamma oscillations in schizophrenia: insights from computational modeling. Front. Hum. Neurosci. 3, 33.

Starc, M., Murray, J.D., Santamauro, N., Savic, A., Diehl, C., Cho, Y.T., Srihari, V., Morgan, P.T., Krystal, J.H., Wang, X.-J., Repovs, G., Anticevic, A., 2017. Schizophrenia is associated with a pattern of spatial working memory deficits consistent with cortical disinhibition. Schizophr. Res. 181, 107–116.

Teller, D.Y., 1984. Linking propositions. Vis. Res. 24, 1233–1246.

Uhlhaas, P.J., 2013. Dysconnectivity, large-scale networks and neuronal dynamics in schizophrenia. Curr. Opin. Neurobiol. 23, 283–290.

Uhlhaas, P.J., Singer, W., 2010. Abnormal neural oscillations and synchrony in schizophrenia. Nat. Rev. Neurosci. 11, 100–113.

Vierling-Claassen, D., Siekmeier, P., Stufflebeam, S., Kopell, N., 2008. Modeling GABA alterations in schizophrenia: a link between impaired inhibition and altered gamma and beta range auditory entrainment. J. Neurophysiol. 99, 2656–2671.

Volman, V., Behrens, M.M., Sejnowski, T.J., 2011. Downregulation of parvalbumin at cortical GABA synapses reduces network gamma oscillatory activity. J. Neurosci. 31, 18137–18148.

Wang, M., Yang, Y., Wang, C.J., Gamo, N.J., Jin, L.E., Mazer, J.A., Morrison, J.H., Wang, X.-J., Arnsten, A.F.T., 2013. NMDA receptors subserve persistent neuronal firing during working memory in dorsolateral prefrontal cortex. Neuron 77, 736–749.

Wang, X.-J., 1999. Synaptic basis of cortical persistent activity: the importance of NMDA receptors to working memory. J. Neurosci. 19, 9587–9603.

Wang, X.-J., 2001. Synaptic reverberation underlying mnemonic persistent activity. Trends Neurosci. 24, 455–463.

Wang, X.-J., 2002. Probabilistic decision making by slow reverberation in cortical circuits. Neuron 36, 955–968.

Wang, X.-J., 2008. Decision making in recurrent neuronal circuits. Neuron 60, 215–234.

Wang, X.-J., 2010. Neurophysiological and computational principles of cortical rhythms in cognition. Physiol. Rev. 90, 1195–1268.

Wang, X.-J., 2013. The prefrontal cortex as a quintessential "cognitive-type" neural circuit: working memory and decision making. In: Stuss, D.T., Knight, R.T. (Eds.), Principles of Frontal Lobe Function, second ed. Oxford University Press, New York, pp. 226–248 (Chapter 15).

Wang, X.-J., Krystal, J.H., 2014. Computational psychiatry. Neuron 84, 638–654.

Wang, X.-J., Tegnér, J., Constantinidis, C., Goldman-Rakic, P.S., 2004. Division of labor among distinct subtypes of inhibitory neurons in a cortical microcircuit of working memory. Proc. Natl. Acad. Sci. U.S.A. 101, 1368–1373.

Wei, Z., Wang, X.-J., Wang, D.H., 2012. From distributed resources to limited slots in multiple-item working memory: a spiking network model with normalization. J. Neurosci. 32, 11228–11240.

Wong, K.F., Huk, A.C., Shadlen, M.N., Wang, X.-J., 2007. Neural circuit dynamics underlying accumulation of time-varying evidence during perceptual decision making. Front. Comput. Neurosci. 1, 6.

Wong, K.F., Wang, X.-J., 2006. A recurrent network mechanism of time integration in perceptual decisions. J. Neurosci. 26, 1314–1328.

Yang, G.J., Murray, J.D., Repovs, G., Cole, M.W., Savic, A., Glasser, M.F., Pittenger, C., Krystal, J.H., Wang, X.-J., Pearlson, G.D., Glahn, D.C., Anticevic, A., 2014. Altered global brain signal in schizophrenia. Proc. Natl. Acad. Sci. U.S.A. 111, 7438–7443.

Yang, G.J., Murray, J.D., Wang, X.-J., Glahn, D.C., Pearlson, G.D., Repovs, G., Krystal, J.H., Anticevic, A., 2016a. Functional hierarchy underlies preferential connectivity disturbances in schizophrenia. Proc. Natl. Acad. Sci. U.S.A. 113, E219–E228.

Yang, G.R., Murray, J.D., Wang, X.-J., 2016b. A dendritic disinhibitory circuit mechanism for pathway-specific gating. Nat. Commun. 7, 12815.

Zorumski, C.F., Izumi, Y., Mennerick, S., 2016. Ketamine: NMDA receptors and beyond. J. Neurosci. 36, 11158–11164.

# 2

# Serotonergic Modulation of Cognition in Prefrontal Cortical Circuits in Major Depression

*Juan P. Ramirez-Mahaluf[1], Albert Compte[2]*

[1] P. Universidad Católica de Chile, Santiago, Chile;
[2] Institut d'Investigacions Biomèdiques August Pi i Sunyer (IDIBAPS), Barcelona, Spain

Major depression disease (MDD) is a highly prevalent and debilitating psychiatric disease characterized primarily by emotional symptoms, such as persistent negative mood or anhedonia. These main symptoms often go together with vegetative dysfunctions and with cognitive deficits, such as impaired memory, attention, and executive function (Snyder, 2013; American Psychiatric Association, 2013). Such cognitive impairments are thought to reflect the altered functioning of circuits in the prefrontal cortex (PFC) as a result of the disease (Levin et al., 2007), and they may provide clues on the functional neuropathology of depression disorders (Austin et al., 2001) and on the interaction of emotional and cognitive processing in the brain (Pessoa, 2008; Ramirez-Mahaluf et al., 2017a).

Genetic studies of MDD have failed to specify a single robust neuropathology mechanism, pointing at the heterogeneity of the disorder on the basis of multiple biological mechanisms that converge on similar symptoms (Flint and Kendler, 2014). Several biological mechanisms have been associated with MDD (Manji et al., 2001; Belmaker and Agam, 2008). Monoaminergic neurotransmitter systems, and especially the serotonergic system, have received much attention based on the fact that they are targeted by most effective pharmacological treatments (Maes and Meltzer, 1995). A lowered metabolism of serotonin (5-hydroxytryptamine, 5-HT) in depression is thus thought to play an important role in the pathophysiology of MDD, through widespread serotonergic projections to limbic, striatal, and prefrontal cortical circuits. However, the idea that depression is simply the deficiency of central monoamines is now viewed as simplistic, and a more complex chain of pathophysiological events is considered to be causing the disease (Belmaker and Agam, 2008; Krishnan and Nestler, 2008).

It has been proposed that regional neuroanatomical alterations may underlie the disease (Drevets, 2001; Price and Drevets, 2012), possibly by virtue of stress-induced atrophy in vulnerable brain regions (Manji et al., 2001). Indeed, MDD is associated with changes in the functional structure and activity of some specific brain areas, including the ventral anterior cingulate cortex (vACC), the amygdala, the fronto-polar cortex, and the dorsolateral prefrontal cortex (dlPFC) (Mayberg, 1997). Prolonged stress is often a trigger for MDD (Kendler et al., 2001). Both chronic stress and MDD have been associated with neuronal and synaptic atrophy in the PFC and hippocampus as a result of excitotoxicity following excessive glutamate release and glial death, especially in vACC (Choudary et al., 2005; Arnsten, 2009; Rajkowska et al., 1999; Cotter et al., 2001; Ongür et al., 1998). Also, alterations of glutamate reuptake correlate with vACC hyperactivity during resting state and depressive symptoms (Walter et al., 2009; Horn et al., 2010). This evidence leads to a glutamatergic hypothesis for the pathophysiology of MDD, concerning a set of cortical areas, in particular the vACC (Sanacora et al., 2012). Conversely, limbic areas such as amygdala or nucleus accumbens undergo changes consistent with synaptic potentiation. Adaptive mechanisms for dealing with stress, including interactions among homeostatic mechanisms (Turrigiano, 2011), inflammatory signals (Miller et al., 2009), and raphe—raphe interactions (Jasinska et al., 2012), may become maladaptive when engaged chronically and eventually lead to a systemic failure in MDD. The circuitry involved in this system collapse is likely to involve the dorsal raphe nucleus (DRN), as its interactions with vACC determine differential responses to controllable and uncontrollable stressors (Maier, 2015). In addition, recent evidence associates the short allele of the serotonin transporter gene with greater susceptibility to MDD following a stressor. This allele causes faster serotonin reuptake from the synapse, thus reducing serotonergic impact in the system, and it has been argued that serotonin-mediated inhibitions within the DRN may be at the core of its association with MDD (Jasinska et al., 2012). Finally, the PFC and the midbrain raphe nuclei are reciprocally connected and exert a mutual control that has been implicated in the response to behavioral challenges, and enhanced serotonergic activity in this circuit has been associated to stress resilience (Franklin et al., 2012). This evidence suggests that the interplay between serotonergic and glutamatergic mechanisms underlies the various aspects of MDD heterogeneous symptomatology, and it raises the question about what are the specific primary causes of the disease.

These different biological mechanisms may concurrently operate in patients, or they may characterize different subcategories of the disease within this disorder, but the specific way by which they induce cognitive processing deficits as observed clinically is currently unknown. A consistent mechanistic framework spanning from neurochemical to network activity and behavioral function is required to formulate informed hypotheses that advance toward this understanding.

Recently, efforts are being applied to formulating mechanistically explicit models to converge toward computational theories of network dysfunction in psychiatric diseases (Huys et al., 2016; Wang and Krystal, 2014; Siekmeier, 2015). We have recently proposed a computational network model to understand the dynamics of the dlPFC–vACC network during the processing of emotional and cognitive information, its alterations in MDD and its response to treatments (Ramirez-Mahaluf et al., 2017b). Here, we extend this model to include an additional area that supports spatial working memory function (posterior parietal cortex, PPC) to address the role of serotonergic mechanisms in modulating cognitive dysfunction in MDD.

## 2.1 METHODS

We built a computational model of three interconnected cortical areas: vACC, dlPFC, and PPC, each composed of recurrently coupled excitatory and inhibitory spiking neurons (Fig. 2.1A). We accomplished this by adding a vACC network to the model of areas dlPFC and PPC in Edin et al. (2009). This new area vACC was identical to the previously existent dlPFC network, and they were mutually coupled through symmetric excitatory projections to target inhibitory neurons (Ramirez-Mahaluf et al., 2017b). Apart from this network addition, the simulation was identical to the one

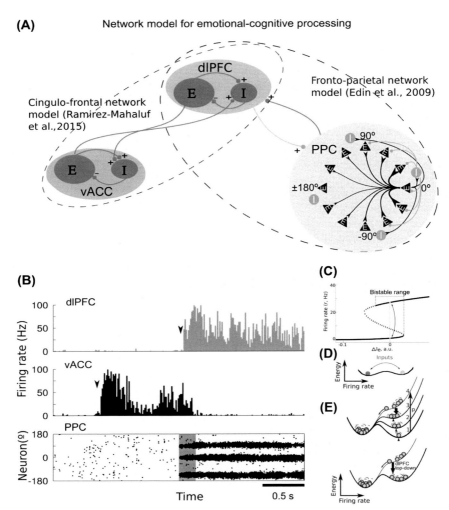

**FIGURE 2.1** Effective mutual inhibition between hubs in emotional and cognitive brain networks can explain the behavioral flexibility for alternating rapidly between emotional and cognitive states. (A) Schematic diagram of the computational network model. Recurrently coupled excitatory (E) and inhibitory (I) neurons comprise each of the three cortical areas simulated (*dlPFC*, dorsolateral prefrontal cortex; *PPC*, posterior parietal cortex; *vACC*, ventral anterior cingulate cortex). The PPC network is organized as a ring model, with neurons selective to specific visual locations and coupled according to the difference in selectivity of pre- and postsynaptic neurons (width of schematic connections). The network model is the joint simulation of two computational models previously published (see Ramirez-Mahaluf et al., 2017b; Edin et al., 2009). For full details see Section 2.1. (B) Network simulation of concatenated emotional and cognitive tasks (see Section 2.1). An initial neutral period (500 ms) is followed by brief inputs to vACC, which signal an emotional situation (downward arrow in vACC panel). This triggers vACC activation, which is stopped via disynaptic inhibition from dlPFC when a cognitive task engages the two

described in Edin et al. (2009) except for the parameter changes indicated below. In brief, neurons were simulated with the integrate-and-fire formalism, whereby a voltage variable describes the excitability of model neurons and integrates synaptic currents resulting from incoming action potentials from presynaptic neurons. Whenever the voltage reaches a predefined threshold, neurons fire an action potential that travels to all other neurons in the network (all-to-all connectivity). Spikes trigger membrane voltage deflections in postsynaptic neurons given by differential equations specific to AMPA ($\alpha$-amino-3-hydroxy-5-methyl-4-isoxazolepropionic acid), NMDA (N-methyl-D-aspartic acid), and GABA$_A$ (gamma-Aminobutyric acid) receptor mediated synaptic transmission, depending on whether presynaptic neurons were excitatory (AMPA and NMDA synapses) or inhibitory (GABA$_A$ synapses). The specific equations and parameters for membrane and synaptic dynamics can be retrieved from Edin et al. (2009). Different areas in our model were characterized by their number of neurons ($N_E = 64$, $N_I = 16$ for vACC and dlPFC; $N_E = 512$, $N_I = 128$ for PPC) and by the patterns of connection strengths: while all connections between matched types of neurons (excitatory or inhibitory) were the same in vACC and dlPFC (Connection strength G, in ns, E $\rightarrow$ E AMPA 0.03, E $\rightarrow$ E NMDA 0.983, E $\rightarrow$ I AMPA 0.01, E $\rightarrow$ I NMDA 0.74, I $\rightarrow$ E 0.937, I $\rightarrow$ I 0.725, X $\rightarrow$ E 3.3, X $\rightarrow$ I 2.5, notations as in Edin et al., 2009), the connections between neurons in the PPC network depended on the distance between spatial location preference of the presynaptic and postsynaptic neuron according to a circular Gaussian with parameters: Connection curve width $\sigma$, in degrees, E $\rightarrow$ E 9.4, E $\rightarrow$ I 32.4, I $\rightarrow$ E 32.4;

◀

cognitive areas (dlPFC and PPC). Top-down projections from dlPFC onto PPC increase the capacity of the PPC storage network as described in Edin et al. (2009), so that the three items presented initially (*shaded area* in PPC panel) can be maintained autonomously by the dlPFC–PPC network system. In the absence of dlPFC top-down input, the PPC network can only maintain one item on its own (not shown). (C) Reduced graphical representation to interpret the conditions for bistable function in the three areas of our network. The combination of cellular and network mechanisms in each area achieves a network bistable regime as depicted by the set of network firing rates achieved at each specific cellular excitability level $I_e$. The range of $I_e$ values for which there are two possible stable network states (*solid line*) is called the bistable range. Network states that are unstable are marked with *dashed lines*. The specific level of excitability $I_e$ at which the network operates is the operating regime (*vertical gray line*). (D) Equivalent representation of the dynamics in the operating regime depicted in panel (C) using an energy landscape. Two wells in the landscape indicate the two stable states available, a low-rate (left well) and a high-rate state (right well). Brief external inputs impose transitions between the two states, which are then persistent in the circuit. (E) The working memory capacity boosting mechanism illustrated with the energy landscape representation (Edin et al., 2009). As more and more items are loaded onto the high-rate state (balls on the right well), the energy landscape deforms and the high-rate well progressively loses its stability (here at load $p = 4$, top scheme). A nonspecific top-down input from dlPFC in its turn promotes the stability of the high-rate state and allows the storage of additional items (bottom scheme).

Connection curve height $J^+$, unitless, E → E 6.5, E → I 1.4, I → E 1.4; Connection strength G, in ns, E → E AMPA 0.0, E → E NMDA 0.77, E → I AMPA 0.0, E → I NMDA 0.479, I → E 3.9, I → I 2.896, X → E 7.2, X → I 5.8; notations as in Edin et al. (2009). The three networks were then connected in the following way: excitatory neurons in the vACC projected to inhibitory neurons in the dlPFC and excitatory neurons in the dlPFC projected to inhibitory neurons in the vACC through AMPA-mediated synapses of strength 0.0056 ns; and excitatory neurons in the dlPFC network projected to excitatory neurons in the PPC circuit uniformly with strength 0.01 ns. To illustrate the operation of our network of areas, we simulated three concatenated tasks that engaged our networks in specific ways. First, the network was simulated in a neutral condition, with neither cognitive nor emotional content (Fig. 2.1B, first 500 ms). In a second task, the network received excitatory inputs to the vACC network (Fig. 2.1B, first arrowhead, 500–800 ms), simulating an emotionally intense task, such as sadness provocation (Ramirez-Mahaluf et al., 2017a). Finally, we injected the same excitatory inputs to the dlPFC network (Fig. 2.1B, second arrowhead, 1500–1800 ms), while we simultaneously activated specifically three subgroups of excitatory neurons in the PPC network (Fig. 2.1B, gray area in bottom panel). This last task represents cognitive engagement in a demanding working memory condition. The whole simulated time was 3000 ms.

We simulated the glutamatergic hypothesis for MDD as an increase in the time constant of NMDA and AMPA synaptic transmission within the vACC network (5% increase in Fig. 2.2A, 10% increase in Fig. 2.2B). We used a custom C++ code, with integration time 0.02 ms and a simulation of 3 s took ~10 min of computation time.

## 2.2 RESULTS

Our model assumes that dlPFC and vACC are two hub nodes of larger cognitive and emotional networks of areas, respectively (as in Ramirez-Mahaluf et al., 2017a). Here, we model explicitly one area within the cognitive network being controlled by dlPFC, the PPC network. The vACC and dlPFC exert effective inhibitory influences on each other based on direct excitatory projections to inhibitory neurons in the target area (Medalla and Barbas, 2010), or via some interposed area (Fig. 2.1A). This organization ensures complementary activation of these areas in tasks with a range of emotional and cognitive loads (Bush et al., 2000), and it implements a competition between emotional and cognitive networks to determine the system's response to stimuli, so that cognitive processes are diminished when the system is engaged in strongly emotional processing (Ikeda et al., 1996; Shackman et al., 2006) and, conversely, strongly focused

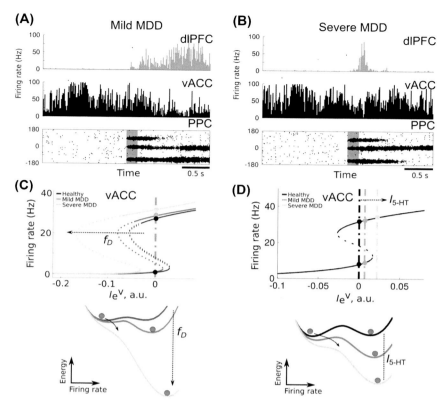

**FIGURE 2.2** Hyperactivity of ventral anterior cingulate cortex (vACC), either through glutamatergic or serotonergic mechanisms, causes network dynamics consistent with cognitive and emotional symptoms of major depression disease (MDD). (A) Slowdown of glutamate reuptake (5% increase in the time constant of all glutamate-mediated synaptic processes in vACC) induced hyperactivity in vACC that limited the response of dorsolateral prefrontal cortex (dlPFC) to cognitive inputs, and as a result reduced the amount of top-down input onto posterior parietal cortex (PPC), which could only store two out of the three items initially presented (gray area). (B) Further slowdown of glutamate processes in vACC (10% increase in time constant) increased vACC hyperactivity and precluded dlPFC activation, eliminating all top-down boosting signals onto PPC. Working memory capacity in these conditions reduced further marking additional working memory impairment. (C) Representation of network dynamics in vACC as a result of progressive glutamate slowdown (parameter $f_D$, see Ramirez-Mahaluf et al., 2017b), mimicking the progressive stages of the disease (healthy to mild and severe MDD in shades of gray). As $f_D$ increases, the network in its operating regime increases firing rates in the active state and eventually loses the stability of the low-rate state, resulting in permanent aberrant activity in vACC. This is depicted both in the $r$-$I_e$ plane (top) and using an energy landscape representation (bottom). Notice the narrowing of the bistable range as $f_D$ increases (top). (D) Graphical representation of the progressive depolarization of vACC neurons, as a result of less serotonergic tone in vACC and reduced 5-HT$_{1A}$ receptor activity. This alteration of vACC has the same functional impact as shown above (not shown, similar simulations as panels A and B). The progression of the disease is simulated here with a progressive displacement of the operating regime ($I_{5\text{-HT}}$, vertical lines in shades of gray, top), which also modifies the energy landscape by increasing the high-rate stability and losing the stability of the low-rate state (bottom). Notice the lack of bistable range narrowing in this mechanistic hypothesis.

cognitive processing reduces the impact of emotional aspects on the system's behavior. In our computational simulations, this competitive dynamics was reflected in the ability of one of the two hub areas, the dlPFC or the vACC, to disrupt and prevent further sustained activation of the other area when it was itself activated by specific external stimuli signaling the prevalence of either cognitive or emotional processing, respectively (Fig. 2.1B). Although sustained nonselective activation of neurons in each of these circuits can only be seen as the most elementary possible computation, and hardly able to achieve the complexity of human cognitive and emotional processing, such tonic activation in their hub function (Cole et al., 2012, 2013) may have important consequences as a source of top-down control over other areas performing more specific, selective computations, such as working memory. Specifically, it has been shown that such tonic activation can be the source of the dynamical control of working memory capacity, by virtue of its top-down effects on storage circuits in the PPC (Edin et al., 2009). We have integrated this effect in our model through the unspecific top-down connection of dlPFC excitatory neurons onto neurons in the PPC network (see Section 2.1). Without tonic dlPFC activity, the PPC network would sustain just a limited number of items in neural activity bumps (Edin et al., 2009), however, the conjoint activation of dlPFC in cognitive demanding situations generates a nonselective top-down input to the PPC network and it increases its capacity limit to store a few more items and thus improve cognitive processing (Fig. 2.1B). This modulatory role over specific computational processes (here illustrated by multiitem working memory) taking place in distributed networks (here demonstrated with the PPC network) is what our model proposes for these two competing hub areas, the dlPFC and the vACC.

The operation of all these circuits can be understood conceptually with the aid of low-dimensional graphic representations. On the one hand, the set of stable network states (characterized by a mean rate, $r$) available to each one of the areas at a given level of excitability $I_e$ can be represented with a solid-line curve in the $r$-$I_e$ plane (Fig. 2.1C). For the network regime of interest, this curve presents an S-shape folding that defines a *bistable range* as the range of $I_e$ values for which the system can reside stably in two very different network states, a low-rate state and a high-rate state. This regime is important, because it allows the network to maintain a memory of brief inputs that impose transitions between each of these stable states. This computational principle is what underlies both the switching behavior of dlPFC and vACC, and the sustained bump activations in the PPC network in Fig. 2.1B. Another way to represent this dynamics graphically is by making use of an energy landscape representation, whereby the system can stabilize at one of the two local minima of the energy curve that describes the network (Fig. 2.1D), and that correspond to the two stable firing rate regimes (low- and high-rate solutions). If these two energy wells are approximately as deep, whether the system resides in one of these two states is largely determined by the history of transient external inputs (Fig. 2.1D). For networks that can store a variety of items (in our case, the PPC network that stores stimulus locations, Fig. 2.1A) this energy representation allows to understand conceptually the storage capacity limitations in the network, and the role of top-down inputs. Indeed, the loading of each new item in memory has an impact in the energy landscape, so that the high-rate state becomes progressively less stable as more and more items are loaded in memory to the point that a limit is reached where only by allowing one memory to die away can the other memories remain active (Fig. 2.1E, top). This sets the capacity limit in the network, and it is primarily a function of the strength of long-range inhibition in the network (Edin et al., 2009). Sustained, unspecific top-down inputs on this circuit, however, can increase the intrinsic capacity limit by enhancing the stability of high-rate states even in high memory load conditions (Fig. 2.1E, bottom). This mechanism defines working memory capacity as the combined action of two different areas, one involved in storage per se, and another one in capacity boosting for dealing with cognitively demanding tasks (Edin et al., 2009).

When these mechanisms are all in place and well balanced, the system operates as expected, by allowing the flexible transitions between emotional and cognitive processing modes that allow healthy social interactions and effective cognitive functioning. However, alterations that disrupt these balances will have profound impact in both emotional and cognitive processing. In the context of MDD, we will expose here the impact of two alterations related to this disease (Belmaker and Agam, 2008). On the one hand, a system-wide reduction in 5-HT metabolism may affect differentially emotional and cognitive networks based on the differential expression levels of 5-HT receptors (Palomero-Gallagher et al., 2009; Santana et al., 2004), causing an imbalance that favors emotional over cognitive processing. On the other hand, a regionally specific deficit of glutamate reuptake in the vACC, as described in the literature (Choudary et al., 2005; Walter et al., 2009; Horn et al., 2010; Portella et al., 2011), would also cause a network-wide imbalance with widespread consequences even if originally caused by a very focal dysfunction in the network (Ramirez-Mahaluf et al., 2017a). In the conceptual framework defined in Fig. 2.1C, changes in cellular excitability ($I_e$) and/or in synaptic network interactions (that affect the S-shaped curve) will move the networks outside of the bistable range, and have important computational impact in their function.

Several lines of evidence have identified glutamatergic hyperactivity in vACC in depression. On the one hand, the death of glial cell in vACC (Rajkowska et al., 1999; Cotter et al., 2001; Ongür et al., 1998) reduces the concentration of glutamine synthetase and glial high-affinity glutamate transporters (Choudary et al., 2005) leading to increased glutamatergic transmission in MDD patients. The recurrence and chronicity of MDD was also correlated with progressive alteration of glutamate metabolism in vACC (Portella et al., 2011), glutamate alterations are correlated with vACC aberrant activity during resting state and depressive symptoms (Walter et al., 2009; Horn et al., 2010). An alternative hypothesis suggests alterations in GABAergic function (Price and Drevets, 2012; Northoff and Sibille, 2014), deficits in inhibitory transmission that could lead to an E/I imbalance similar to that generated by glutamatergic hyperactivity.

In the computational model, when excess glutamatergic transmission is simulated in synapses of the vACC, this hub area becomes incessantly active and cognitive commands from the dlPFC are unable to suppress its activity, which in turn prevents tonic activations in dlPFC (Fig. 2.2A and B). The absence of tonic activation in dlPFC does not prevent working memory function altogether because other working memory circuits in frontal and parietal cortex can still operate; but because these other circuits now do not get their capacity boosted by control signals from dlPFC (Edin et al., 2009), there is an overall reduction of cognitive performance (Fig. 2.2B). Notice that this cognitive effect is a result of a focal dysfunction solely in the emotional circuit of our network (vACC), showing how regionally limited alterations can have multimodal, widespread behavioral consequences by virtue of long-range connections between specialized but interdependent cortical areas. The specific alteration of function that affects the vACC circuit in our network can be conceptually understood with the aid of our graphical representations (Fig. 2.2C). Thus, the progressive slowdown of glutamate reuptake (increasing $f_D$, see Ramirez-Mahaluf et al., 2017b) causes an excess of glutamate availability at the synapse that changes the S-shaped $r$-$I_e$ curve in vACC so that it moves away from the level of cellular excitability in the network ($\Delta I_e^v = 0$) and the network loses its bistability, being only able to reside in the high-rate regime of activity (Fig. 2.2C, top). In addition, the bistable range becomes narrower as $f_D$ increases (Fig. 2.2C), due to an oscillatory instability in the high-rate network state (dotted lines, for details see Ramirez-Mahaluf et al., 2017b). Based on the energy representation, the high-rate state becomes progressively more and more stable (deeper energy well) to the point that the low-rate state loses its quality of local minimum and cannot remain stable (Fig. 2.2C, bottom). Given this persistent activation in vACC, the dlPFC network is now constantly inhibited by virtue of effectively inhibitory projections from vACC, and this sets its own $I_e$ also outside the bistable range, leaving only the low-rate state stable. Finally, this also has implications for the PPC network in that the dlPFC top-down input is now absent, and the capacity boosting of Fig. 2.1E (bottom) is now abolished. This chain of interdependencies propagates the focal dysfunction of the vACC network to fine cognitive processes dependent on other cortical circuits, like PPC.

However, our mechanistic model will also show unbalanced behavioral biases based on a diffuse, general reduction of 5-HT metabolism, considering the regional specificities of 5-HT receptors. The two primary 5-HT receptors in the neocortex are the $5\text{-HT}_{1A}$ and the $5\text{-HT}_{2A}$ receptors (Artigas, 2013), and they exert opposing influences on postsynaptic neurons, while $5\text{-HT}_{1A}$ receptors result in postsynaptic inhibition, $5\text{-HT}_{2A}$ receptors excite postsynaptic neurons. Interestingly, vACC has a higher incidence of $5\text{-HT}_{1A}$ receptors relative to other prefrontal regions (Palomero-Gallagher et al., 2009; Santana et al., 2004), suggesting that a general reduction of serotonergic activity will have a comparatively higher excitatory effect in vACC. Such increase in vACC excitability in our model (Fig. 2.2D) will generate network dynamics, and behavioral effects, analogous to the ones described above for glutamatergic hyperactivity in vACC (Fig. 2.2A and B). Specifically, reduced 5-HT activity would reduce $5\text{-HT}_{1A}$ inhibition in vACC and increase its tonic activity, thus suppressing activation in dlPFC, which would then be unable to enhance the capacity of working memory storage circuits in PPC. Note, however, that despite the general similarity in functional effects of these two mechanisms (5-HT hypofunction and glutamatergic hyperactivity in vACC), the underlying network dynamics are different. This is best illustrated using the graphical representation. Indeed, the S-shaped $r$-$I_e$ curve in vACC is altered following glutamate transmission increase, with a narrowing of its bistable range (Fig. 2.2C), while 5-HT hypofunction does not affect the bistable range of the S-shaped $r$-$I_e$ curve and only changes the general level of excitability of the network (Fig. 2.2D). This difference will be relevant to interpret treatment response (see below).

In parallel, however, a general alteration of 5-HT metabolism would now also impact the functioning of the working memory storage circuits in the PPC network. Indeed, it has been shown that a computational network model maintaining information in persistent neural activity is sensitive to serotonergic neuromodulation through $5\text{-HT}_{1A}$ and $5\text{-HT}_{2A}$ receptors (Cano-Colino et al., 2013, 2014). In particular, a well-balanced concentration of 5-HT is necessary for maintaining the information correctly in the PPC network (Fig. 2.3A). This is because both a reduction and an increase in 5-HT concentration induce spatial working memory errors in the model, primarily through the action of $5\text{-HT}_{1A}$ receptors (Cano-Colino et al., 2014). The specific nature of the working memory errors committed, however, depends on the specific alteration in 5-HT concentration. Thus, increased 5-HT concentration leads to an increase of low-confidence behavioral errors, due to the inability of the network to sustain activity bumps through the working memory delay (Fig. 2.3B). In

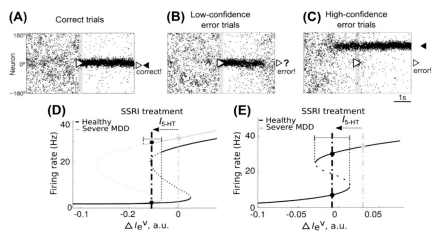

FIGURE 2.3    Serotonergic neuromodulation of network dynamics. (A–C) Alterations of serotonin concentration in the posterior parietal cortex network also alter working memory processes directly (Cano-Colino et al., 2014). In the normally operating circuit (panel A), a stimulus briefly presented at a given location (*empty right-pointing triangle*) is maintained via sustained network activity through the delay period and the location decoded at the end of the delay (*filled right-pointing triangle*) matches the correct stimulus location, resulting in a successful working memory trial. When activity of serotonergic receptors is increased in the network, stimulus-related delay activity often fails to bridge the whole delay period and it fades away before being decoded (panel B). This results in behavioral errors of low-confidence, given the lack of neural information on the stimulus location at the end of the delay. Reduced serotonergic receptor activity, instead, results in high-confidence errors (panel C), where a strong signal is decoded at the end of the delay, but it falsely recapitulates the stimulus location due to spontaneous network activation during the baseline period. (D) Serotonergic treatment of major depression disease (MDD) networks under the glutamatergic hypothesis of the disease (Fig. 2.2C) can recover ventral anterior cingulate cortex (vACC) bistable function, but within a narrower bistable range. Treatment is modeled as hyperpolarization of vACC excitatory neurons ($I_{5\text{-HT}}$), displacing the operating regime from the *thick vertical gray line to the thick vertical black line*. (E) For MDD networks under the serotonergic hypothesis (Fig. 2.2D), serotonergic treatment ($I_{5\text{-HT}}$) reinstates vACC bistable function within the bistable range of healthy operation. SSRI, selective serotonin reuptake inhibitors. *(A–C) Adapted from Cano-Colino, M., Almeida, R., Gomez-Cabrero, D., Artigas, F., Compte, A., 2014. Serotonin regulates performance nonmonotonically in a spatial working memory network. Cereb. Cortex 24 (9), 2449–2463. http://dx.doi.org/10.1093/cercor/bht096, by permission of Oxford University Press.*

contrast, reductions in 5-HT lead to an increase of high-confidence working memory errors (or "false memories" see Mayer and Park, 2012; Lee et al., 2008), which reflect the emergence of spontaneous bumps of activity in the network that interfere with the task (Fig. 2.3C). As a result, under the hypothesis of low 5-HT metabolism for MDD, our computational model predicts a primary effect in the increase of high-confidence errors in trials with low memory load. Notice that the possible reduction in working memory capacity in high-load memory trials due to the hyperexcitability of vACC described above might be compensated by the increase in excitability of the working memory storage network, depending on the relative incidence of these two effects.

Serotonergic modulations of cognitive function in MDD may also occur through the pharmacological treatments that patients undergo. Indeed, current antidepressant medications typically target the serotonergic system, and may have per se cognitive implications that our network model can help to identify and conceptualize. We have previously shown that we can simulate the action of selective serotonin reuptake inhibitors (SSRIs) in our network model as a specific hyperpolarization of vACC neurons (based on the higher incidence of 5-HT$_{1A}$ receptors in pyramidal neurons of vACC, Palomero-Gallagher et al., 2009) and partially restore the unbalanced dynamics of the dlPFC−vACC system (Fig. 2.3D and E). Nevertheless, depending on the primary dysfunction of the MDD model, either serotonergic or glutamatergic, the predicted outcome could be different.

In the glutamatergic MDD model the bistable range becomes narrower with the progressive slowing down of glutamate decay (Figs. 2.2C and 2.3D). The narrower bistable range has implications for serotonergic treatment outcome in the model. In the mild MDD model, the simulated SSRI restores the operating regime within the bistable range, recovering network function close to normal (Ramirez-Mahaluf et al., 2017b). In contrast, the severe glutamatergic MDD model presents a markedly narrower bistable range (Figs. 2.2C and 2.3D), so that even a dose-optimized serotonergic treatment that sets the network excitability within the bistable range (Fig. 2.3D) is unable to fully stabilize the inactive vACC network state, and random fluctuations restart aberrant vACC activations in the absence of emotional trigger stimuli (for details see Ramirez-Mahaluf et al., 2017b). Such imperfect stabilization of vACC inactive state inhibits the proper activation of dlPFC during working memory tasks, and this will have an impact on the working memory capacity of the PPC network. Thus, in the glutamatergic MDD model the progression of MDD is linked to a worsening in the outcome of SSRI treatment, both in relation to emotional and cognitive symptoms.

In contrast, in the serotonergic MDD model the bistable range is not affected by the reduction of the 5-HT levels (Fig. 2.2D). As a result, the simulated SSRI at a sufficient dose can restore the operating regime within the bistable range independently of the severity of the MDD model, thus recovering the normal function of the network (Fig. 2.3E), both in terms of emotional and cognitive symptoms. This difference in the prognosis of SSRI treatment in severe MDD patients may thus be a behavioral marker of different primary causes of the disease.

The difference in the response to SSRI treatment follows from the fact that in the glutamatergic MDD model, SSRI treatment acts by compensating the E/I balance with a mechanism different from the primary disturbance. In the serotonergic MDD model, the SSRI treatment recovers normal function of the network because it targets the primary dysfunction specifically. This difference could explain why a percentage of patients respond to treatment, while another group of patients present multiple recurrences of depressive episodes in spite of serotonergic treatment (Kendler et al., 2001).

## 2.3 DISCUSSION

In this chapter we integrated the findings of complementary computational models (Ramirez-Mahaluf et al., 2017b; Edin et al., 2009; Cano-Colino et al., 2013, 2014) to review pathophysiological hypotheses of MDD in relation to serotonergic neuromodulation of cognitive processes. We focused on cognitive networks in prefrontal and parietal cortex, and we analyzed mechanistically the impact of serotonergic and glutamatergic alterations on a simulated long-range network that includes the anterior cingulate cortex (Fig. 2.1A).

We provide a biophysical computational model for three areas: vACC, dlPFC, and PPC. In healthy function, the model accomplishes two complementary functions: for one, cingulo-frontal networks alternate between cognitive and emotional processing according to task demands. Secondly, top-down inputs from dlPFC onto PPC increase working memory capacity during demanding cognitive tasks. The healthy coordination of these two functions is impaired when either of the participating circuits is affected. This occurs when simulating MDD alterations, either glutamatergic hyperactivity in vACC or serotonergic hypoactivity. For the glutamatergic hypothesis, slower glutamate reuptake in vACC affects the capacity to switch from emotional to cognitive processing in the vACC−dlPFC loop, and thus it compromises the working memory capacity boosting function of dlPFC projections onto PPC, thus resulting in impaired cognitive function (specifically lower working memory capacity). This model provides a mechanistic framework to understand how slower glutamate reuptake in vACC (Choudary et al., 2005; Cotter et al., 2001; Ongür et al., 1998) disrupts the E/I balance (Walter et al., 2009; Horn et al., 2010), causes hyperactivity in vACC (Seminowicz et al., 2004; Mayberg et al., 1999; Mayberg et al., 2005), and ends up generating sadness and anhedonia through the inability to disengage from emotional processing during cognitive tasks (Gohier et al., 2009; Rose and Ebmeier, 2006; Watts and Sharrock, 1985; Paelecke-Habermann et al., 2005; Disner et al., 2011). Such hyperactivity of vACC generates dorsal frontal hypoactivity and impairs working memory capacity (Mayberg, 1997; Oda et al., 2003; Bench et al., 1992; Videbech et al., 2002; Kennedy et al., 2001).

For the serotonergic hypothesis the model generates the same functional dysfunctions as described above for the glutamatergic hypothesis, only that the primary mechanism now is the reduced activation of 5-$HT_{1A}$ receptors in cortical areas enriched with this receptor, such as vACC. Thus, a decrease in serotonergic tone would generate vACC hyperactivity (Seminowicz et al., 2004; Mayberg et al., 1999; Mayberg et al., 2005), and the effective inhibition between vACC and dlPFC would cause dlPFC hypoactivity, limited capacity boosting in PPC, and cognitive dysfunctions (Mayberg, 1997; Oda et al., 2003; Bench et al., 1992; Videbech et al., 2002; Kennedy et al., 2001). However, the reduction in serotonergic tone is systemic and not only generates hyperactivity in the vACC through 5-$HT_{1A}$ receptors, but it also affects directly the working memory storage circuit, leading to an increase in high-confidence behavioral errors in a spatial working memory task (Cano-Colino et al., 2014). From this result, our model formulates a specific feature that distinguishes glutamatergic from serotonergic mechanistic bases of MDD. In MDD patients where the pathophysiology is a decrease in 5-HT release, a spatial delayed-response working memory task would result in increased high-confidence errors (Cano-Colino et al., 2014). No specific change in the proportion of high-confidence errors is expected in MDD patients with a primary pathophysiology based on slow glutamate reuptake in vACC.

An additional distinction between these two pathophysiology hypotheses for MDD emerges from our models in relation to the treatment outcome of SSRIs. This is related to the differential impact of the glutamatergic and serotonergic dysfunctions on the bistable range of the vACC. Following the glutamatergic vACC dysfunction, the bistable range in the model becomes narrower and this decreases the treatment outcome even if the SSRI dose is optimized (Fig. 2.3D). This could be related to experimental evidence relating glutamate dysfunction with the progression of MDD (Portella et al., 2011) and weaker response to SSRI treatment as the disease progresses (Kendler et al., 2000, 2001; Keller, 1992). This effect, however, is absent if we simulate MDD in our model according to reduced 5-HT levels, in which case SSRI treatment always reverses the behavioral symptoms, provided the optimal treatment dose can be achieved (Fig. 2.3E). In cases in which pharmacologically acceptable doses of SSRIs are unable to restore network function, other treatments such as deep brain stimulation may be capable of achieving larger levels of vACC hyperpolarization and thus recover the bistable range and reduce MDD symptoms (Ramirez-Mahaluf et al., 2017b).

A different perspective may be taken on this confrontation of two alternative pathophysiology mechanisms for MDD. Indeed, MDD is a heterogeneous illness and there have been attempts to classify it in different subtypes, for instance melancholic and atypical depression, based on the patient's symptomatology, severity, course, and response to treatment (Harald and Gordon, 2012). These different subtypes could be related to a different pathophysiology in these patients. In particular, it is suggestive that the clinical course of atypical depression is characterized by a tendency to chronicity (Stewart et al., 2009; McGinn et al., 1996; Posternak and Zimmerman, 2002), which our modeling approach would relate to the reduction in the bistable range and subsequently poor treatment outcome and higher recurrence. This would link atypical depression with a glutamatergic origin of the disease according to our model, and this resonates with other clinical features of this depression subtype (Stewart et al., 2009; McGinn et al., 1996; Posternak and Zimmerman, 2002), in particular its association with bipolar disorder (Mitchell and Malhi, 2004; Benazzi, 2002). Mood stabilizers typically used to treat bipolar depression target glutamatergic and GABAergic transmission (Krystal et al., 2002), which are effective to decrease the recurrence of depressive episodes. In our model, the normalization of glutamate transmission in the vACC network will lead to the stabilization of the active network states in vACC, an increase of the bistable range, and a better treatment outcome. This suggests that the glutamatergic dysfunction simulated in our model may underlie atypical depression symptomatology. Instead, melancholic depression has been related to a decrease in plasma serotonin levels (Sarrias et al., 1987), suggesting a possible relation with our serotonergic network model of MDD, although our model's predictions of a better treatment outcome in this scenario is not specifically supported clinically (Uher et al., 2011). The difficulty in achieving a consistent classification of depression subtypes that account for its heterogeneous course and treatment outcome (Harald and Gordon, 2012; Uher et al., 2011; Hadzi-Pavlovic and Boyce, 2012; Lamers et al., 2016) may be partly addressed by considering the specific implications of the different primary mechanisms that could underlie depression symptoms, as predicted by computational network models of this disease.

## Acknowledgments

This work was funded by the Spanish Ministry of Economy and Competitiveness and the European Regional Development Fund (Ref: BFU2012-34838); by AGAUR of the Generalitat de Catalunya (Ref: SGR14-1265); and by CONICYT PIA ACT 1414. The work was carried out at the Esther Koplowitz Centre, Barcelona, and Psychiatry department of P. Universidad Católica de Chile, Santiago.

# References

American Psychiatric Association, 2013. DSM 5. American Psychiatric Association.

Arnsten, A.F.T., 2009. Stress signalling pathways that impair prefrontal cortex structure and function. Nat. Rev. Neurosci. 10 (6), 410–422. http://dx.doi.org/10.1038/nrn2648.

Artigas, F., 2013. Serotonin receptors involved in antidepressant effects. Pharmacol. Ther. 137 (1), 119–131. http://dx.doi.org/10.1016/j.pharmthera.2012.09.006.

Austin, M.P., Mitchell, P., Goodwin, G.M., 2001. Cognitive deficits in depression: possible implications for functional neuropathology. Br. J. Psychiatry 178, 200–206. http://dx.doi.org/10.1192/bjp.178.3.200.

Belmaker, R.H., Agam, G., 2008. Major depressive disorder. N. Engl. J. Med. 358 (1), 55–68. http://dx.doi.org/10.1056/NEJMra073096.

Benazzi, F., 2002. Psychomotor changes in melancholic and atypical depression: unipolar and bipolar-II subtypes. Psychiatry Res. 112 (3), 211–220.

Bench, C.J., Friston, K.J., Brown, R.G., Scott, L.C., Frackowiak, R.S., Dolan, R.J., 1992. The anatomy of melancholia—focal abnormalities of cerebral blood flow in major depression. Psychol. Med. 22 (3), 607–615. http://dx.doi.org/10.1017/S003329170003806X.

Bush, G., Luu, P., Posner, M., 2000. Cognitive and emotional influences in anterior cingulate cortex. Trends Cogn. Sci. (Regul. Ed.) 4 (6), 215–222. http://dx.doi.org/10.1016/S1364-6613(00)01483-2.

Cano-Colino, M., Almeida, R., Compte, A., 2013. Serotonergic modulation of spatial working memory: predictions from a computational network model. Front. Integr. Neurosci. 7, 71. http://dx.doi.org/10.3389/fnint.2013.00071.

Cano-Colino, M., Almeida, R., Gomez-Cabrero, D., Artigas, F., Compte, A., 2014. Serotonin regulates performance nonmonotonically in a spatial working memory network. Cereb. Cortex 24 (9), 2449–2463. http://dx.doi.org/10.1093/cercor/bht096.

Choudary, P.V., Molnar, M., Evans, S.J., et al., 2005. Altered cortical glutamatergic and GABAergic signal transmission with glial involvement in depression. Proc. Natl. Acad. Sci. U.S.A. 102 (43), 15653–15658. http://dx.doi.org/10.1073/pnas.0507901102.

Cole, M.W., Yarkoni, T., Repovs, G., Anticevic, A., Braver, T.S., 2012. Global connectivity of prefrontal cortex predicts cognitive control and intelligence. J. Neurosci. 32 (26), 8988–8999. http://dx.doi.org/10.1523/JNEUROSCI.0536-12.2012.

Cole, M.W., Reynolds, J.R., Power, J.D., Repovs, G., Anticevic, A., Braver, T.S., 2013. Multitask connectivity reveals flexible hubs for adaptive task control. Nat. Neurosci. 16 (9), 1348–1355. http://dx.doi.org/10.1038/nn.3470.

Cotter, D., Mackay, D., Landau, S., Kerwin, R., Everall, I., 2001. Reduced glial cell density and neuronal size in the anterior cingulate cortex in major depressive disorder. Arch. Gen. Psychiatry 58 (6), 545–553.

Disner, S.G., Beevers, C.G., Haigh, E.A.P., Beck, A.T., 2011. Neural mechanisms of the cognitive model of depression. Nat. Rev. Neurosci. 12 (8), 467–477. http://dx.doi.org/10.1038/nrn3027.

Drevets, W.C., 2001. Neuroimaging and neuropathological studies of depression: implications for the cognitive-emotional features of mood disorders. Curr. Opin. Neurobiol. 11 (2), 240–249. http://dx.doi.org/10.1016/S0959-4388(00)00203-8.

Edin, F., Klingberg, T., Johansson, P., McNab, F., Tegnér, J., Compte, A., 2009. Mechanism for top-down control of working memory capacity. Proc. Natl. Acad. Sci. U.S.A. 106 (16), 6802–6807. http://dx.doi.org/10.1073/pnas.0901894106.

Flint, J., Kendler, K.S., 2014. The genetics of major depression. Neuron 81 (3), 484–503. http://dx.doi.org/10.1016/j.neuron.2014.01.027.

Franklin, T.B., Saab, B.J., Mansuy, I.M., 2012. Neural mechanisms of stress resilience and vulnerability. Neuron 75 (5), 741–761. http://dx.doi.org/10.1016/j.neuron.2012.08.016.

Gohier, B., Ferracci, L., Surguladze, S.A., et al., 2009. Cognitive inhibition and working memory in unipolar depression. J. Affect. Disord. 116 (1–2), 100–105. http://dx.doi.org/10.1016/j.jad.2008.10.028.

Hadzi-Pavlovic, D., Boyce, P., 2012. Melancholia. Curr. Opin. Psychiatry 25 (1), 14–18. http://dx.doi.org/10.1097/YCO.0b013e32834dc147.

Harald, B., Gordon, P., 2012. Meta-review of depressive subtyping models. J. Affect. Disord. 139 (2), 126–140. http://dx.doi.org/10.1016/j.jad.2011.07.015.

Horn, D.I., Yu, C., Steiner, J., et al., 2010. Glutamatergic and resting-state functional connectivity correlates of severity in major depression – the role of pregenual anterior cingulate cortex and anterior insula. Front. Syst. Neurosci. 4. http://dx.doi.org/10.3389/fnsys.2010.00033.

Huys, Q.J.M., Maia, T.V., Frank, M.J., 2016. Computational psychiatry as a bridge from neuroscience to clinical applications. Nat. Neurosci. 19 (3), 404–413. http://dx.doi.org/10.1038/nn.4238.

Ikeda, M., Iwanaga, M., Seiwa, H., 1996. Test anxiety and working memory system. Percept. Mot. Skills 82 (3 Pt 2), 1223–1231. http://dx.doi.org/10.2466/pms.1996.82.3c.1223.

Jasinska, A.J., Lowry, C.A., Burmeister, M., 2012. Serotonin transporter gene, stress and raphe-raphe interactions: a molecular mechanism of depression. Trends Neurosci. 35 (7), 395–402. http://dx.doi.org/10.1016/j.tins.2012.01.001.

Keller, M.B., 1992. Time to recovery, chronicity, and levels of psychopathology in major depression. Arch. Gen. Psychiatry 49 (10), 809. http://dx.doi.org/10.1001/archpsyc.1992.01820100053010.

Kendler, K.S., Thornton, L.M., Gardner, C.O., 2000. Stressful life events and previous episodes in the etiology of major depression in women: an evaluation of the "kindling" hypothesis. Am. J. Psychiatry 157 (8), 1243–1251. http://dx.doi.org/10.1176/appi.ajp.157.8.1243.

Kendler, K.S., Thornton, L.M., Gardner, C.O., 2001. Genetic risk, number of previous depressive episodes, and stressful life events in predicting onset of major depression. Am. J. Psychiatry 158 (4), 582–586. http://dx.doi.org/10.1176/appi.ajp.158.4.582.

Kennedy, S.H., Evans, K.R., Krüger, S., et al., 2001. Changes in regional brain glucose metabolism measured with positron emission tomography after paroxetine treatment of major depression. Am. J. Psychiatry 158 (6), 899–905. http://dx.doi.org/10.1176/appi.ajp.158.6.899.

Krishnan, V., Nestler, E.J., 2008. The molecular neurobiology of depression. Nature 455 (7215), 894–902. http://dx.doi.org/10.1038/nature07455.

Krystal, J.H., Sanacora, G., Blumberg, H., et al., 2002. Glutamate and GABA systems as targets for novel antidepressant and mood-stabilizing treatments. Mol. Psychiatry 7 (Suppl. 1), S71–S80. http://dx.doi.org/10.1038/sj.mp.4001021.

Lamers, F., Beekman, A.T.F., van Hemert, A.M., Schoevers, R.A., Penninx, B.W., 2016. Six-year longitudinal course and outcomes of subtypes of depression. Br. J. Psychiatry 208 (1), 62–68. http://dx.doi.org/10.1192/bjp.bp.114.153098.

Lee, J., Folley, B.S., Gore, J., Park, S., 2008. Origins of spatial working memory deficits in schizophrenia: an event-related FMRI and near-infrared spectroscopy study. PLoS One 3 (3), e1760. http://dx.doi.org/10.1371/journal.pone.0001760.

Levin, R.L., Heller, W., Mohanty, A., Herrington, J.D., Miller, G.A., 2007. Cognitive deficits in depression and functional specificity of regional brain activity. Cogn. Ther. Res. 31 (2), 211–233. http://dx.doi.org/10.1007/s10608-007-9128-z.

Maes, M., Meltzer, H., 1995. The serotonin hypothesis of major depression. In: Psychopharmacology: The Fourth Generation of Progress, vol. 10, pp. 933–934.

Maier, S.F., 2015. Behavioral control blunts reactions to contemporaneous and future adverse events: medial prefrontal cortex plasticity and a corticostriatal network. Neurobiol. Stress 1, 12–22. http://dx.doi.org/10.1016/j.ynstr.2014.09.003.

Manji, H.K., Drevets, W.C., Charney, D.S., 2001. The cellular neurobiology of depression. Nat. Med. 7 (5), 541–547. http://dx.doi.org/10.1038/87865.

Mayberg, H.S., 1997. Limbic-cortical dysregulation: a proposed model of depression. J. Neuropsychiatry Clin. Neurosci. 9 (3), 471–481. http://dx.doi.org/10.1176/jnp.9.3.471.

Mayberg, H.S., Liotti, M., Brannan, S.K., et al., 1999. Reciprocal limbic-cortical function and negative mood: converging PET findings in depression and normal sadness. Am. J. Psychiatry 156 (5), 675–682. http://dx.doi.org/10.1176/ajp.156.5.675.

Mayberg, H.S., Lozano, A.M., Voon, V., et al., 2005. Deep brain stimulation for treatment-resistant depression. Neuron 45 (5), 651–660. http://dx.doi.org/10.1016/j.neuron.2005.02.014.

Mayer, J.S., Park, S., 2012. Working memory encoding and false memory in schizophrenia and bipolar disorder in a spatial delayed response task. J. Abnorm. Psychol. 121 (3), 784–794. http://dx.doi.org/10.1037/a0028836.

McGinn, L.K., Asnis, G.M., Rubinson, E., 1996. Biological and clinical validation of atypical depression. Psychiatry Res. 60 (2–3), 191–198.

Medalla, M., Barbas, H., 2010. Anterior cingulate synapses in prefrontal areas 10 and 46 suggest differential influence in cognitive control. J. Neurosci. 30 (48), 16068–16081. http://dx.doi.org/10.1523/JNEUROSCI.1773-10.2010.

Miller, A.H., Maletic, V., Raison, C.L., 2009. Inflammation and its discontents: the role of cytokines in the pathophysiology of major depression. Biol. Psychiatry 65 (9), 732–741. http://dx.doi.org/10.1016/j.biopsych.2008.11.029.

Mitchell, P.B., Malhi, G.S., 2004. Bipolar depression: phenomenological overview and clinical characteristics. Bipolar Disord. 6 (6), 530–539. http://dx.doi.org/10.1111/j.1399-5618.2004.00137.x.

Northoff, G., Sibille, E., 2014. Why are cortical GABA neurons relevant to internal focus in depression? A cross-level model linking cellular, biochemical and neural network findings. Mol. Psychiatry 19 (9), 966–977. http://dx.doi.org/10.1038/mp.2014.68.

Oda, K., Okubo, Y., Ishida, R., et al., 2003. Regional cerebral blood flow in depressed patients with white matter magnetic resonance hyperintensity. Biol. Psychiatry 53 (2), 150–156. http://dx.doi.org/10.1016/S0006-3223(02)01548-2.

Ongür, D., Drevets, W.C., Price, J.L., 1998. Glial reduction in the subgenual prefrontal cortex in mood disorders. Proc. Natl. Acad. Sci. U.S.A. 95 (22), 13290–13295.

Paelecke-Habermann, Y., Pohl, J., Leplow, B., 2005. Attention and executive functions in remitted major depression patients. J. Affect. Disord. 89 (1–3), 125–135. http://dx.doi.org/10.1016/j.jad.2005.09.006.

Palomero-Gallagher, N., Vogt, B.A., Schleicher, A., Mayberg, H.S., Zilles, K., 2009. Receptor architecture of human cingulate cortex: evaluation of the four-region neurobiological model. Hum. Brain Mapp. 30 (8), 2336–2355. http://dx.doi.org/10.1002/hbm.20667.

Pessoa, L., 2008. On the relationship between emotion and cognition. Nat. Rev. Neurosci. 9 (2), 148–158. http://dx.doi.org/10.1038/nrn2317.

Portella, M.J., de Diego-Adeliño, J., Gómez-Ansón, B., et al., 2011. Ventromedial prefrontal spectroscopic abnormalities over the course of depression: a comparison among first episode, remitted recurrent and chronic patients. J. Psychiatr. Res. 45 (4), 427–434. http://dx.doi.org/10.1016/j.jpsychires.2010.08.010.

Posternak, M.A., Zimmerman, M., 2002. Partial validation of the atypical features subtype of major depressive disorder. Arch. Gen. Psychiatry 59 (1), 70–76.

Price, J.L., Drevets, W.C., 2012. Neural circuits underlying the pathophysiology of mood disorders. Trends Cogn. Sci. (Regul. Ed.) 16 (1), 61–71. http://dx.doi.org/10.1016/j.tics.2011.12.011.

Rajkowska, G., Miguel-Hidalgo, J.J., Wei, J., et al., 1999. Morphometric evidence for neuronal and glial prefrontal cell pathology in major depression. Biol. Psychiatry 45 (9), 1085–1098.

Ramirez-Mahaluf, J.P., Perramon, J., Otal, B., Villoslada, P., Compte, A., 2017a. Subgenual anterior cingulate cortex controls sadness-induced modulations of cognitive and emotional network hubs. Biorxiv. http://dx.doi.org/10.1101/163709.

Ramirez-Mahaluf, J.P., Roxin, A., Mayberg, H.S., Compte, A., 2017b. A computational model of major depression: the role of glutamate dysfunction on cingulo-frontal network dynamics. Cereb. Cortex 27 (1), 660−679. http://dx.doi.org/10.1093/cercor/bhv249.

Rose, E.J., Ebmeier, K.P., 2006. Pattern of impaired working memory during major depression. J. Affect. Disord. 90 (2−3), 149−161. http://dx.doi.org/10.1016/j.jad.2005.11.003.

Sanacora, G., Treccani, G., Popoli, M., 2012. Towards a glutamate hypothesis of depression: an emerging frontier of neuropsychopharmacology for mood disorders. Neuropharmacology 62 (1), 63−77. http://dx.doi.org/10.1016/j.neuropharm.2011.07.036.

Santana, N., Bortolozzi, A., Serrats, J., Mengod, G., Artigas, F., 2004. Expression of serotonin1A and serotonin2A receptors in pyramidal and GABAergic neurons of the rat prefrontal cortex. Cereb. Cortex 14 (10), 1100−1109. http://dx.doi.org/10.1093/cercor/bhh070.

Sarrias, M.J., Artigas, F., Martínez, E., et al., 1987. Decreased plasma serotonin in melancholic patients: a study with clomipramine. Biol. Psychiatry 22 (12), 1429−1438.

Seminowicz, D.A., Mayberg, H.S., McIntosh, A.R., et al., 2004. Limbic-frontal circuitry in major depression: a path modeling metanalysis. Neuroimage 22 (1), 409−418. http://dx.doi.org/10.1016/j.neuroimage.2004.01.015.

Shackman, A.J., Sarinopoulos, I., Maxwell, J.S., Pizzagalli, D.A., Lavric, A., Davidson, R.J., 2006. Anxiety selectively disrupts visuospatial working memory. Emotion 6 (1), 40−61. http://dx.doi.org/10.1037/1528-3542.6.1.40.

Siekmeier, P.J., 2015. Computational modeling of psychiatric illnesses via well-defined neurophysiological and neurocognitive biomarkers. Neurosci. Biobehav. Rev. 57, 365−380. http://dx.doi.org/10.1016/j.neubiorev.2015.09.014.

Snyder, H.R., 2013. Major depressive disorder is associated with broad impairments on neuropsychological measures of executive function: a meta-analysis and review. Psychol. Bull. 139 (1), 81−132. http://dx.doi.org/10.1037/a0028727.

Stewart, J.W., McGrath, P.J., Quitkin, F.M., Klein, D.F., 2009. DSM-IV depression with atypical features: is it valid? Neuropsychopharmacology 34 (13), 2625−2632.

Turrigiano, G., 2011. Too many cooks? Intrinsic and synaptic homeostatic mechanisms in cortical circuit refinement. Annu. Rev. Neurosci. 34, 89−103. http://dx.doi.org/10.1146/annurev-neuro-060909-153238.

Uher, R., Dernovsek, M.Z., Mors, O., et al., 2011. Melancholic, atypical and anxious depression subtypes and outcome of treatment with escitalopram and nortriptyline. J. Affect. Disord. 132 (1−2), 112−120. http://dx.doi.org/10.1016/j.jad.2011.02.014.

Videbech, P., Ravnkilde, B., Pedersen, T.H., et al., 2002. The Danish PET/depression project: clinical symptoms and cerebral blood flow. A regions-of-interest analysis. Acta Psychiatr. Scand. 106 (1), 35−44. http://dx.doi.org/10.1034/j.1600-0447.2002.02245.x.

Walter, M., Henning, A., Grimm, S., et al., 2009. The relationship between aberrant neuronal activation in the pregenual anterior cingulate, altered glutamatergic metabolism, and anhedonia in major depression. Arch. Gen. Psychiatry 66 (5), 478−486. http://dx.doi.org/10.1001/archgenpsychiatry.2009.39.

Wang, X.-J., Krystal, J.H., 2014. Computational psychiatry. Neuron 84 (3), 638−654. http://dx.doi.org/10.1016/j.neuron.2014.10.018.

Watts, F.N., Sharrock, R., 1985. Description and measurement of concentration problems in depressed patients. Psychol. Med. 15 (2), 317−326.

# 3

# Dopaminergic Neurons in the Ventral Tegmental Area and Their Dysregulation in Nicotine Addiction

*Gregory Dumont[1,a], Reinoud Maex[1,a], Boris Gutkin[1,2]*

[1] Ecole Normale Superieure PSL Research University, Paris, France;
[2] National Research University Higher School of Economics, Moscow, Russia

[a] Equal contribution.

*Computational Psychiatry*
http://dx.doi.org/10.1016/B978-0-12-809825-7.00003-1

**47**

# 3.1 NICOTINE, DOPAMINE, AND ADDICTION

The connection between addiction and dopamine started to be established in the 1950s with the observation by Olds and Milner of the rewarding effect of electrical self-stimulation in certain midline brain structures, and with the discovery by Carlsson of dopamine as a central neurotransmitter. The circuit involved is the mesolimbic system and it comprises of the mesencaphalic ventral tegmental area (VTA) and its major projection areas: the nucleus accumbens or ventral striatum (part of the telencephalic basal ganglia) and the prefrontal cortex (PFC) (Lindvall et al., 1974).

Phylogenetically, dopaminergic innervation of basal ganglia structures is already found in precursors of the vertebrata (such as the lamprey) that evolved more than 500-million-years ago (Smeets et al., 2000; Yamamoto and Vernier, 2011). It is not surprising then that dopamine dysregulation is associated with severe pathology. For instance, insufficient dopamine production in the nearby substantia nigra causes Parkinson's disease, and depletion of dopamine in the PFC was found in monkeys to be as devastating as a surgical cortical ablation (Brozoski et al., 1979). Likewise, addiction is widely considered a dopamine disease: all drugs of abuse acutely enhance the release of dopamine (Volkow et al., 2004), whereas chronic abuse and addiction are associated with blunted dopamine responses (van de Giessen et al., 2016; Perez et al., 2015). The VTA itself has recently been implicated in the addiction to benzodiazepines (Tan et al., 2010) and morphine (Jalabert et al., 2011) as well as vast majority of other drugs of abuse (e.g., nicotine, alcohol, cocaine, etc.).

After the discovery of central nicotinic acetylcholine receptors (nAChRs) in the 1980s, these receptors were found to show the same distribution as dopamine receptors, with the highest density located in the tegmentum, striatum, and PFC (Zhou et al., 2001). This colocalization of dopamine and acetylcholine (ACh) signaling offers not only a substrate for nicotine addiction but also for the nicotinic control of dopamine release, and hence also a rationale for the use of nicotinic compounds as a potential treatment for Parkinson's disease (Quik and Wonnacott, 2011) and schizophrenia.

While nicotinic receptors are expressed throughout the central and peripheral nervous system, experimental data using genetic manipulations of receptors indicated that the effects of nicotine on the dopamine neurons in the VTA are central to acquisition of nicotine self-administration. It is clear that nicotine affects the dopaminergic neurotransmission by acting on the nAChRs, yet it is not completely clear where is the exact site of action that is responsible for the motivational effects of nicotine within the VTA. Previous data indeed show that the VTA is a neuronal microcircuit, containing $\sim 80\%$ dopamine neurons targeted locally by the GABAergic cells (20%) (Lacey et al., 1989; Johnson and North, 1992; Ikemoto et al., 1997; Sugita et al., 1992). This local circuit receives glutamatergic (Glu) afferents from the PFC (Tong et al., 1996; Sesack and Pickel, 1992; Christie et al., 1985; Steffensen et al., 1998) and is furthermore innervated with Glu (Clements and Grant, 1990; Cornwall et al., 1990; Forster and Blaha, 2000) and cholinergic projections (Oakman et al., 1995) from the brainstem. In turn dopamine and GABAergic projections from the VTA target numerous areas of the brain including the PFC (Thierry et al., 1973; Berger et al., 1976; Swanson, 1982; Steffensen et al., 1998; Carr and Sesack, 2000) and limbic/striatal structures (Anden et al., 1966; Ungerstedt, 1971; Oades and Halliday, 1987; Van Bockstaele and Pickel, 1995). Thus the VTA dopamine and GABAergic neurons send projections throughout the brain generating dopamine and GABAergic signals in response to cortical (Taber and Fibiger, 1995; Tong et al., 1996) and subcortical inputs (Floresco et al., 2003; Lodge and Grace, 2006) as well as to nicotine.

Release of the endogenous ligand ACh into the VTA causes nearly synchronous activation of nAChRs (Dani et al., 2001). The rapid delivery and breakdown of ACh by acetylcholinesterase precludes significant nAChR desensitization (Feldberg, 1945). Nicotine activates and then desensitizes nAChRs within seconds to minutes (Katz and Thesleff, 1957; Pidoplichko et al., 1997; Fenster et al., 1997) since it remains elevated in the blood of smokers during and after smoking (Henningfield et al., 1993). Importantly, the various subtypes of nAChRs have distinct affinities for ACh as well as nicotine, exhibit markedly different activation/desensitization kinetics (Changeux et al., 1998), and have different expression targets:

1. Low-affinity $\alpha7$ containing nAChRs desensitize rapidly (Champtiaux et al., 2003) and are found on Glu terminals;
2. High-affinity, slowly desensitizing $\alpha4\beta2$ containing nAChR on GABAergic cells;
3. $\alpha4$-and $\alpha6$-containing nAChRs on dopamine neurons (Calabresi et al., 1989; Klink et al., 2001)

In this chapter we will focus mainly on the first two types. Experimental results have suggested that in the VTA most of the nAChR-mediated currents and the reinforcing properties in response to nicotine are mediated by the $\alpha4\beta2$ nAChR subtype (Picciotto et al., 1998), whereas $\alpha7$ nAChRs play an important role in neuroplasticity of Glu afferents to the VTA. On a longer time scale nicotine is known to lead to upregulation of the $\alpha4\beta2$ nAChRs. This increases the density of receptors, boosting their number, while the receptors remain fully functional (Champtiaux et al., 2003). Interestingly, the receptor upregulation appears to be differentially targeted to the GABAergic neurons (Nashmi et al., 2007); this will be discussed in detail in this chapter below. In contrast, the $\alpha7$ nAChRs have therefore been suggested to contribute to long-term potentiation of Glu afferents onto dopamine neurons (Bonci and Malenka, 1999; Mansvelder and McGehee, 2000).

Despite the vast array of gathered knowledge (Benowitz, 2009; Taly et al., 2009), the pathogenesis of nicotine addiction is still incompletely understood. Complicating factors are the great diversity of receptor subtypes with their typical nonlinear (bell-shaped) dose-response curves, the flexibility of the expression of different receptor subtypes by different neuron classes, an incomplete understanding of central cholinergic physiology and of the mechanisms of cholinergic and dopaminergic homeostasis, and finally the still elusive mechanism underlying different patterns (tonic vs. phasic) of dopamine release.

While there have been a number of computational models suggesting how altered phasic dopamine release evoked by the drugs of addiction might translate into a pathological reinforcement and learned drug-seeking behavior (e.g., see Redish, 2004; Dezfouli et al., 2009; Keramati and Gutkin, 2002), these models treated almost exclusively cocaine as the paradigmatic example, and were not geared to address explicitly the link between drug action at identified receptors, their adaptations, and the behavioral outcome. As an initial approach to address the latter issue and to focus on nicotine specifically we introduced a large-scale neuro-computational framework (Gutkin et al., 2006). This framework integrated nicotine effects on a generic dopaminergic (DAergic) neuron population at the receptor level (signaling the reward-related information), together with a simple model of action selection. This model also incorporated a dopamine-dependent learning rule that gives distinct roles to the phasic and tonic dopamine neurotransmission. The model was a proposal on the differential roles of the positive (rewarding) and opponent processes in the acquisition and maintenance of drug taking behavior, and the development of such behavior into a rigid habit.

The major hypothesis for this large-scale approach was that the nicotine effects on dopamine signaling initiate a cascade of neuroadaptations that in turn bias Glu learning processes in the dorsal striatum-related structures that are responsible for behavioral choice, leading to the onset of stable self-administration. This initial modeling effort focused on nicotine as an antagonist of the nicotinic receptors and showed how through activation and upregulation nicotine dynamically changes the gain of the dopaminergic signaling. Hence, nicotine in that model both potentiated the phasic dopamine response to rewarding stimuli and evoked such a signal by itself (Dani and Heinemann, 1996; Changeux et al., 1998). The phasic dopamine in turn instructed the learning and plasticity in a Glu action selection circuit that was modeled as a stochastic winner-take-all network (e.g, Usher and McClelland, 2001; Wang, 2012). Since data show that dopamine and nicotine potentiate Glu plasticity in the dorsal striatum (Reynolds and Wickens, 2002), the learning in the decision-making circuit proposed that tonic dopamine gates Hebbian learning in the excitatory (cortico–striatal–cortical) synapses. Persistent nicotine-dependent depression in tonic dopamine removed the gate necessary for learning and stamped in the already learned drug-seeking action selection. In other words, a slow-onset opponent process on the nicotinic receptor-mediated function disrupts dopamine neurotransmission to the point that extinction learning or response unlearning is impaired.

Simulations of this general framework showed that drug-induced neuroadaptations in the dopaminergic circuitry and drug-modulated learning in the action selection system are sufficient to account for the development and maintenance of self-administration. Importantly, the positive rewarding effect of the drug is translated into biased action selection and choice making, whereas the slow opponent process plays a key role in cementing the drug-associated behavior by removing the dopamine signal from the range where learning (and unlearning) can take place. Interestingly, the computational framework implied that the sensitization of behavior by nicotine through dopamine-dependent processes may be disassociated from the acquisition of self-administration. At low doses/short duration, nicotine may lead to apparent behavioral sensitization, but not self-administration.

In summary this general and large-scale model suggested how, in the long run, processes that oppose the primary reward of nicotine ingrain the drug-related behavior making it independent of the motivation state and value of various action choices and difficult to modify in the face of changing contingencies. While providing a synthetic view of how the various positive and negative effects of nicotine in the combined dopa-mine—glutamate system may results in nicotine seeking, the global model was not designed to pinpoint the specific local mechanisms by which nicotine may bias the dopamine signaling.

This chapter presents a computational modeling strategy in the study of nicotine addiction, using models of varying complexity for receptors, neurons, and circuits. We build beyond the global approach that was initially taken (see above) and present two progressively detailed modeling approaches. The first is geared toward seeing the VTA as a local circuit governed by the nicotinic receptors and helps us to identify the potential specific targets for acute and chronic nicotine. The circuit model shows how receptor adaptations to chronic nicotine modify the dopamine responses to endogenous Ach in such a way as to have them blunted unless nicotine is present. Hence, the model allows us to propose that nicotine acts to renormalize endogenous motivational signaling in an otherwise pathological addicted individual. The second approach goes even deeper in detail, to the level of a single spiking dopamine neuron, and attempts to show exactly how nicotine-induced changes in the input structure to such a neuron may promote burst firing. In turn this burst firing leads to increased dopamine output, and thereby potentially me-diates the motivational properties of nicotine.

## 3.2 MODELING RECEPTOR KINETICS

It is probably fair to say that many neuroscientists and pharmacologists have too simple an idea of the interaction between pharmacological or addictive compounds and their receptors. The goal of the present section is to illustrate the rich dynamics among nicotinic compounds, nicotinic receptors, and the endogenous neurotransmitter ACh. We deal with this topic in some detail, because the ability to disentangle changes in receptor, neuron, and circuit dynamics is one of the strengths of computational models. The effects arising from the differential expression of receptor subtypes by different neuron classes in the VTA is the subject of the next section.

Nicotinic receptors are pentameric ion channels of the same family as the ionotropic γ-amino butyric acid (GABA)-gated channels and glycine-gated channels (Hille, 2001). Natural nicotinic agents codevel-oped with their receptors and were produced by plants and animals to paralyze or intoxicate their predator or prey (Albuquerque et al., 2009). Typically, after having crossed the blood—brain barrier, these agents are neither taken up by transporters and recycled nor rapidly broken down by acetylcholinesterase, and their binding to receptors may even slow down their clearance from the brain (a process called buffered diffusion), giving them plenty of time to desensitize their receptors, which is the major difference that distinguishes them from synaptically released ACh (Quick and Lester, 2002).

This desensitization may lead to apparent paradoxes. For instance succinylcholine, which is an agonist at the muscular nAChR because it depolarizes the end plate, is used as a muscle relaxant in general anesthesia (Giniatullin et al., 2005). As for centrally acting nicotinic compounds (or nicotine itself), the relative contributions in vivo of receptor activation and desensitization are much more difficult to assess, and they are still a matter of debate.

### 3.2.1 Ligand−Receptor Interaction

We deal here only with the kinetics of ligand binding, based on the law of mass action,

$$R + S \underset{k_b}{\overset{k_f}{\rightleftharpoons}} RS + S \underset{k_b}{\overset{k_f}{\rightleftharpoons}} RS_2 \tag{3.1}$$

where $k_f$ and $k_b$ are the rate constants of the forward and backward reactions, respectively.

The fraction of fully liganded receptors $RS_2/R_{total}$ is a nonlinear function of the substrate concentration $[S]$ for two reasons. First, at high concentrations the reaction saturates because there is only a limited number of receptors available within a relative abundance of ligand. Secondly, toward very low ligand concentrations little $RS_2$ is produced as it becomes increasingly improbable for the receptor to hit the two molecules required to become fully liganded.

Both these phenomena are captured by the Hill formula that readers with a biochemical background will recognize as a variant of the Michaelis−Menten formula for enzyme kinetics (Segel, 1975, 1984),

$$r = \frac{R_{active}}{R_{total}} = \frac{S^n}{S^n + K_S^n} = \frac{1}{1 + \left(\frac{K_S}{S}\right)^n} = \frac{\left(\frac{S}{K_S}\right)^n}{1 + \left(\frac{S}{K_S}\right)^n} \tag{3.2}$$

where $K_S$ is the dissociation constant $k_b/k_f$ and $n$ the Hill exponent.

In this formula, which calculates the steady-state fraction $r$ of fully liganded receptor $RS_2$, it has been tacitly assumed that the intermediary state $RS$, in which only a single ligand is bound, is inactive and transitory, and hence can be neglected. A Hill exponent $n = 2$ reflects the two binding sites of nAChRs; a greater value would indicate cooperation between the receptor's subunits in binding ligand, a lower value competition.

This formula would suffice, as a first approximation, to describe the binding of ACh that is immediately hydrolyzed after release and does not appreciably desensitize the receptor. It would be insufficient, however, to describe the effect of systemically taken substances such as nicotine, or the endogenous effect of ACh in the presence of inhibitors of acetylcholinesterase.

### 3.2.1.1 A Two-Gate Receptor Model

Graupner and Gutkin (Graupner and Gutkin, 2009; Graupner et al., 2013) developed a two-gate receptor model with independent activation and desensitization gates, akin to the activation and inactivation gates of voltage-gated ion channels, but with ligand concentration instead of voltage as the independent variable. This separation into two gates is justified by the observed dissociation between the activating and desensitizing capacities of many nicotinic agents (M. Bencherif, personal communication).

The two-gate model is a simplification, for use in circuit models, of the cyclic desensitization model of Katz and Thesleff (1957), further called the KT model, in which $R$ and $D$ denoted the receptor in its sensitive and desensitized state, respectively:

$$S + R \; \rightleftarrows \; SR$$
$$\uparrow \qquad \downarrow$$
$$S + D \; \rightleftarrows \; SD$$

The main new assumptions of the two-gate model are

1. the reactions of activation and (de)sensitization are separable, and
2. at each ligand concentration [$S$] the steady-state fraction of fully liganded receptors (in the above scheme denoted by $SR$ and $SD$, respectively) can be calculated from the dissociation constant and Hill coefficient derived from Eq. (3.2). This is more convenient than using the rate constants of the cyclic reaction scheme, because these Hill functions are precisely the format pharmacologists use to characterize the binding properties of new compounds.

Hence, in the presence of substrate $S$, receptors can be activated ($R_0 \rightarrow R_1$) and/or desensitized ($D_1 \rightarrow D_0$) through the reactions

$$S + R_0 \overset{K_S^a}{\rightleftarrows} SR_1 \qquad S + D_1 \overset{K_S^d}{\rightleftarrows} SD_0$$

The four possible receptor states are then:

1. inactive ($R_0$) but sensitive ($D_1$) ($R$ in the cyclic KT model, a0s1 in the two-gate model),
2. active ($R_1$) and sensitive ($D_1$) ($SR$, or a1s1),
3. active ($R_1$) but desensitized ($D_0$) ($SD$, or a1s0),
4. inactive ($R_0$) and desensitized ($D_0$) ($D$ or a0s0).

Only in state 2 is the channel open and conductive, and in the following we will denote by $a$ and $s$ the fraction of channels in the active and sensitive state, respectively. The fraction of open channels is then the product of $a$ and $s$. Note that although the two gates are independent, they are not uncorrelated because their dependencies on ligand concentration greatly overlap. Hence desensitized receptors will mostly be ligand bound but closed and nonconducting.

The two-gate model departs from the KT model at two points: (1) resensitization cannot be faster than desensitization in the two-gate model, and (2) the spontaneous transitions from the unbound sensitive to the unbound desensitized state (or unbound active and inactive states) is vanishingly small as opposed to the theoretical presence of unliganded open channels in the KT model and other allosteric receptor formulations (Galzi et al., 1996; Shelley and Cull-Candy, 2010).

### 3.2.1.2 Steady-State Receptor Current

The two gates have first-order kinetics:

$$\tau_a \frac{da}{dt} = -a + a_\infty$$

and

$$\tau_s \frac{ds}{dt} = -s + s_\infty$$

where the time-constants $\tau_a$ and $\tau_s$ and the steady-state values $a_\infty$ and $s_\infty$ depend on the substrate concentration [$S$]. The values of $a_\infty$ and $s_\infty$, which are Hill functions as described in Eq. (3.2), are shown in Fig. 3.1 for the heteromeric $\alpha 4 \beta 2$ subtype of nAChRs (A) and the homomeric $\alpha 7$ nAChRs (B).

The clearance of nicotinic compounds from the brain is so slow that ligand concentrations can often be considered approximately constant. In such a case the two most important factors determining their effect are (1) the size of the window current they generate, on the positive side, and (2) on the negative side, the endogenous current they prevent ACh from generating by desensitizing the receptor (and which is equivalent to a reduction of the pool of available receptors). The window current can be

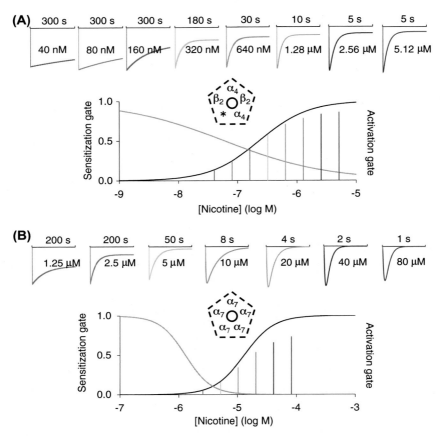

FIGURE 3.1    Kinetics of the heteromeric $\alpha 4\beta 2$ nicotinic acetylcholine receptors (nAChRs) (A) and the homomeric $\alpha 7$ nAChRs (B) at varying nicotine concentrations. The left and right vertical axes plot the steady-state values $s_\infty$ and $a_\infty$ of the desensitization and activation gates, respectively. The upper traces plot the current, or time-course of channel opening, at the indicated nicotine concentrations. Note different timescales above each trace; the peak of each trace (the fraction of open channels) is indicated by the height of the corresponding vertical bar on the Hill plot. Parameters as in Maex et al. (2014), based mainly on Fenster et al. (1997) and Buisson and Bertrand (2001).

appreciated from the degree of overlap of the $s_\infty$ and $a_\infty$ Hill functions in Fig. 3.1, and corresponds, for the $\alpha 4\beta 2$ nAChRs, to area a1s1 in Fig. 3.2.

### 3.2.1.3 Temporal Dynamics of the Receptor Current

In the present model, activation of the receptor was taken fast ($\tau = $ constant $= 5$ ms), whereas its desensitization was slow (between 0.5 and 600 s for $\alpha 4\beta 2$ nAChRs) and varying with concentration according to

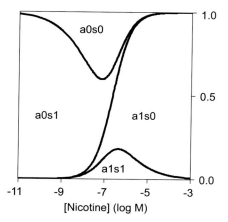

FIGURE 3.2 Fraction of $\alpha4\beta2$ nicotinic acetylcholine receptors in each of the four possible receptor states at steady nicotine concentrations. The label a1 indicates that the receptor is activated, s1 that it is sensitive (or not desensitized). Only in state a1s1 is the channel open.

the same Hill function as used for $s_\infty$ (Grady et al., 1997). The enhanced speed of desensitization at high nicotine concentrations can be appreciated from the time-course of channel opening during steady nicotine application in Fig. 3.1.

The fast and deep desensitization at high-nicotine concentrations also leads to a bell-shaped (or inverted-U-shaped) dose-response curve. This effect, which is typical of many neuropharmacological compounds, is illustrated in Fig. 3.3, where varying doses of nicotine were delivered as a

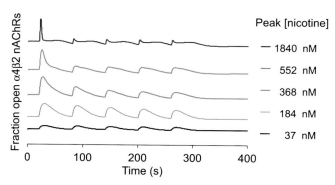

FIGURE 3.3 Dose dependency of channel opening. Five nicotine pulses were applied with 60 s interval, each 1 s pulse being convolved with an alpha function of 10 s time constant to account for the pharmacokinetics of systemic nicotine administration. The applied dose of nicotine increased from bottom to top, resulting in the peak concentrations indicated in the legend. Each trace plots the fraction of open $\alpha4\beta2$ nicotinic acetylcholine receptor (nAChR), which in this simple model can be taken as a proxy for dopamine release. Parameters as in Maex et al. (2014).

series of pulses. At low doses of nicotine (each pulse generating a peak concentration of 37 nM), the $\alpha4\beta2$ nAChR opens only slightly, whereas with high doses (1840 nM) the receptor already completely desensitizes after the first pulse. Hence the optimal dose for receptor opening (and cumulative dopamine release) is the one that generates peak nicotine concentrations of about 300 nM (middle trace), which, interestingly, is about the concentration obtained after smoking of a single cigarette (357 nM in Rose et al., 2010).

## 3.2.2 Competition and Cooperation Between Ligands, or Between Ligand and the Endogenous Transmitter

As mentioned in Section 3.2.1.2 the effect of a systemically applied substance will not only depend on the size of the window current it generates, but also on the degree to which it impedes transmission by the endogenous transmitter. When the nicotinic window current of Fig. 3.2 is corrected by subtracting the lost ACh current, a more nuanced picture arises (Fig. 3.4). In the presence of increasing levels of cholinergic tone, the current gained by nicotine is dwarfed, and even made negative, by the apparent loss of nAChRs that have been desensitized by the continual presence of nicotine.

Apart from this overtly dual action of nicotinic compounds (activation vs. desensitization of the receptor), some other actions have been invoked to explain more subtle effects.

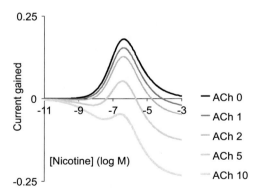

FIGURE 3.4    Window current of the $\alpha4\beta2$ nicotinic acetylcholine receptors generated by nicotine in the presence of a background of cholinergic tone. The window current of Fig. 3.2 has been corrected by subtracting from it the current that is being lost when endogenous acetylcholine (ACh) cannot open the receptors that nicotine has desensitized. The level of cholinergic tone is expressed as an equivalent synaptic concentration of ACh in units of μM.

### 3.2.2.1 *Competitive Inhibition*

In Eq. (3.1) only a single substrate was assumed to be available to bind the receptor. When multiple substrates are present, the Hill function must be extended to include all compounds:

$$a_\infty = \sum_i w_i \frac{\left(\frac{x_i}{K_i}\right)^{n_i}}{1 + \sum_j \left(\frac{x_j}{K_j}\right)^{n_j}} \tag{3.3}$$

where for brevity $x_i$ denotes the concentration of compound $i$, $K_i$ is its dissociation constant, and $w_i$ its efficacy (the strength of its effect relative to that of ACh, which is taken 1). A similar expression holds at the desensitization gate, except for the omission of ACh which, because of its very transitory presence in the synaptic cleft, is assumed not to desensitize the receptors.

Competitive antagonism arises when several compounds compete for the same (so-called homosteric) binding site, for instance ACh and nicotine, or both of them in the presence of varenicline, a medicine used to quit smoking. Typically such different compounds have different efficacies $w_i$, and when a compound with low efficacy is present in sufficient concentration to occupy binding sites that would more effectively be occupied by a ligand present at a lower concentration, then the less effective compound antagonizes the more effective one. Competitive inhibition can always be overcome by increasing the concentration of the compound with stronger efficacy. It arises because the coefficient $w_i$ is present only in the numerator of Eq. (3.3), and not in the denominator, which leads to a kind of divisive inhibition.

Typical competitive antagonists have zero efficacy. Substrates with nonzero efficacy are called partial agonists. They act as agonists when present alone, but at high ACh concentrations their main effect is competition with ACh (or any other more effective agonist). Hence, a potential pharmacological advantage of partial agonists is that they can stabilize receptor function, enhancing its function when the endogenous transmitter is depleted, suppressing its function in case of overactivation.

Fig. 3.5 illustrates this effect for the interaction between nicotine and varenicline at the activation gate of the $\alpha 4\beta 2$ nAChR, using the parameters tabulated in Rollema et al. (2010). Varenicline has an efficacy that is only 25% that of nicotine, and hence is a partial agonist at the $\alpha 4\beta 2$ nAChR (but it is also a full agonist at the $\alpha 7$ nAChR). Note that it is unknown whether this competitive inhibition contributes to the neuropharmacological effect of varenicline.

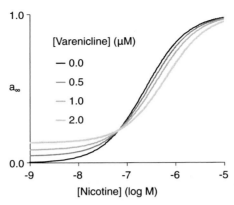

FIGURE 3.5   Illustration of competitive antagonism between high concentrations of varenicline (a partial agonist) and nicotine (a full agonist). The vertical axis plots the steady-state activation of the $\alpha 4 \beta 2$ nicotinic acetylcholine receptor for varying concentrations of nicotine (horizontal axis) in the absence (*black curve*) or the presence (gray) of three different fixed concentrations of varenicline.

### 3.2.2.2 *Coagonism*

In Eqs. (3.1)–(3.3) only the diliganded receptor could have the open-channel configuration, and individual receptors always bound pairs of ligands of the same species. To add to the complexity of the ligand–receptor interaction, it has been observed that low concentrations of some muscarinic compounds (for instance atropine) can activate nicotinic receptors in the presence of ACh (Cachelin and Rust, 1994; Zwart and Vijverberg, 1997). At high concentrations, however, they act as antagonists. The putative mechanism is coagonism of the muscarinic compound with ACh, in which the receptor liganded to one molecule of each can go into the open state. In the formulae, this implies that the cross terms among different ligands can no longer be neglected. For a Hill exponent of two, the formula with cross-terms reads (Cachelin and Rust, 1994; Zwart and Vijverberg, 1997):

$$a_\infty = \frac{w_{1,1}\left(\dfrac{x_1}{K_1}\right)^2 + 2w_{1,2}\dfrac{x_1}{K_1}\dfrac{x_2}{K_2} + w_{2,2}\left(\dfrac{x_2}{K_2}\right)^2}{1 + \left(\dfrac{x_1}{K_1}\right)^2 + 2\dfrac{x_1}{K_1}\dfrac{x_2}{K_2} + \left(\dfrac{x_2}{K_2}\right)^2} \tag{3.4}$$

Typically the muscarinic term would have zero efficacy on its own ($w_{2,2} = 0$), and the cross term would have a nonvanishing efficacy (for instance $w_{1,2} = 0.2$). Positive cognitive effects evoked by nanomolar concentrations of nicotinic compounds have been attributed to this mechanism (see Maex et al., 2014 for a simulation of this effect and references).

### 3.2.2.3 Positive Allosteric Modulation

Positive allosteric modulators bind the receptor at a different (allosteric) binding site and have zero efficacy when applied alone. Nevertheless, they can potentiate the effect of the endogenous transmitter (or any other, even silent agonist) by preventing desensitization and enhancing activation (such effects can be modeled by shifting the $K_d$ values, as in Maex et al., 2014). The best known example within the same receptor family are the benzodiazepines that potentiate GABA$_A$ receptor channels, but similar effects have been observed with some nicotinic agents in vitro. These compounds are actively researched because they would not interfere with the physiological pattern of endogenous receptor activation, but only enhance its effect (Williams et al., 2011). Negative allosteric modulators are also called noncompetitive antagonists.

## 3.2.3 Miscellaneous and Secondary Receptor Effects

Besides receptor desensitization, which itself recruits several processes on timescales from milliseconds to tens of minutes, many other long-term effects of nicotinic agents have been observed. These include presumed metabotropic effects at the closed $\alpha 7$ nAChRs, downregulation of $\alpha 4\beta 2$ nAChRs following the activation of $\alpha 7$ receptors (Melis et al., 2013), and long-term potentiation of Glu synapses by activation of axonal (presynaptic) $\alpha 7$ nAChRs.

# 3.3 CIRCUIT MODELS OF THE VENTRAL TEGMENTAL AREA

Histochemical and electrophysiological studies (reviewed in Oliva and Wanat, 2016) have demonstrated that most neurons in the VTA synthesize either dopamine or GABA. Both these neuron classes express nAChRs of the $\alpha 4\beta 2$ subtype on their soma and dendrite (Wu and Lukas, 2011). These receptors are activated by ACh released by afferents from the pedunculopontine (PPTN) and laterodorsal tegmental nuclei (LDTN) (Forster and Blaha, 2000, 2003).

In addition to controlling the excitability of dopaminergic and GABAergic neurons, nAChRs also gate both the inputs to, and outputs from, the VTA. At the input stage, $\alpha 7$ receptors are found not only on corticofugal Glu terminals to both dopamine (Jones and Wonnacott, 2004) and GABA neurons (Taylor et al., 2013), but also on terminals from the PPTN (Good and Lupica, 2009; Forster and Blaha, 2003) and on neurons in the LDTN (McDaid et al., 2016). Although the homomeric $\alpha 7$

nAChRs have a low affinity for ACh and nicotine (Fig. 3.1B), this does not make them less important physiologically, as they have been suggested to be located, at least in cortex, closer to the ACh release sites (Bennett et al., 2012). In addition, $\alpha$7s have a Ca permeation of 12%, which is comparable to that of N-methyl-d-aspartate receptors, and this Ca influx is involved in the regulation of glutamate release and in synaptic plasticity.

At the output stage, dopamine release in the nucleus accumbens is gated by nAChRs of the $\alpha6\beta2$ subtype located on the dopaminergic axons (Quik et al., 2011). ACh, released by spontaneously active giant cholinergic interneurons in the nucleus accumbens, can stimulate dopamine release either directly or indirectly: directly by activating the $\alpha6\beta2$ receptors on the terminal dopamine axon, indirectly by activating $\alpha$7 receptors on the glutaminergic terminals to the medium-sized spiny neurons (the principal neurons of the striatum), stimulating glutamate release that secondarily promotes dopamine release by activating glutamate receptors on the dopamine terminal (Cachope et al., 2012; Maex et al., 2014; Glowinski et al., 1988).

### 3.3.1 Reorganization of the Ventral Tegmental Area in Addiction

Nicotinic receptors are exceptional in that nicotine treatment induces a rapid upregulation of their numbers, whereas other receptors are classically downregulated in the continual presence of their agonist (Wonnacott, 1990; Buisson and Bertrand, 2002).

Although short-term intermittent nicotine exposure (1 week) leads to locomotor sensitization and an upregulation of nAChRs on the dopamine neurons (Baker et al., 2013; Walsh et al., 2008; Henderson and Lester, 2015), a chronic treatment of 4 weeks leads to a blunted dopamine response (Perez et al., 2015; Koranda et al., 2014) as is the case with most other dopamine-enhancing substances (van de Giessen et al., 2016). This blunted dopamine response is associated with addiction and craving. At this stage there is a considerable (40%) upregulation of $\alpha4\beta2$s on the GABA neurons (Nashmi et al., 2007), but no longer on dopamine neurons. Hence, after a nicotine challenge, the cholinergic system seems to maintain its homeostasis by enhanced receptor expression on the GABA neurons, rather than downregulating the receptors on the effector (dopamine) neurons. A less conspicuous upregulation has also been observed for the $\alpha$7 subtype on Glu terminals (Pakkanen et al., 2005). There may also be on upregulation of $\alpha4\beta2$s on the dopamine axon terminal (Xiao et al., 2009) and a downregulation of $\alpha6\beta2$s (Marks et al., 2014).

## 3.3.2 Circuit Simulations of the Normal and Reorganized Ventral Tegmental Area

We conducted computer simulations of a minimal VTA circuit model to examine how the reorganized VTA, as compared to the naive VTA, responds (1) to an external load of nicotine, and (2) to a series of endogenous ACh peaks. The model comprised both dopamine and GABA neurons, but because many details are still lacking, we calculated only their presumed population response. The firing rate of the population of dopamine neurons is described as:

$$\tau \frac{dV_{dop}}{dt} = -V_{dop} - [I_{gab}] + I_0 + sI_{\alpha7} + rI_{\alpha4\beta2} \tag{3.5}$$

with a similar equation describing the GABA neurons,

$$\tau \frac{dV_{gab}}{dt} = -V_{gab} - I_0 + (1-s)I_{\alpha7} + (1-r)I_{\alpha4\beta2}$$

where $I_{\alpha7}$ and $I_{\alpha4\beta2}$ are excitatory input currents controlled by their respective nAChRs (and hence dependent on the concentration of ACh, nicotine, and any other nicotinic agent, as described in Section 3.2), the $I_0$ are nAChR-independent basal levels of excitation, and the [] (square brackets in Eq. 3.5) indicate half-wave rectification. Importantly, $s$ and $r$ are fractional parameters that determine the relative distribution of receptor subtypes over the two neuron classes.

The output of the model is dopamine efflux in the nucleus accumbens. Since dopamine homeostasis depends also on nonmodeled components such as feedback to dopamine receptors, all output is presented here as fluctuations around the same baseline dopamine level of about 50 nM. For more details the reader can refer to Graupner et al. (2013) and Maex et al. (2014), and to the legend of Fig. 3.6.

In the simulations of naive animals, a nicotine injection (*green traces* in Fig. 3.6) evokes a strong phasic dopamine efflux (*blue trace* in left panel of 3.6B). After chronic nicotine exposure, the response reverses in polarity (right panel), although a positive peak can still be evoked with higher doses of nicotine as are commonly used in in vitro experiments (*oblique arrow* in panel C). The reason for the negative dopamine response in B is the much stronger response of the enriched GABA neurons (*red trace*). (In C this trough is much narrower because the upregulated receptors almost immediately desensitize, as in the upper trace of Fig. 3.3). When the upregulation of $\alpha4\beta2s$ on the GABA neurons is less pronounced, a flat trace can be obtained (not shown). It could therefore be concluded that the VTA has successfully managed to counter the chronic nicotine challenge by upregulating the $\alpha4\beta2s$ on the GABA neurons.

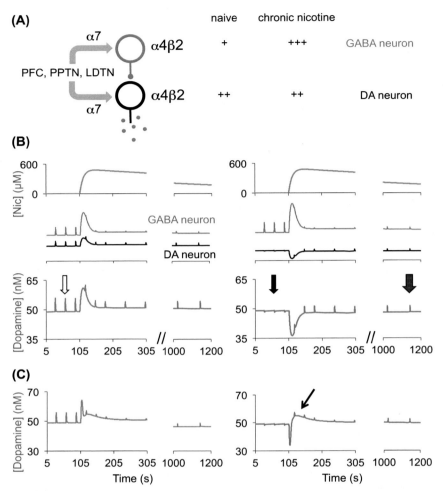

FIGURE 3.6    Circuit simulations of the naive (B,C *left*) and reorganized ventral tegmental area (B,C *right*). (A) Cartoon of the two major neuron populations, with α4/β2 nicotinic acetylcholine receptors (nAChRs) located somatodendritically. The Glu inputs from the prefrontal cortex (PFC) and the pedunculopontine (PPTN) and laterodorsal (LDTN) tegmental nuclei are gated by presynaptic α7 nAChRs. Chronic exposure to nicotine enhances by 40% the expression of α4β2s on the γ-amino butyric acid (GABA) neurons. (B) The predicted dopamine efflux (*blue traces*) to a nicotine load (green) in the naive (*left*) and reorganized circuit (*right*). The population responses of the GABAergic and dopaminergic neurons are plotted (in unspecified units) as *red* and *black traces*, respectively. The sharp peaks are the responses to brief (1 s) pulses of acetylcholine (ACh) of 12 μM amplitude. (C) Dopamine response as in (B), but to a four times higher nicotine dose. Model parameters: $r = 0.65$ for naive animals, and $r = 0.55$ after chronic nicotine; $s = 0.8$ throughout; there was further assumed to be a baseline cholinergic activity producing an equivalent baseline ACh concentration of 5 μM (see Fig. 3.4).

The responses to the pulses of endogenous ACh, however, reveal a serious side effect. These responses are much smaller in the reorganized VTA so as to vanish almost completely (compare the first three responses, before the nicotine is applied, indicated in the left and right panel by the open and *black arrow*, respectively). Paradoxically, the responses to ACh partly recover in the reorganized VTA provided nicotine is acutely applied (compare *narrow* and *broad black arrows* in right panel), because desensitization of the $\alpha 4\beta 2$s on the GABA neurons now disinhibits the dopamine neuron. A similar phenomenon has been found in slice experiments, where the dopamine neurons were much more sensitive to disinhibition by gabazine in nicotine-treated animals (Xiao et al., 2009). Hence this example suggests that nicotine may act to regularize the endogenous ACh response in nicotine addiction, and that smoking may be regarded as a kind of self-medication.

# 3.4 MODELING TONIC VERSUS PHASIC DOPAMINE RELEASE

The modeling framework we laid out above (Eqs. 3.1−3.5) is designed to understand the influence of receptor adaptations evoked by nicotine on the overall levels of activity in the VTA local circuit. However, it is well appreciated that dopamine neurons can sustain activation in a different way than a mere increase in firing rate. Their activity pattern, and in particular burst firing as opposed to tonic pace making, also matters for dopamine release (Goto et al., 2007). Therefore a more complex model has to be designed to fully restitute the emergent properties of dopamine neuron activity patterns as they are modified by acute and chronic nicotine. Here we go beyond the firing rate description of the DAergic and GABAergic populations in the VTA to focus on spiking models of dopamine neuron and how their activity is controlled by the patterns of afferent inputs that are in turn modified by nicotine.

VTA dopamine neurons display two distinct firing modes: tonic and phasic, as schematically indicated in Fig. 3.7. Tonic firing refers to spontaneously occurring baseline spikes (1−4 Hz) and is believed to be driven by pacemaker-like membrane currents (Grace and Onn, 1989; Grace and Bunney, 1984). In contrast, phasic activation is characterized by transiently occurring multispike firing sequences (15−30 Hz) called bursts. A crucial association between the firing pattern and the extracellular concentration of dopamine has been made (Dreyer et al., 2010; Goto et al., 2007). While tonic activity of VTA dopamine neurons sets a low background dopamine level in downstream regions and therefore underlies the baseline of dopamine concentration (Floresco et al., 2003; Smith and Grace, 1992), dopamine efflux is significantly elevated by burst firing.

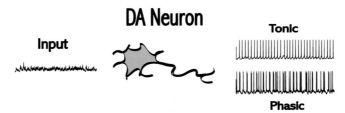

FIGURE 3.7    Illustration of the tonic versus phasic firing pattern in dopamine neurons. In the upper right panel, the neuron fires regularly (tonic mode). In the bottom right panel, burst firing patterns are observable (phasic mode).

Bursts trigger a high-amplitude, transient, phasic dopamine release at the synapse (Floresco et al., 2003; Grace, 1991).

The fact that all drugs of abuse have been inherently related to an increase of dopamine neurotransmission within the limbic system could be explained on the basis of the tonic–phasic dopaminergic activity (Grace, 2000). Addictive substances are known to increase dopaminergic signaling, and dopamine neurons suffer vigorous drug-induced internal/external changes as a consequence of chronic drug exposure. The duality between tonic and phasic dopamine systems might sustain a framework in which we can formulate a theory of addiction (Grace, 2000). Consequently, a deep understanding of the mechanisms underlying the electrical dynamics of dopamine neurons is of crucial importance to achieve a new level of insight.

Nonetheless, the way by which it is possible to tune the firing properties of dopamine cells has not been systematically explored. The ion channels involved in each mode as well as in the transition between modes remain unresolved. Because burst firing patterns are absent in slice preparations, afferent inputs onto dopamine neurons have been suggested to be critical (Grace and Onn, 1989; Overton and Clark, 1997). Because, in addition, nicotine can affect the firing pattern of dopamine neurons (Mameli-Engvall et al., 2006), we can broadly ask the following question: How does nicotine change the bursting properties of dopamine neurons? To address this issue, a modeling approach is used as aid to reasoning.

## 3.4.1 Modeling the Firing Pattern of Dopamine Neurons

From way back to the pioneering work of Hodgkin and Huxley in 1952, for which they received the 1963 Nobel Prize, it has become natural to represent neurons by electrical circuits. It offers a picture to describe their electrical properties and helps us to understand how neurons react when a current is applied. Just like an electrical circuit, applying a voltage pulse across the membrane generates an abrupt current. The amount of current that flows through these resistors is given by Ohm's law, which allows us to write:

$$C_m \frac{\mathrm{d}V(t)}{\mathrm{d}t} = I(t),$$

where $I$ represents the current, $C_m$ the membrane capacitance, and $\mathrm{d}V/\mathrm{d}t$ the rate of voltage change with respect to time (Dayan and Abbott, 2001). For the dopamine neuron, there is a minimal set of ionic currents which is necessary to reproduce the peculiar tonic/phasic firing pattern (Fig. 3.7). The ionic currents are written using the Hodgkin and Huxley formalism (Hodgkin and Huxley, 1952), and their detailed expressions can be found in Appendix A. The cell membrane is then modeled as a resistor and capacitor in parallel (Fig. 3.8). These all are some basic electronic components, which will make the model behave like actual cells and allow to reproduce the pattern of activity so specific to dopamine cells (see Fig. 3.7). Note that the previous modeling framework (see Eq. 3.5) did not have access to this, because rate models only capture the firing rate of dopamine neurons.

The advantage of a modeling approach is that causal explanations can be provided to figure out why dopamine neurons exhibit those distinct firing modes (Drion, 2013). The tonic activity pattern of dopamine neurons can be understood via a bifurcation diagram (Ermentrout, 2007). In applied mathematics, bifurcation theory is a standard tool introduced by Henri Poincaré to study how the behavior of dynamical systems changes with respect to variations in their parameters.

Fig. 3.9 illustrates this concept by showing how the membrane potential equilibrium changes as a function of calcium concentration. Drion et al. (2011) were the first to note the two values of calcium (low value and high value) present in the bifurcation diagram. These two values delimit the range in which there are two stable possible states indicated by the *solid red line*. The appearance of two stable states gives rise to the

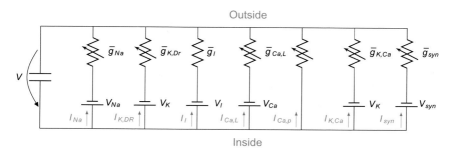

FIGURE 3.8  Schematic representation of the neuron equivalent circuit. The ionic currents are written using the Hodgkin–Huxley formalism, see Appendix A.

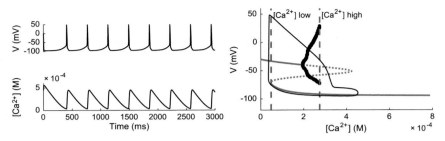

FIGURE 3.9  Mechanism of tonic firing pattern of the dopamine neuron model (Drion et al., 2011). On the left is shown the time evolution of the membrane potential (upper panel), as well as the time variation of the calcium concentration (bottom panel). The right plot illustrates the bifurcation diagram with $[Ca^{2+}]$ as the bifurcation parameter. The red line indicates fixed points, which can be stable (*solid line*) or unstable (*dashed*), and the black dots the presence of a stable limit cycle. The *vertical dashed lines* delimit two values of calcium (low and high) within which a range with two stable states exists (Drion et al., 2011). The *black line* indicates the trajectory to the cell. The bifurcation diagram was made using XPPAUT (Ermentrout, 2007). The full model is given in Appendix A with parameters of the simulation listed in Table 3.1.

phenomenon of hysteresis. This hysteresis in the $[Ca^{2+}]$-$V$ curve is essential for neurons and one can see how it is then possible to get a tonic firing pattern. Indeed, the neuron dynamics is such that with high-calcium concentration the voltage decreases, while the calcium concentration increases as soon as the voltage gets high, see Appendix A. As a consequence, the voltage will oscillate making up and down jumps between these limit states (see left panels of Fig. 3.9).

We can now push the analysis further to learn more about the conditions responsible for phasic activity patterns. The mechanism able to induce a transition toward the phasic mode is of particular interest because it accounts for a substantially increase of dopamine neurotransmission. Knowing that midbrain dopamine neurons exhibit bursting when exposed to pharmacological blockade of the Ca-activated small-conductance (SK) potassium channel, we designed a set of experiments in the spirit of Drion (2013) to test the reaction of the cell under two conditions: SK activated and SK blocked (Fig. 3.10). As expected, applying a step current induces an increase in firing rate, but the presence of SK channels keeps the response to the stimulus under control (Fig. 3.10 *left*). In contrast, once SK is blocked, the dopamine neuron becomes highly responsive and the increase in firing rate is brisk (Fig. 3.10 *right*).

Of course, in a real setting, a neuron's stimulus would be the result of a set of afferent neurons activating its synapses. To gain in realism, we model the input as a stochastic arrival of Glu pulses. This will cause fluctuations and brief increases of the membrane potential. The stochastically generated spike train of 50 Glu neurons, the resulting synaptic

TABLE 3.1    Main Notations Used Throughout This Paper and Their Biophysical Interpretations

| Parameter | Value |
|---|---|
| $C_m$ | $1\mu F/cm^2$ |
| $V_{Na}$ | $50\,mV$ |
| $V_K$ | $-95\,mV$ |
| $V_l$ | $-54.3\,mV$ |
| $V_{Ca}$ | $120\,mV$ |
| $\bar{g}_{Na}$ | $0.16\,S/cm^2$ |
| $\bar{g}_{K,DR}$ | $0.024\,S/cm^2$ |
| $\bar{g}_l$ | $0.0003\,S/cm^2$ |
| $\bar{g}_{Ca,L}$ | $0.0031\,S/cm^2$ |
| $\bar{g}_{syn}$ | $0.0001\,S/cm^2$ |
| $\bar{g}_{K,Ca}$ | $0.005\,S/cm^2$ |
| $I_{Ca,pump,max}$ | $0.0156\,mA/cm^2$ |
| $K_{M,P}$ | $0.0001\,mM$ |
| $K_{M,L}$ | $0.00018\,mM$ |
| $k_1$ | $0.0001375$ |
| $k_C$ | $0$ |
| $k_D$ | $0.0004\,mM$ |
| $k_2$ | $0.0000018$ |

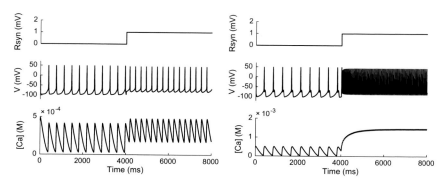

FIGURE 3.10    Influence of the small-conductance (SK) channel on the response of the dopamine neuron to current injection (Drion, 2013). In the left panels the SK channels are present, in the right panels they have been blocked. From top to bottom, the panels show the injected current (step function), membrane voltage, and calcium concentration.

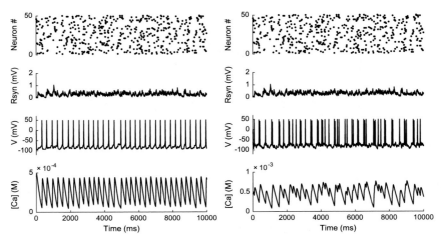

FIGURE 3.11    Tonic versus phasic firing in a stochastically stimulated dopamine neuron (Drion, 2013). Simulation of the dopamine neuron model receiving stochastically generated spike trains under two different conditions: small-conductance (SK) activated (left) and SK blocked (right). From top to bottom, the panels plot the stochastically generated spike trains, the resulting synaptic input, membrane voltage, and calcium concentration.

input, and the neural response are depicted in Fig. 3.11. Note that when SK is activated, the large fluctuations of the stimulus have little influence on the dynamics. The dopamine neuron keeps firing in a regular fashion. In contrast, when SK is blocked, the cell reacts to fluctuations by fast spiking transients. As a consequence, the SK channel appears to be a key player in switching between firing modes (see Fig. 3.12). By preserving the tonic activity, the SK channel might be seen as a regulator of the neural response (Drion, 2013).

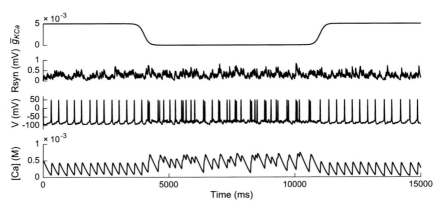

FIGURE 3.12    A switching mechanism (Drion, 2013). Simulation of the dopamine neuron model receiving a stochastic spike train. From top to bottom, the panels show the varied small-conductance, the stochastic synaptic input, membrane voltage, and calcium concentration.

This provides a clear explanation why bursts have not been observed in in vitro brain preparations (Grace and Onn, 1989). Afferent input is an indispensable ingredient for burst generation. The burst pattern can be seen as a response to stimulus variations. This is an important remark as it is thought that transitions between these two modes encode the context of unexpected rewarding stimuli and ultimately changes in dopamine concentration help to encode saliency to stimuli (Schultz, 2007; Redgrave et al., 2008).

## 3.4.2 Effects of Nicotine on the Dopamine Cell's Spike Pattern

Nicotine has been related to increases in dopamine release (Mameli-Engvall et al., 2006). By facilitating the emergence of irregularities in the spiking activity of dopamine neurons, nicotine increases bursting and ultimately enhances the dopamine transmission (Mameli-Engvall et al., 2006). On the other hand, experimental observations also suggest that nicotine enhances the synchrony between Glu neurons. So far, no link between the two phenomena has been established. In this section, we ask if a causal relation can be drawn between the enhancement in Glu synchrony and the enhanced bursting of dopamine neurons. To be more precise, we address the following question: Can the synchronization of afferent Glu neurons affect the bursting of dopamine neurons?

We report in Fig. 3.13 the effect of synchronization on the bursting level of the dopamine neurons. In a follow-up series of experiments, we have

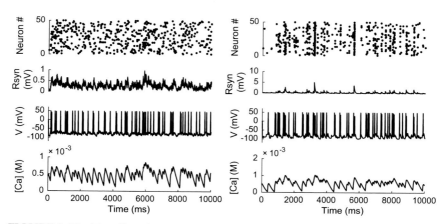

FIGURE 3.13    The effect of glutamatergic synchrony on bursting. Simulation of the dopamine neuron receiving a stochastically generated spike train under two different conditions: no correlation between spikes (*left*) and high correlation (*right*). The small-conductance channel had been downregulated to induce bursting more frequently. From top to bottom, the panels show the stochastically generated spike train, the resulting synaptic input, membrane voltage, and calcium concentration.

modified the amount of synchronization between the afferent Glu neurons (Macke et al., 2009). The stochastically generated spike trains of 50 Glu neurons are depicted in the upper panels of Fig. 3.13. Note the difference in the level of synchrony between the raster plots. Let us emphasize that the individual firing rates, as well as the population activity of the Glu afferent cells were kept constant, only the level of correlation between spikes was changed. In other words, the number of afferent spikes in the two scenarios is the same, only their temporal structure is modified. The consequence on the neural response can be seen on the voltage trace, the bursting aspect of the dopamine neuron seems enhanced.

The apparently enhanced bursting needs to be quantified in a more objective manner. Although several metrics have been suggested, there is no generally accepted statistic (or set of statistics) to quantify bursting activity (Ko et al., 2012; van Elburg and Ooyen, 2004; Robina et al., 2009). Bimodal interspike interval (ISI) histograms can be indicative of burst responses (Izhikevich, 2006). The rationale behind such a statement is that short ISIs occur more frequently when induced by burst firing, whereas longer ISIs correspond to pauses between bursts and indicate the time scale of burst separation (Fig. 3.14). This remark defines by itself a method for burst quantification and is at the core of most existing metrics that are built on a separation between the higher and lower activity regimes (Ko et al., 2012). As a first approach, we may simply quantify the bursting by looking at the coefficient of variation (CV) of the ISI histogram (Robina et al., 2009).

Fig. 3.15 plots the CV of the distribution of ISIs between the spikes of a single dopamine neuron for varying degrees of synchrony between its Glu afferents (Macke et al., 2009). The level of synchrony has been varied between the two regimes shown in Fig. 3.13. In other words, the left and right panels of Fig. 3.13 correspond to the lowest and highest levels of input covariance, respectively, simulated in Fig. 3.15. We observe an interesting linear relationship between synchrony of the afferents and

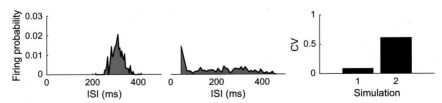

FIGURE 3.14   Distributions of interspike intervals (ISIs) for different spiking modes of the dopamine neuron. Left panel: the neuron fires in a tonic fashion and its ISI distribution is unimodal. Middle panel: the neuron fires in phasic way and its associated ISI histogram is bimodal. The right panel shows the corresponding coefficient of variation (CV). The ISIs (simulation 1–2) were obtained from the spiking activity of the simulations presented in Fig. 3.11.

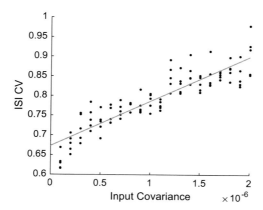

FIGURE 3.15 Effect of glutamatergic synchrony on bursting of the dopamine neuron. The coefficient of variation (CV) of the interspike interval (ISI) distribution is plotted against the input covariance. The red line shows the best linear fit. The input synchrony was varied between the two levels used in Fig. 3.13, and modified according to a procedure used in Macke et al. (2009).

burstiness of the dopamine neuron. This illustrates how synchronization of Glu afferents might be responsible for the increased bursting, and as a consequence also for the rise in the amount of dopamine released.

## 3.5 SUMMARY

Our aim in this chapter was to show how computational approaches may shed light on the mechanisms by which nicotine exerts its control over motivational signaling in the brain. Although both acute and chronic effects of nicotine have been seen across a wide range of brain circuits, and the receptors responding to nicotine are expressed throughout the brain, we focused on the dopaminergic signaling and notably on the VTA. The reasons for this are threefold: first, the high-affinity nAChRs show a particularly dense pattern of expression in the VTA, making it a perfect candidate for nicotine action. Second, much research over the past decades has pointed to the VTA and the dopamine neurotransmission emanating from this structure as the hub of signals that drive and instruct motivated behaviors. Third, last but not the least, genetic manipulations of the high-affinity receptors in the VTA supported the notion that this structure is indeed the key to the motivational effects of nicotine. In fact, Maskos et al. (2005) showed that mice that lack the nAChRs in the VTA do not acquire nicotine self-administration. Hence, one could argue that without these receptors, nicotine is not seen as rewarding. Even more interestingly, reexpression of these receptors only in the VTA was sufficient to rescue nicotine self-administration (Maskos et al., 2005).

In our initial global approach to the question of why nicotine might be addictive, while not particularly hedonically powerful (no "high" due to nicotine, and even wide anecdotal reports of initially experienced negative effects such as nausea), we lumped the VTA into a single population of dopaminergic neurons and examined how the dopamine signals, that instruct action selection learning, are dysregulated by nicotine. Interestingly, over the past decade the VTA has started to be seen as a heterogeneous structure with dopamine and local inhibitory interneurons, as well as endogenous Glu neurotransmission. Beyond that, recent data (Tolu et al., 2013) showed that reexpression of nAChRs on dopamine neurons by themselves is not sufficient to account neither for the changes in dopamine outflow, nor for the nicotine-evoked changes in dopamine cell firing patterns, and more crucially not for stable self-administration of the nicotine. Coexpression of the receptors on the local VTA GABA interneurons was also necessary (interestingly in GABA-reexpressed animals nicotine was aversive). These data clearly pointed that to understand how exactly nicotine exerts its effect on dopamine, a detailed circuit approach is necessary. We show in this chapter two complementary ways we have started to tackle these questions.

The first shows how we constructed a circuit level model of the VTA firing-rate dynamics that also included the dynamics of nicotine-driven (and endogenous ACh) effects. Of course these effects are mediated by the nAChRs. We are greatly aided in modeling the effect of these receptors by the lucky fact that they are in fact the classical ligand-gated ion channel that when open lead to depolarizing currents. While it is certain that modeling the receptors in detail is a complex enterprise, a classical simple kinetic scheme for the receptor-mediated currents (Katz and Thesleff, 1957) can be readily adapted to be used in the circuit models (Graupner et al., 2013). In the initial sections of the chapter we showed how this modeling framework can be used to describe the nicotine- and Ach-evoked current dynamics, and how to model different subtypes of nAChR-mediated currents. We further show that such a simple modeling scheme is sufficient to describe actions of competition and cooperation between ligands and allosteric modulation. The model in particular shows that the desensitization of the receptors is responsible for the descending shoulder of the prototypical U-shape dose-response curve to nicotine (whereas endogenous Ach, being rapidly broken down by the cholinesterase, does not lead to desensitization). This implies that there is an optimal dose for receptor opening and maximal current. One could speculate that this optimal dose may be related to doses of nicotine that are most efficient to evoke a behavioral response. The model further can profile how competitive agonists may in fact induce a suppression of response to nicotine, with an example of Varenicline (a compound marketed as a smoking reduction drug) taken to illustrate. The receptor models are then incorporated in a local circuit model of the VTA that includes dopamine and GABAergic neuronal populations. These populations are modeled at the level of their firing rates, and technically speaking the model is very similar to rate models that have been previously used for cortical circuits (e.g., the Wilson–Cowan model). Both dopamine and GABA populations receive Glu input and the dopamine population provides the dopamine output. Compared to the cortical circuitry, the connectivity is much simplified—GABA populations projects to the dopamine population. Because there is no evidence of dopamine receptors on the GABA neurons, we did not include feedback, and so the model presents the VTA as a feed-forward inhibition circuit. The crux of the model is that we explicitly include the simple receptor dynamic model that we can parametrically express on the dopamine and the GABA populations (for the model parameterized for the high-affinity $\alpha4\beta2$ receptors) and on the glutamate inputs (for the $\alpha7$ receptors). We have previously validated this approach by showing that it can account for responses to tonic nicotine and its dependence on the endogenous Ach (Graupner et al., 2013), responses to nicotine puffs (Tolu et al., 2013), and agonist induced dopamine outflow suppression (Maex et al., 2014).

In this chapter, we explicitly turned our attention to the interplay between the acute effects of nicotine and endogenous Ach and circuit reorganization induced by chronic nicotine. In particular exposure to chronic nicotine in rats showed that high-affinity nAChRs are increased in number differentially on GABA neurons in the VTA. We mimicked this increase in our circuit model. Our simulations showed that acute nicotine at doses that caused dopamine increase in control conditions, now cause a dopamine suppression in the reorganized circuit. Yet increased nicotine dose can still evoke a strong dopamine increase transient. Moreover, response of dopamine to endogenous Ach was suppressed by the nicotine-induced neuroadaptation, but when acute nicotine was present in the "addicted" circuit, Ach responding was recovered. In summary the simulations showed that the upregulation of nAChRs on the GABA neurons was sufficient to account qualitatively for (1) blunted dopamine response to moderate nicotine dose and (2) requirement for increased nicotine to evoke dopamine transient. Intriguingly, we also were able to show that nicotine in the addicted state may remediate responses to endogenous Ach. The latter point, we might speculate, could indeed be a key to the addictive liability of nicotine: In an addicted organism the presence of nicotine may restore "normal" function of the VTA circuitry, and if reward information in the environment is signaled by Ach (or some other variable attributable to the rewarding quality), restore motivational expectations. Arguably this squares up with heuristic reports of smokers that in the absence of nicotine they feel "out of sorts" and need a smoke to "make themselves feel normal." Furthermore, we might note that based on our previous work, the changes in dopamine response to nicotine and phasic Ach are dependent on the general levels of the endogenous Ach, hence it will be interesting to profile how the response in the reorganized "addicted" circuit may change if the Ach tone is increased or decreased.

Of course, the reorganization of receptors in the VTA is only part of the neuroadaptations induced by nicotine. The structure of the dopamine firing has also been observed to change, favoring bursts as opposed to tonic firing in the dopamine neurons. Also, as mentioned above, nicotine induces increased coherence of activity in the brain circuits that furnish Glu input to the VTA (e.g., the PFC and the hippocampal formation). The circuit model discussed above is too general to account for changes in the firing modes of dopamine neurons, as it represents the mean firing rate dynamics of the dopamine neuron population. To understand how the change in the input structure may promote burst firing in nicotine conditions, we turned to a more detailed modeling approach where we focus on a spiking model of a prototypical dopamine neuron. We showed how modulation on the intrinsic current composition of the neuron, notably downregulating the slow potassium SK current, may lead to increased bursting. We further showed how slow decreases in the intracellular calcium may lead to a decrease in irregularity and burstiness of the dopamine neuron. We then subjected our model to dynamic excitatory input whose statistics reflected the nicotine-induced changes seen. Our simulation showed that in dopamine neurons with a downregulated SK, an increase in coherence (partial synchrony) of the inputs is sufficient to induce a significant increase in both irregularity and bursting.

Burst firing in dopamine neurons has been previously linked to an increase on dopamine outflow, and specifically in dopamine transients. It would be natural to expect that coupling our spiking dopamine model with a dynamic model of dopamine release would lead to the observed effect of nicotine on dopamine release. Furthermore, a natural extension for this second approach is to incorporate the nAChR effects that we explored in the first part of the chapter. Here the challenge is to not only understand how activation and desensitization of the receptors on the dopamine neuron itself changes its excitability and responding to excitatory inputs, but also to understand what would be the role of altered VTA inhibitory dynamics on dopamine firing and outflow. In fact, our recent modeling showed that increased GABA neuron synchrony in the VTA in consort with dynamic excitatory inputs can increase high-frequency firing in dopamine neurons, and potentially bursting.

Incorporating biophysically realistic spiking models of dopamine and GABA neurons into a VTA circuit, taking into account the nAChR effect models, is indeed the next step. We believe that beyond understanding why nAChRs on both dopamine and GABA neurons are necessary to explain increased dopamine outflow under acute nicotine (Tolu et al., 2013), the spiking circuit/receptor modeling will enable us to predict what might happen after chronic nicotine. Furthermore, we may be able to tease out in the model circuit why certain drugs of addiction, such as nicotine and alcohol, show a strong positive interaction in both dopamine release and comorbidity to addiction. Last but not least two aspects of our circuit modeling framework should lead to new insights. First, we can use the nicotine receptor models to go beyond the VTA and to examine what might be the contributions of the various cellular types to generating the effects of nicotine (both pathological and therapeutic) in other brain regions, e.g., the PFC, hippocampus, etc. Second, we can use our circuit modeling approach to understand the cellular mechanisms of reward signaling in the dopamine circuitry, the effect of dopamine cell heterogeneity, the relative roles of local GABA and dopamine neurons, as well as inputs to constructing functional variables such as the reward prediction error. This approach should provide direct links with larger scale cognitive models, and may indeed allow us to tease apart how nicotine may lead to the onset of addiction.

# APPENDIX A: THE DOPAMINE NEURON MODEL

To assess how nicotine can subserve an adequate basis for the release of dopamine we have used a neural model first introduced in Drion et al. (2011). The cell is described by a single-compartment conductance-based model and obeys to

$$C_m \frac{d}{dt} V = -I_{Na} - I_{K,DR} - I_L - I_{Ca,L} - I_{Ca,pump} - I_{syn} - I_{K,Ca}.$$

The ionic currents are written using the Hodgkin–Huxley formalism.

$$I_{Na} = \bar{g}_{Na} m_\infty^3 h (V - V_{Na}),$$

$$I_K = \bar{g}_{K,DR} n_\infty^4 (V - V_K),$$

$$I_L = \bar{g}_L (V - V_L),$$

$$I_{Ca,L} = \bar{g}_{Ca,L} d_L f_L (V - V_{Ca}),$$

$$I_{Ca,pump} = I_{Ca,pump,max} \left( 1 + \frac{K_{M,P}}{[Ca^{2+}]_{in}} \right)^{-1},$$

$$I_{syn} = R_{syn} \bar{g}_{syn} (V - 0),$$

$$I_{K,Ca} = \bar{g}_{K,Ca} \left( \frac{[Ca^{2+}]_{in}}{K_D + [Ca^{2+}]_{in}} \right)^2 (V - V_K),$$

with the gating variable following

$$\frac{d}{dt}m = \alpha_m(1-m) - \beta_m m,$$

$$\frac{d}{dt}h = \alpha_h(1-h) - \beta_h h,$$

$$\frac{d}{dt}n = \alpha_n(1-n) - \beta_n n,$$

$$\tau_{d_L}\frac{d}{dt}d_L = d_{L,\inf}(v) - d_L,$$

$$\frac{d}{dt}\left[Ca^{2+}\right]_{in} = k_1\left(I_{Ca,L} - I_{Ca,pump} - k_c\left[Ca^{2+}\right]_{in} + k_2 I_{Na}\right),$$

where

$$\alpha_m = \frac{-0.025 * (V+40)}{\exp(-0.1(V+28))+1},$$

$$\beta_m = \exp(-(V+65)/18),$$

$$\alpha_h = 0.0175 \exp(-(V+65)/20),$$

$$\beta_h = \frac{0.25}{(1+\exp(-(V+35)/10))},$$

$$\alpha_n = \frac{0.0025(V+55)}{(1-\exp(-(V+55)/10))},$$

$$\beta_n = 0.03125 \exp(-((V+65)/80)).$$

$$d_{L,\inf} = \frac{1}{1+\exp(-(V+55)/3)}$$

$$\tau_{d_L} = 72 \exp\left(-(V+45)^2\big/400\right) + 6$$

$$f_L = \frac{K_{M,L}}{K_{M,L} + [Ca^{2+}]_{in}}.$$

# References

Albuquerque, E.X., Pereira, E.F., Alkondon, M., Rogers, S.W., 2009. Mammalian nicotinic acetylcholine receptors: from structure to function. Physiol. Rev. 89 (1), 73–120.

Anden, N.E., Dahlstrom, A., Fuxe, K., Larsson, K., Olson, L., Ungerstedt, U., 1966. Ascending monoamine neurons to the telencephalon and diencephalon. Acta Physiol. 67 (3), 313–326.

Baker, L.K., Mao, D., Chi, H., Govind, A.P., Vallejo, Y.F., Iacoviello, M., Herrera, S., Cortright, J.J., Green, W.N., McGehee, D.S., Vezina, P., 2013. Intermittent nicotine exposure upregulates nAChRs in VTA dopamine neurons and sensitises locomotor responding to the drug. Eur. J. Neurosci. 37 (6), 1004–1011.

Bennett, C., Arroyo, S., Berns, D., Hestrin, S., 2012. Mechanisms generating dual-component nicotinic EPSCs in cortical interneurons. J. Neurosci. 32 (48), 17287–17296.

Benowitz, N.L., 2009. Pharmacology of nicotine: addiction, smoking-induced disease, and therapeutics. Annu. Rev. Pharmacol. Toxicol. 49, 57–71.

Berger, B., Thierry, A.M., Tassin, J.P., Moyne, M.A., 1976. Dopaminergic innervation of the rat prefrontal cortex: a fluorescence histochemical study. Brain Res. 106 (1), 133–145.

Bonci, A., Malenka, R.C., 1999. Properties and plasticity of excitatory synapses on dopaminergic and gabaergic cells in the ventral tegmental area. J. Neurosci. 19 (10), 3723–3730.

Brozoski, T.J., Brown, R.M., Rosvold, H.E., Goldman, P.S., 1979. Cognitive deficit caused by regional depletion of dopamine in prefrontal cortex of rhesus monkey. Science 205 (4409), 929–932.

Buisson, B., Bertrand, D., 2001. Chronic exposure to nicotine upregulates the human $\alpha 4\beta 2$ nicotinic acetylcholine receptor function. J. Neurosci. 21 (6), 1819–1829.

Buisson, B., Bertrand, D., 2002. Nicotine addiction: the possible role of functional upregulation. Trends Pharmacol. Sci. 23 (3), 130–136.

Cachelin, A.B., Rust, G., 1994. Unusual pharmacology of (+)-tubocurarine with rat neuronal nicotinic acetylcholine receptors containing beta 4 subunits. Mol. Pharmacol. 46 (6), 1168–1174.

Cachope, R., Mateo, Y., Mathur, B.N., Irving, J., Wang, H.L., Morales, M., Lovinger, D.M., Cheer, J.F., 2012. Selective activation of cholinergic interneurons enhances accumbal phasic dopamine release: setting the tone for reward processing. Cell Rep. 2 (1), 33–41.

Calabresi, P., Lacey, M.G., North, R.A., 1989. Nicotinic excitation of rat ventral tegmental neurones in vitro studied by intracellular recording. Br. J. Pharmacol. 98 (1), 135–140.

Carr, D.B., Sesack, S.R., 2000. GABA-containing neurons in the rat ventral tegmental area project to the prefrontal cortex. Synapse 38 (2), 114–123.

Champtiaux, N., Gotti, C., Cordero-Erausquin, M., David, D.J., Przy- bylski, C., Lena, C., Clementi, F., Moretti, M., Rossi, F.M., Le Novere, N., McIntosh, J.M., Gardier, A.M., Changeux, J.P., 2003. Subunit composition of functional nicotinic receptors in dopaminergic neurons investigated with knock-out mice. J. Neurosci. 23 (21), 7820–7829.

Changeux, J.P., Bertrand, D., Corringer, J.P., Dehaene, S., Edelstein, S., Lena, C., Le Novere, N., Marubio, L., Picciotto, M., Zoli, M., 1998. Brain nicotinic receptors: structure and regulation, role in learning and reinforcement. Brain Res. Rev. 26 (2), 198–216.

Christie, M.J., Bridge, S., James, L.B., Beart, P.M., 1985. Excitotoxin lesions suggest an aspartatergic projection from rat medial prefrontal cortex to ventral tegmental area. Brain Res. 333 (1), 169–172.

Clements, J.R., Grant, S., 1990. Glutamate-like immunoreactivity in neurons of the laterodorsal tegmental and pedunculopontine nuclei in the rat. Neurosci. Lett. 120 (1), 70–73.

Cornwall, J., Cooper, J.D., Phillipson, O.T., 1990. Afferent and efferent connections of the laterodorsal tegmental nucleus in the rat. Brain Res. Bull. 25 (2), 271–284.

Dani, J.A., Heinemann, S., 1996. Molecular and cellular aspects of nicotine abuse. Neuron 16, 905–908.

Dani, J.A., Ji, D., Zhou, Fu-M., 2001. Synaptic plasticity and nicotine addiction. Neuron 31 (3), 349–352.

Dayan, P., Abbott, L., 2001. Theoretical Neuroscience. MIT Press.

Dezfouli, A., Piray, P., Keramati, M.M., Ekhtiari, H., Lucas, C., Mokri, A., 2009. A neurocomputational model for cocaine addiction. Neural Comput. 21 (10), 2869–2893.

Dreyer, J.K., Herrik, K.F., Berg, R.W., Hounsgaard, J.D., 2010. Influence of phasic and tonic dopamine release on receptor activation. J. Neurosci. 30 (42), 14273–14283.

Drion, G., Massotte, L., Sepulchre, R., Seutin, V., 2011. How modeling can reconcile apparently discrepant experimental results: the case of pacemaking in dopaminergic neurons. PLoS Comput. Biol. 7, e1002050.

Drion, G., 2013. Regulation of Excitability, Pacemaking and Bursting: Insights from Dopamine Neuron Electrophysiology. Master's thesis. University of Liege.

Ermentrout, B., 2007. Xppaut. Scholarpedia 2 (1), 1399.

Feldberg, W., 1945. Recent views on the mode of action of acetylcholine in the central nervous system. Physiol. Rev. 25, 596−642.

Fenster, C.P., Rains, M.F., Noerager, B., Quick, M.W., Lester, R.A., 1997. Influence of subunit composition on desensitization of neuronal acetylcholine receptors at low concentrations of nicotine. J. Neurosci. 17 (15), 5747−5759.

Floresco, S.B., West, A.R., Ash, B., Moore, H., Grace, A.A., 2003. Afferent modulation of dopamine neuron firing differentially regulates tonic and phasic dopamine transmission. Nat. Neurosci. 6, 968−973.

Forster, G.L., Blaha, C.D., 2000. Laterodorsal tegmental stimulation elicits dopamine efflux in the rat nucleus accumbens by activation of acetylcholine and glutamate receptors in the ventral tegmental area. Eur. J. Neurosci. 12 (10), 3596−3604.

Forster, G.L., Blaha, C.D., 2003. Pedunculopontine tegmental stimulation evokes striatal dopamine efflux by activation of acetylcholine and glutamate receptors in the midbrain and pons of the rat. Eur. J. Neurosci. 17 (4), 751−762.

Galzi, J.L., Edelstein, S.J., Changeux, J., 1996. The multiple phenotypes of allosteric receptor mutants. Proc. Natl. Acad. Sci. U.S.A. 93 (5), 1853−1858.

Giniatullin, R., Nistri, A., Yakel, J.L., 2005. Desensitization of nicotinic ach receptors: shaping cholinergic signaling. Trends Neurosci. 28 (7), 371−378.

Glowinski, J., Cheramy, A., Romo, R., Barbeito, L., 1988. Presynaptic regulation of dopaminergic transmission in the striatum. Cell Mol. Neurobiol. 8 (1), 7−17.

Good, C.H., Lupica, C.R., 2009. Properties of distinct ventral tegmental area synapses activated via pedunculopontine or ventral tegmental area stimulation in vitro. J. Physiol. 587 (Pt 6), 1233−1247.

Goto, Y., Otani, S., Grace, A., 2007. The yin and yang of dopamine release: a new perspective. Neuropharmacology 53, 583−587.

Grace, A.A., Bunney, B.S., 1984. The control of firing pattern in nigral dopamine neurons: burst firing. J. Neurosci. 4 (11), 2877−2890.

Grace, A.A., Onn, S.P., 1989. Morphology and electrophysiological properties of immunocytochemically identified rat dopamine neurons recorded in vitro. J. Neurosci. 9 (10), 3463−3481.

Grace, A., 1991. Phasic versus tonic dopamine release and the modulation of dopamine system responsivity: a hypothesis for the etiology of schizophrenia. Neuroscience 41 (1), 1−24.

Grace, A., 2000. The tonic/phasic model of dopamine system regulation and its implications for understanding alcohol and psychostimulant craving. Addiction 95, 119−128.

Grady, S.R., Grun, E.U., Marks, M.J., Collins, A.C., 1997. Pharmacological comparison of transient and persistent [$^3$H]dopamine release from mouse striatal synaptosomes and response to chronic ʟ-nicotine treatment. J. Pharmacol. Exp. Ther. 282 (1), 32−43.

Graupner, M., Gutkin, B., 2009. Modeling nicotinic neuromodulation from global functional and network levels to nAChR based mechanisms. Acta Pharmacol. Sin. 30 (6), 681−693.

Graupner, M., Maex, R., Gutkin, B., 2013. Endogenous cholinergic inputs and local circuit mechanisms govern the phasic mesolimbic dopamine response to nicotine. PLoS Comput. Biol. 9 (8), e1003183.

Gutkin, B., Dehaene, S., Changeux, J.P., 2006. A neurocomputational hypothesis for nicotine addiction. PNAS 103 (4), 1106−1111.

Henderson, B.J., Lester, H.A., 2015. Inside-out neuropharmacology of nicotinic drugs. Neuropharmacology 96 (Pt B), 178−193.

Henningfield, J.E., Stapleton, J.M., Benowitz, N.L., Grayson, R.F., London, E.D., 1993. Higher levels of nicotine in arterial than in venous blood after cigarette smoking. Drug Alcohol Depend. 33 (1), 23–29.

Hille, B., 2001. Ion Channels of Excitable Membranes. Sinauer, Sunderland, MA.

Hodgkin, A.L., Huxley, A.F., 1952. A quantitative description of membrane current and its application to conduction and excitation in nerve. J. Neurophysiol. 117 (4), 500–544.

Ikemoto, S., Kohl, R.R., McBride, W.J., 1997. GABA(A) receptor blockade in the anterior ventral tegmental area increases extracellular levels of dopamine in the nucleus accumbens of rats. J. Neurochem. 69 (1), 137–143.

Izhikevich, E.M., 2006. Bursting. Scholarpedia 1 (3), 1300.

Jalabert, M., Bourdy, R., Courtin, J., Veinante, P., Manzoni, O.J., Barrot, M., Georges, F., 2011. Neuronal circuits underlying acute morphine action on dopamine neurons. Proc. Natl. Acad. Sci. U.S.A. 108 (39), 16446–16450.

Johnson, S.W., North, R.A., 1992. Two types of neurone in the rat ventral tegmental area and their synaptic inputs. J. Physiol. 450, 455–468.

Jones, I.W., Wonnacott, S., 2004. Precise localization of $\alpha7$ nicotinic acetylcholine receptors on glutamatergic axon terminals in the rat ventral tegmental area. J. Neurosci. 24 (50), 11244–11252.

Katz, B., Thesleff, S., 1957. A study of the desensitization produced by acetylcholine at the motor end-plate. J. Physiol. 138 (1), 63–80.

Keramati, M., Gutkin, B.S., 2002. Imbalanced decision hierarchy in addicts emerging from drug-hijacked dopamine spiraling circuit. PLoS One 8 (4), e61489.

Klink, R., de Kerchove d'Exaerde, A., Zoli, M., Changeux, J.P., 2001. Molecular and physiological diversity of nicotinic acetylcholine receptors in the midbrain dopaminergic nuclei. J. Neurosci. 21 (5), 1452–1463.

Ko, D., Wilson, C.J., Lobb, C.J., Paladini, C.A., 2012. Detection of bursts and pauses in spike trains. J. Neurosci. Methods 15 (211), 145–158.

Koranda, J.L., Cone, J.J., McGehee, D.S., Roitman, M.F., Beeler, J.A., Zhuang, X., 2014. Nicotinic receptors regulate the dynamic range of dopamine release in vivo. J. Neurophysiol. 111 (1), 103–111.

Lacey, M.G., Mercuri, N.B., North, R.A., 1989. Two cell types in rat substantia nigra zona compacta distinguished by membrane properties and the actions of dopamine and opioids. J. Neurosci. 9 (4), 1233–1241.

Lindvall, O., Björklund, A., Moore, R.Y., Stenevi, U., 1974. Mesencephalic dopamine neurons projecting to neocortex. Brain Res. 81 (2), 325–331.

Lodge, D.J., Grace, A.A., 2006. The laterodorsal tegmentum is essential for burst firing of ventral tegmental area dopamine neurons. PNAS 103 (13), 5167–5172.

Macke, J.H., Berens, P., Ecker, A.S., Tolias, A.S., Bethge, M., 2009. Generating spike trains with specified correlation coefficients. Neural Comput. 21 (2), 397–423.

Maex, R., Grinevich, V.P., Grinevich, V., Budygin, E., Bencherif, M., Gutkin, B., 2014. Understanding the role $\alpha7$ nicotinic receptors play in dopamine efflux in nucleus accumbens. ACS Chem. Neurosci. 5 (10), 1032–1040.

Mameli-Engvall, M., Evrard, A., Pons, S., Maskos, U., Svensson, T.H., Changeux, J.P., Faure, P., 2006. Hierarchical control of dopamine neuron-firing patterns by nicotinic receptors. Neuron 50 (6), 911–921.

Mansvelder, H.D., McGehee, D.S., 2000. Long-term potentiation of excitatory inputs to brain reward areas by nicotine. Neuron 27 (2), 349–357.

Marks, M.J., Grady, S.R., Salminen, O., Paley, M.A., Wageman, C.R., Mcintosh, J.M., Whiteaker, P., 2014. $\alpha6\beta2*$-subtype nicotinic acetylcholine receptors are more sensitive than $\alpha4\beta2*$-subtype receptors to regulation by chronic nicotine administration. J. Neurochem. 130 (2), 185–198.

Maskos, U., Molles, B.E., Pons, S., Besson, M., Guigard, B.P., Guilloux, J.P., Evrard, A., Cazala, P., Cormier, A., Mameli-Engvall, M., Dufour, N., Cloez-Tayarani, I., Bemelmans, A.P., Mallet, J., Gardier, A.M., David, V., Faure, P., Granon, S., Changeux, J.P., 2005. Nicotine reinforcement and cognition restored by targeted expression of nicotinic receptors. Nature 27 (436), 103–107.

McDaid, J., Abburi, C., Wolfman, S.L., Gallagher, K., McGehee, D.S., 2016. Ethanol-induced motor impairment mediated by inhibition of $\alpha7$ nicotinic receptors. J. Neurosci. 36 (29), 7768–7778.

Melis, M., Scheggi, S., Carta, G., Madeddu, C., Lecca, S., Luchicchi, A., Cadeddu, F., Frau, R., Fattore, L., Fadda, P., Ennas, M.G., Castelli, M.P., Fratta, W., Schilstrom, B., Banni, S., De Montis, M.G., Pistis, M., 2013. PPAR$\alpha$ regulates cholinergic-driven activity of midbrain dopamine neurons via a novel mechanism involving $\alpha7$ nicotinic acetylcholine receptors. J. Neurosci. 33 (14), 6203–6211.

Nashmi, R., Xiao, C., Deshpande, P., McKinney, S., Grady, S.R., Whiteaker, P., Huang, Q., McClure-Begley, T., Lindstrom, J.M., Labarca, C., Collins, A.C., Marks, M.J., Lester, H.A., 2007. Chronic nicotine cell specifically upregulates functional $\alpha4^*$ nicotinic receptors: basis for both tolerance in midbrain and enhanced long-term potentiation in perforant path. J. Neurosci. 27 (31), 8202–8218.

Oades, R.D., Halliday, G.M., 1987. Ventral tegmental (A10) system: neurobiology. 1. Anatomy and connectivity. Brain Res. 434 (2), 117–165.

Oakman, S.A., Faris, P.L., Kerr, P.E., Cozzari, C., Hartman, B.K., 1995. Distribution of ponto-mesencephalic cholinergic neurons projecting to substantia nigra differs significantly from those projecting to ventral tegmental area. J. Neurosci. 15 (9), 5859–5869.

Oliva, I., Wanat, M.J., 2016. Ventral tegmental area afferents and drug-dependent behaviors. Front. Psychiatry 7, 30.

Overton, P.G., Clark, D., 1997. Burst firing in midbrain dopaminergic neurons. Brain Res. Rev. 25 (3), 312–334.

Pakkanen, J.S., Jokitalo, E., Tuominen, R.K., 2005. Up-regulation of $\beta2$ and $\alpha7$ subunit containing nicotinic acetylcholine receptors in mouse striatum at cellular level. Eur. J. Neurosci. 21 (10), 2681–2691.

Perez, X.A., Khroyan, T.V., McIntosh, J.M., Quik, M., 2015. Varenicline enhances dopamine release facilitation more than nicotine after long-term nicotine treatment and withdrawal. Pharmacol. Res. Perspect. 3 (1), e00105.

Picciotto, M.R., Zoli, M., Rimondini, R., Lena, C., Marubio, L.M., Pich, E.M., Fuxe, K., Changeux, J.P., 1998. Acetylcholine receptors containing the beta2 subunit are involved in the reinforcing properties of nicotine. Nature 391, 173–177.

Pidoplichko, V.I., DeBiasi, M., Williams, J.T., Dani, J.A., 1997. Nicotine activates and desensitizes midbrain dopamine neurons. Nature 390, 401–404.

Quick, M.W., Lester, R.A., 2002. Desensitization of neuronal nicotinic receptors. J. Neurobiol. 53 (4), 457–478.

Quik, M., Wonnacott, S., 2011. $\alpha6\beta2^*$ and $\alpha4\beta2^*$ nicotinic acetylcholine receptors as drug targets for parkinson's disease. Pharmacol. Rev. 63 (4), 938–966.

Quik, M., Perez, X.A., Grady, S.R., 2011. Role of $\alpha6$ nicotinic receptors in cns dopaminergic function: relevance to addiction and neurological disorders. Biochem. Pharmacol. 82 (8), 873–882.

Redgrave, P., Gurney, K., Reynolds, J., 2008. What is reinforced by phasic dopamine signals? Brain Res. Rev. 322–339.

Redish, A.D., 2004. Addiction as a computational process gone awry. Science 306 (7503), 1944–1997.

Reynolds, J.N., Wickens, J.R., 2002. Dopamine-dependent plasticity of corti-costriatal synapses. Neural Netw. 15 (4), 507–521.

Robina, K., Mauriceb, N., Degosb, B., Deniau, J.-M., Martineriec, J., Pezard, L., 2009. Assessment of bursting activity and interspike intervals variability: a case study for methodological comparison. J. Neurosci. Methods 179 (142−149).

Rollema, H., Shrikhande, A., Ward 3rd, K.M., Tingley, F.D., Coe, J.W., O'Neill, B.T., Tseng, E., Wang, E.Q., Mather, R.J., Hurst, R.S., Williams, K.E., de Vries, M., Cremers, T., Bertrand, S., Bertrand, D., 2010. Preclinical properties of the $\alpha 4\beta 2$ nicotinic acetylcholine receptor partial agonists varenicline, cytisine and dianicline translate to clinical efficacy for nicotine dependence. Br. J. Pharmacol. 160 (2), 334−345.

Rose, J.E., Mukhin, A.G., Lokitz, S.J., Turkington, T.G., Herskovic, J., Behm, F.M., Garg, S., Garg, P.K., 2010. Kinetics of brain nicotine accumulation in dependent and nondependent smokers assessed with pet and cigarettes containing 11c-nicotine. Proc. Natl. Acad. Sci. U.S.A. 107 (11), 5190−5195.

Schultz, W., 2007. Behavioral dopamine signals. Trends Neurosci. 30 (5), 203−210.

Segel, I.H., 1975. Enzyme Kinetics. Wiley & Sons, New York.

Segel, L.A., 1984. Modeling Dynamic Phenomena in Molecular and Cellular Biology. Cambridge University Press, Cambridge, UK.

Sesack, S.R., Pickel, V.M., 1992. Prefrontal cortical efferents in the rat synapse on unlabeled neuronal targets of catecholamine terminals in the nucleus accumbens septi and on dopamine neurons in the ventral tegmental area. J. Comp. Neurol. 320 (2), 145−160.

Shelley, C., Cull-Candy, S.G., 2010. Desensitization and models of receptor-channel activation. J. Physiol. 588 (Pt 9), 1395−1397.

Smeets, W.J., Marin, O., Gonzalez, A., 2000. Evolution of the basal ganglia: new perspectives through a comparative approach. J. Anat. 196 (Pt 4), 501−517.

Smith, I.D., Grace, A.A., 1992. Role of the subthalamic nucleus in the regulation of nigral dopamine neuron activity. Synapse 12 (4), 287−303.

Steffensen, S.C., Svingos, A.L., Pickel, V.M., Henriksen, S.J., 1998. Electrophysiological characterization of GABAergic neurons in the ventral tegmental area. J. Neurophysiol. 1 (19), 8003−8015.

Sugita, S., Johnson, S.W., Alan North, R., 1992. Synaptic inputs to GABAA and GABAB receptors originate from discrete afferent neurons. Neurosci. Lett. 134 (2), 207−211.

Swanson, L.W., 1982. The projections of the ventral tegmental area and adjacent regions: a combined fluorescent retrograde tracer and immunofluorescence study in the rat. Brain Res. Bull. 1 (6), 321−353.

Taber, M.T., Fibiger, H.C., 1995. Electrical stimulation of the prefrontal cortex increases dopamine release in the nucleus accumbens of the rat: modulation by metabotropic glutamate receptors. J. Neurophysiol. 15 (5), 3896−3904.

Taly, A., Corringer, P.J., Guedin, D., Lestage, P., Changeux, J.P., 2009. Nicotinic receptors: allosteric transitions and therapeutic targets in the nervous system. Nat. Rev. Drug Discov. 8 (9), 733−750.

Tan, K.R., Brown, M., Labouebe, G., Yvon, C., Creton, C., Fritschy, J.M., Rudolph, U., Luscher, C., 2010. Neural bases for addictive properties of benzodiazepines. Nature 463 (7282), 769−774.

Taylor, D.H., Burman, P.N., Hansen, M.D., Wlcox, R.S., Larsen, B.R., Blanchard, J.K., Merrill, C.B., Edwards, J.G., Sudweeks, S.N., Jie Wu, M.D., Arias, H.R., Steffensen, S.C., 2013. Nicotine enhances the excitability of GABA neurons in the ventral tegmental area via activation of alpha 7 nicotinic receptors on glutamate terminals. Biochem. Pharmacol. S1. http://dx.doi.org/10.4172/2167-0501. S1-002.

Thierry, A.M., Blanc, G., Sobel, A., Stinus, L., Glowinski, J., 1973. Dopaminergic terminals in the rat cortex. Science 182 (4111), 499−501.

Tolu, S., Eddine, R., Marti, F., David, V., Graupner, M., Pons, S., Baudonnat, M., Husson, M., Besson, M., Reperant, C., Zemdegs, J., PagÃŔs, C., Hay, Y.A., Lambolez, B., Caboche, J., Gutkin, B., Gardier, A.M., Changeux, J.P., Faure, P., Maskos, U., 2013. Co-activation of VTA DA and GABA neurons mediates nicotine reinforcement. Mol. Psychiatry 27 (18), 382–393.

Tong, Z.Y., Overton, P.G., Clark, D., 1996. Stimulation of the prefrontal cortex in the rat induces patterns of activity in midbrain dopaminergic neurons which resemble natural burst events. Synapse 22 (3), 195–208.

Ungerstedt, U., 1971. Stereotaxic mapping of the monoamine pathways in the rat brain. Acta Physiol. Scand. Suppl. 367, 1–48.

Usher, M., McClelland, J.L., 2001. The time course of perceptual choice: the leaky, competing accumulator model. Psychol. Rev. 10 (3), 550–592.

Van Bockstaele, E.J., Pickel, V.M., 1995. GABA-containing neurons in the ventral tegmental area project to the nucleus accumbens in rat brain. Brain Res. 682 (1), 215–221.

van de Giessen, E., Weinstein, J.J., Cassidy, C.M., Haney, M., Dong, Z., Ghazzaoui, R., Ojeil, N., Kegeles, L.S., Xu, X., Vadhan, N.P., Volkow, N.D., Slifstein, M., Abi-Dargham, A., 2016. Deficits in striatal dopamine release in cannabis dependence. Mol. Psychiatry 22, 68–75. http://dx.doi.org/10.1038/mp.2016.21.

van Elburg, R.A.J., Ooyen, A.van, 2004. A new measure for bursting. Neurocomputing 58, 497–502.

Volkow, N.D., Fowler, J.S., Wang, G.J., Swanson, J.M., 2004. Dopamine in drug abuse and addiction: results from imaging studies and treatment implications. Mol. Psychiatry 9 (6), 557–569.

Walsh, H., Govind, A.P., Mastro, R., Hoda, J.C., Bertrand, D., Vallejo, Y., Green, W.N., 2008. Up-regulation of nicotinic receptors by nicotine varies with receptor subtype. J. Biol. Chem. 283 (10), 6022–6032.

Wang, X.J., 2012. Neural dynamics and circuit mechanisms of decision-making. Curr. Opin. Neurobiol. 22 (6), 1039–1046.

Williams, D.K., Wang, J., Papke, R.L., 2011. Positive allosteric modulators as an approach to nicotinic acetylcholine receptor-targeted therapeutics: advantages and limitations. Biochem. Pharmacol. 82 (8), 915–930.

Wonnacott, S., 1990. The paradox of nicotinic acetylcholine receptor upregulation by nicotine. Trends Pharmacol. Sci. 11 (6), 216–219.

Wu, J., Lukas, R.J., 2011. Naturally-expressed nicotinic acetylcholine receptor subtypes. Biochem. Pharmacol. 82 (8), 800–807.

Xiao, C., Nashmi, R., McKinney, S., Cai, H., McIntosh, J.M., Lester, H.A., 2009. Chronic nicotine selectively enhances $\alpha 4\beta 2^*$ nicotinic acetylcholine receptors in the nigrostriatal dopamine pathway. J. Neurosci. 29 (40), 12428–12439.

Yamamoto, K., Vernier, P., 2011. The evolution of dopamine systems in chordates. Front. Neuroanat. 5, 21.

Zhou, F.M., Liang, Y., Dani, J.A., 2001. Endogenous nicotinic cholinergic activity regulates dopamine release in the striatum. Nat. Neurosci. 4 (12), 1224–1229.

Zwart, R., Vijverberg, H.P., 1997. Potentiation and inhibition of neuronal nicotinic receptors by atropine: competitive and noncompetitive effects. Mol. Pharmacol. 52 (5), 886–895.

# MODELING NEURAL SYSTEM DISRUPTIONS IN PSYCHIATRIC ILLNESS

# Computational Models of Dysconnectivity in Large-Scale Resting-State Networks

*Murat Demirtaş*[1,2], *Gustavo Deco*[2,3]

[1] Yale University, New Haven, CT, United States; [2] Universitat Pompeu Fabra, Barcelona, Spain; [3] Institució Catalana de Recerca i Estudis Avançats (ICREA), Barcelona, Spain

OUTLINE

*Computational Psychiatry*
http://dx.doi.org/10.1016/B978-0-12-809825-7.00004-3

87

## 4.1 INTRODUCTION

Phineas Gage, in 1868, survived an occupational accident in which an iron bar passed through the left side of his skull. As a result, the dramatic changes in his personality were attributed to the damage to his brain's left frontal lobe (Harlow, 1868). Seven years earlier, Paul Broca published the autopsy findings of Louis Victor Leborgne, who had experienced a sudden loss of his speech ability, and showed a lesion in the left inferior frontal gyrus (Broca, 1861). In the 20th century, as a result of the removal of bilateral medial temporal lobes in epilepsy surgery, Henry Gustave Molaison (known as H.M.) lost his ability to transfer his short-term memory to long-term memory (Scoville and Milner, 1957). Finally, the behavioral changes due to the removal of interhemispheric connections were reported (Gazzaniga et al., 1963; Myers and Sperry, 1953). These four groundbreaking cases led to the idea of "dysconnexion syndromes," thus emphasizing the role of disconnection between cortical regions in the genesis of clinical disorders (Geschwind, 1965a,b). During the past 2 decades, advances in neuroimaging have made it possible to test disease-related alterations using noninvasive techniques. Recently, researchers were able to reconstruct the lesions in these seminal cases using diffusion weighted imaging (DWI) (de Schotten et al., 2015).

We shall not provide an overview of the literature of neuroimaging studies before we highlight the distinction between the aforementioned cases (i.e., Gage, Leborgne, and Molaison) and widespread psychiatric diseases. As an analogy, we can borrow the definition of "rare diseases," which has important implications in genetic research. Technological progress has allowed researchers to identify genetic markers of most rare diseases, mainly due to their extremely low prevalence and dramatic consequences (Aymé and Schmidtke, 2007). However, despite tremendous efforts the genetic origins of highly prevalent, complex diseases (such as psychiatric and developmental disorders) remain obscure (Motulsky, 2006; McCarthy et al., 2008; Manolio et al., 2009). Noting this, we argue that it is very unlikely that simple loci will be found for most of the psychiatric disorders. Instead, we believe that the overlap, as well as potentially contradictory findings, between different disorders is a natural consequence of the mechanisms behind the working principles of the brain. The seminal cases that suggest the mechanistic link between brain disconnection and function should be treated as very special exceptions. We focus, in a broader context, on the relationship between the large-scale spontaneous coordination in the brain and common psychiatric disorders.

In this chapter, we first review resting-state functional magnetic resonance imaging (fMRI) that is now the most established experimental tool to study large-scale functioning of the brain. We then introduce large-scale computational models and discuss their role in enhancing our current understanding of macroscopic scale neuronal dynamics. By its nature, theoretical studies rely on experimental measures describing various phenomena observed in the brain. Theoretical approaches welcome multifaceted, even controversial phenomena, in the first place because these kinds of phenomena bring opportunity to develop more accurate and unifying models. The purpose of these models is to provide testable hypotheses that can solve difficult scientific problems. For this reason, here we aim to facilitate the communication between experimental and theoretical research in large-scale brain dynamics.

## 4.1.1 The Study of Large-Scale Brain Connectivity

Large-scale brain connectivity can be defined through multiple perspectives. The first one is the direct physical links among brain regions that we observe by the defined anatomical structures. Second, the functional relationship among brain regions can be observed based on measuring the association between direct and indirect measures neuronal activity in the brain. For instance, functional connectivity (FC) can be defined as the statistical correlation of the neural activity of two regions. At the microcircuit level, the relationship between structure and function is extensively studied using recent developments in anatomical and neurophysiological techniques such as optogenetics. These techniques exploit experimental manipulation of neuronal activity in the brain to understand how information is processed. Recent research has been successful in understanding how local neuronal circuits operate to process the information from the environment. This provides a precise but incomplete picture of the inner workings of the brain as the localized focus has to neglect the role of the large-scale interactions among brain areas. How information is integrated across brain networks remains unknown. Some new techniques allow the study of large-scale brain connectivity at much finer temporal and spatial resolution; however, immediate application of high-resolution techniques is limited in humans due to their invasive nature. Here we focus on resting-state fMRI (rs-fMRI) studies.

## 4.2 RESTING-STATE FUNCTIONAL CONNECTIVITY AND NETWORKS IN FUNCTIONAL MAGNETIC RESONANCE IMAGING

The study that is considered the foundation of rs-fMRI is Bharat Biswal's findings on low-frequency temporal correlations in the motor cortex during rest, implying that there is indeed FC among brain networks in the absence of any explicit task (Biswal et al., 1995). However, there was limited interest in rs-fMRI until the review by Gusnard and Raichle that highlighted regional heterogeneity in the baseline activity of the brain and coined the term "default mode" in 2001 (Gusnard and Raichle, 2001). Gusnard and Raichle considered the metabolic cost of baseline activity and pointed out the activations as well as deactivations in the brain across rest and task conditions. Raichle et al. (2001) emphasized that a functionally connected network of regions (comprising posterior cingulate, precuneus, medial prefrontal cortex, and hippocampus) is consistently observed during the baseline state, and defined this particular network as a default mode network (DMN). Another crucial finding was that the two core regions of DMN, the posterior cingulate and ventral anterior cingulate, were activated during rest and attenuated during cognitive processing (Greicius et al., 2003).

Subsequent to the discovery of DMN, other resting-state networks (RSNs) were identified (Fransson, 2005) based on the temporal correlations between the blood-oxygen-level dependent (BOLD) signal in one region (i.e., the seed) and the rest of the brain. Moreover, after Beckmann and Smith (2004) introduced the use of independent component analysis (ICA), many other RSNs were discovered (Beckmann et al., 2005). Another important finding was the inverse relationship between two major RSNs, namely the task-positive network (TPN) and task-negative network (TNN) (Fox et al., 2005; Fox and Raichle, 2007). Despite technical and interpretational criticisms, mounting evidence supported the functional relevance of RSNs (De Luca et al., 2006). Several studies found that learning could modify RSNs (Albert et al., 2009; Lewis et al., 2009). Moreover, graph theoretical analysis of resting-state FC (rs-FC) found a relationship between IQ and efficiency (van den Heuvel et al., 2009a). After developments in anatomical fiber tracking, Greicius et al. (2009) showed that the rs-FC was closely related to the underlying structural connectivity. Later this finding was extended to include the structural origins of RSNs (Damoiseaux and Greicius, 2009; van den Heuvel et al., 2009b). Structural analysis emphasized the role of high-influence regions (or hubs) in the emergence of FC (Achard et al., 2006).

The fundamental concept that uncovered the importance of resting-state fluctuations was FC among brain regions (Friston, 2011). Paradoxically, FC has always been the most controversial concept in resting-state neuroimaging research. Various connectivity measures (such as correlation coefficient, phase coherence, mutual information, etc.) were explored to quantify FC. Among them, Pearson's correlation coefficient has been the most widely used measure for FC (Bandettini et al., 1993; Biswal et al., 1995). The vulnerability of FC measures to the confounding factors such as physiological noise and the validity of the available statistical approaches were frequently addressed in methodological studies (Birn, 2012). Furthermore, relating the structural correlates of connectivity to FC has been considered a highly complex, "multifaceted" problem (Horwitz, 2003; Rogers et al., 2007).

The most frequently discussed concern relates to the impact of a standard preprocessing procedure that removes the global BOLD fluctuations from the signal, i.e., global signal regression (GSR) (Murphy et al., 2009). GSR was intended to remove the fluctuations in the BOLD signal that reflected correlated physiological artifacts rather than being of neuronal origin.

Apart from the preprocessing concerns, seed-based correlations were problematic because they only reflected hypothesis-driven information but ignored the remaining network interactions and introduced bias to the analysis. Furthermore, an increase in the number of seeds would cause additional problems due to difficulties in controlling type-I (and type-II) errors in statistical analyses (multiple comparisons). ICA was proposed as a means to provide a solid statistical framework with which to analyze RSNs (Beckmann and Smith, 2004).

## 4.3 DYNAMIC FUNCTIONAL CONNECTIVITY

Another important debate regarding FC addresses the assumption that the correlation structure between regions is stationary over time. One of the first studies that investigated the temporal dynamics of the resting state introduced a spatiotemporal model of RSNs based on the point process (Vedel Jensen and Thorarinsdottir, 2007). Later, the time-frequency analysis of RSNs based on wavelet analysis and sliding-window analysis revealed that the relationship between two major RSNs (TNN and TPN) changes over time (Chang and Glover, 2010). In brief, they illustrated that the temporal correlations between these network waxes and wanes in time. Nonstationary dynamics in resting-state fluctuations were also observed in animal studies (Majeed et al., 2009, 2011; Thompson et al., 2013). These nonstationary dynamics were further confirmed by comparing the observed time-dependent changes in rs-FC with a stationary null model (Zalesky et al., 2014). The most conventional approach to the estimation of dynamic functional connectivity is the sliding-window analysis approach, which involves the calculation of FC within smaller temporal windows over time (Allen et al., 2012).

Subsequent studies showed alterations in the dynamic FC of schizophrenia patients, such as the dwell times of the states and disruptions in thalamocortical connectivity (Damaraju et al., 2014). Different methods, such as independent vector analysis, also showed altered variability between frontoparietal, cerebellar, and temporal regions in schizophrenia patients (Ma et al., 2014). Furthermore, the temporal changes in graph metrics found differences in schizophrenia patients (Yu et al., 2015). Another study compared dynamic FC networks in schizophrenia and bipolar disorder (Rashid et al., 2014), finding differences in dynamic states in both disorders. Alterations in the dwell time of DMN were also observed between Alzheimer's disease and mild cognitive impairment (Jones et al., 2012). A recent study showed that the variability in dynamic FC can capture additional clinically relevant features in major depressive disorder (Demirtaş et al., 2016).

## 4.4 MEASURING STRUCTURAL CONNECTIVITY

Structural connectivity (SC) refers to the long-range anatomical connections among brain areas through white-matter fiber projections. The idea of fiber tracking based on the bounded diffusion of water molecules in the fibers made it possible to create connectivity maps noninvasively (Conturo et al., 1999; Le Bihan et al., 1986). In the beginning, as an efficient method, diffusion tensor imaging (DTI) was proposed as a means to track the neural fibers (Basser et al., 1994; Le Bihan and Iima, 2015). Later, the poor performance of DTI in distinguishing complex wiring structures, such as crossing fibers, led to more advanced methods such diffusion spectrum imaging (DSI) and DWI that would not compromise the computational efficiency (Fillard et al., 2011; Huisman, 2003; Wedeen et al., 2005). Studies that compared the performance of the noninvasive diffusion-based imaging and invasive techniques showed the validity of these methods (Gigandet et al., 2008; Schmahmann et al., 2007). These developments allowed the emergence of a new field studying the human connectome (Sporns et al., 2005).

Limitations were present in the anatomical tracking of the neural fibers, however. First, despite the improved new techniques that could be used to resolve crossing fibers, diffusion-based imaging still performed poorly in detecting interhemispheric connections. Another problem with anatomical connectivity is the complexity of subcortical structures. For example, the thalamus and cerebellum consist of numerous subregions with complex excitatory and inhibitory connections. Additionally, the available connectivity maps of the human brain coincided with another recently popularized field—graph theory—which has yielded an enormous literature on the topological features of brain connectivity.

## 4.5 EFFECTIVE CONNECTIVITY

Many studies have characterized the relationship between SC and FC. Computational models refined this relationship by demonstrating that the FC among unconnected cortical regions can be explained by indirect connections in SC. However, SC only accounts for the physical characteristics of the fibers. Furthermore, although fiber-tracking methods can measure the weights of the connections, they do not carry any information on the causality of the connections. The issue is further complicated when the biophysical properties of the neural fibers and the cortical regions are considered. The problem becomes even more complicated for FC. FC in rs-fMRI measures overall dependence among the BOLD time series of brain regions. Therefore, besides being undirected, FC between two regions may depend on the whole-brain activity. To address this problem, Friston et al. (1995) introduced the concept of effective connectivity (EC) to describe the modulatory interactions between visual processing regions, later extended as "psychophysiological interaction analysis" (Friston et al., 1997). Likewise, a structural equation modeling approach was proposed based on minimization of predicted and observed dependent variables (Büchel and Friston, 1997). EC has potential implications in uncovering the neural mechanisms behind clinical disorders (Dauwels et al., 2010) and consciousness (Boly et al., 2011; Denis Jordan, 2013).

Briefly, EC refers to a broader definition of structural connectivity, which captures all the features that shape connectivity, such as synaptic strengths, concentration of neurotransmitter, neural excitability, causal structure of the links, etc. Various techniques have been proposed to estimate EC from fMRI BOLD signals, but the applications of these techniques were often overlooked because of the complexity of the methods (Stephan and Roebroeck, 2012). Furthermore, in the review article of Stephan and Roebroeck, they underline the fact that resting-state studies in particular are exploratory in nature and "the richness and complexity of their results can invite unconstrained interpretations and post-hoc injection of meaning" (Stephan and Roebroeck, 2012).

There are two fundamental approaches to estimate EC: model-driven and data-driven. The model-driven approaches rely on the generation of BOLD signals under some assumptions. The most widely used approach is dynamic causal modeling (DCM) (Friston et al., 2003). In DCM the hidden states of neuronal origin are simulated through generative models (i.e., neuronal state equations). Then, the neuronal state equations are linked to the observed data using hemodynamic models.

Finally, given the experimental constraints, the most likely model is inferred from the observed data through Bayesian techniques (Stephan and Roebroeck, 2012).

One of the two important aspects of DCM that were emphasized in the literature was the model for the hemodynamic response function (HRF). The most widely used HRF are based on "Balloon" model (Buxton et al., 1998). Later, various modifications were done for the hemodynamic models (Havlicek et al., 2015; Stephan et al., 2007). However, there are also some regional differences in hemodynamic responses (Huettel and McCarthy, 2001), which increase the free parameters in the DCM model.

Another key aspect of DCM is the selection and verification of the model. The decision regarding an optimal model requires the assessment of the trade-off between fit and complexity. Bayesian model selection provides an ideal framework for this purpose. The estimated parameters are assessed through methodologies such as Akaike information criterion/Bayesian information criterion (Penny et al., 2004) and the more widely used free energy principle (Friston et al., 2007).

The verification of DCMs involves two stages. First, the validity of parameter assessment, namely face validity, is performed using in silico models (Friston et al., 2003; Razi et al., 2015). Specifically, in this approach the model's validity is tested using artificial networks. A more challenging problem for DCM is the predictive validity, which accounts for the performance of the model given different settings. One approach to this question is to test the performance of model for test data given parameter estimates for training data (Smith et al., 2010). Another approach is to compare the model with respect to known independent states, such as the classification of clinical populations (Brodersen et al., 2011). The most convenient approach, however, is to use invasive techniques that assess the validity of the model by simultaneous electrophysiological measurements (Moran et al., 2011).

An alternative approach is to characterize dependencies in the BOLD signals using multivariate (or vector) autoregressive (MAR) models (Harrison et al., 2003). The major difference in MAR models is that they reduce the model's complexity by bypassing the hemodynamic response compromising the accuracy of the estimates given the nonneuronal hemodynamic confounds (Rogers et al., 2010). To address this problem, the MAR approach requires estimation of the model order (Penny and Roberts, 2002).

In contrast to the model-based estimates of EC, data-driven approaches were also considered in the literature. These approaches rely on higher-order statistics, information theoretical measures, or phase relationships in the BOLD signals. The most widely used method by which to extract EC from BOLD signals is Granger causality modeling (Goebel et al., 2003), which quantifies how well the past behavior of a signal can predict the second signal beyond the past of the second one alone. Other approaches include mutual information and transfer entropy (Lizier et al., 2010), Bayesian networks (Ramsey et al., 2010), LiNGAM (Shimizu et al., 2006), Patel's tau (Patel et al., 2006), or nonlinear synchronization (Quian Quiroga et al., 2002). Smith et al. (2011) compared the performance of these measures (Smith et al., 2011) and found that only correlation-based measures and Bayesian networks approach could accurately reflect the presence of the connections. Furthermore, they showed that all of the measures perform poorly in detecting the directionality of the connections, although Patel's tau performed slightly better than the other approaches. Other studies using simultaneous electroencephalography (EEG) recordings showed that regional variations in HRF might confound the causal relationships reflected in data-driven approaches (David et al., 2008). Recent studies proposed an alternative approach to overcome this problem, which involves deconvolution of BOLD signals prior to analysis (Bush et al., 2015). Nevertheless, as suggested by Friston et al. (2014) the EC is naturally model-based.

The techniques inferring that EC performs well are dependent on the context in which the data have been acquired. Crucially, they require experimentally controlled stimulation, such as transmagnetic or deep-brain stimulation. A serious problem arises when the origin of the stimulation is not known, such as rs-fMRI. Extensions of DCM address this problem by using stochastic (Daunizeau et al., 2009), Fourier series (Di and Biswal, 2014), and cross-spectrum-based approaches (Friston et al., 2014).

Another important problem arises as the number of brain regions increases. The current frameworks to infer EC from fMRI BOLD signals are computationally expensive, and the computation time increases exponentially with the number of nodes. For this reason the analysis of EC could cover only small sets of regions in the cortex. Critically, the resting-state research showed that interactions among regions are nontrivial: The entire cortex is coordinated to generate rich, dynamic patterns of activity. Therefore, whole brain-based models are essential for the study of EC.

Inspired by the relationship between structure and function in the resting-state signals revealed by computational models (Deco et al., 2014a) proposed a whole-brain computational model based on dynamic mean-field approximation of a biophysical model to infer EC using observed FC. They illustrated that the analytical approximation of covariance in simulated time series might make it possible to heuristically estimate optimal SC (i.e., EC). The disadvantage of this approach, however, was that the directionality of the connections could not be incorporated. A recent study extended this approach to construct a framework that can account for causal interactions (Gilson et al., 2017).

# 4.6 TOPOLOGICAL ANALYSIS OF THE NETWORKS

Prior to the growing interest in large-scale neuronal interactions, empirical and theoretical research in neuroscience primarily dealt with the analysis of neuronal coding and alterations in brain activity in various brain areas. Recent advances in the field of graph theory provided new ways to study brain connectivity. Principally, these connectivity graphs relied on very simple structures (such as nodes and the edges of a graph), but their complexity grew exponentially with the increasing amount of information. Therefore, it became necessary to analyze and represent such graphs in a comprehensive way. Graph theory fulfilled this need by transforming connectivity information into mathematical objects. Although graph theory was widely used in social sciences during the latter half of the 20th century, its adaptation to brain networks is relatively recent (Felleman and Van Essen, 1991).

In graph theory, some spatially distributed objects are defined as nodes and the links that connects these objects are defined as edges. In regard to brain connectivity, where predefined brain regions are denoted as nodes, the neuronal fibers are considered as edges. The graphs are called binary if the edges take only the values 0 and 1 (connections either exist or do not exist), while the graphs with positive real-value edges are called weighted graphs.

The basic measures of graph connectivity (i.e., the basic graph metrics) describe how the edges are distributed in each graph. Density, the most basic metric, quantifies whether the graph is densely or sparsely con-nected. The path length measures the average number of edges in the shortest path between nodes. At the nodal level, the degree of a node is defined as the sum of the edges that are linked to that specific node. The weighted counterpart of nodal degree is called strength.

These basic metrics are used to generate advanced graph metrics that can capture the more abstract properties of a graph. For example, the clustering coefficient measures the ratio between the number of triangles that a node forms with its neighbors and the number of all possible triangles. As its name suggests, the average clustering coefficient quantifies how densely the nodes of a graph are connected with their neighbors.

It has been long known that many self-organizing, complex networks in nature show a particular property, namely a "small world" that makes the necessary steps to link any two nodes in the graph very efficiently (Milgram, 1967; Travers and Milgram, 1969). Watts and Strogatz (1998) proposed a simple graph metric with which this property can be quan-tified. By drawing the contrast between cyclic networks (i.e., high clus-tering coefficient, long path length) and random networks (i.e., low clustering coefficient, short path length), they showed that small-world networks unite short path lengths and high clustering coefficients. The proposed metric, small worldness quantified this relationship by comparing the ratio between clustering coefficient and path length with the surrogate networks equivalent in connection density.

Centrality, at the nodal level, measures the importance of a particular node in the graph. The betweenness centrality is measured as the number of shortest paths that pass through a particular node. Specifically, betweenness centrality reflects whether a node is a hub or an isolated node having little influence. Many other graph metrics have been proposed and used in the context of brain connectivity (Rubinov and Sporns, 2010).

Following the emergence of diffusion-based tractography, the topological characterization of brain connectivity took place immediately (Iturria-Medina et al., 2007). Shortly afterward, the small-world structure in human brain connectivity was introduced (Hagmann et al., 2007). Furthermore, the structural correlates of RSNs were identified using the topological properties of the brain (Hagmann et al., 2008). This study showed that the regions forming the major RSNs (such as precuneus, posterior cingulate, paracentral lobule, and superior and inferior parietal cortex) were the so-called structural core of the brain with higher degree, strength, efficiency, and centrality.

The topological properties of brain connectivity were immediately echoed in clinical neuroimaging studies. Researchers reported alterations in graph metrics in various clinical disorders (Alaerts et al., 2015).

## 4.7 COMPARING CONNECTIVITY AMONG GROUPS

In resting-state research, particularly in clinical applications, valid statistical approaches with which to make inferences on the differences among groups are essential. However, as the number of measures that can characterize brain connectivity increases, statistical comparison among groups becomes problematic. The relatively trivial problem is that not all of the measures fulfill the assumptions (such as Gaussianity) underlying the parametric statistical methods. The use of nonparametric methods, preferably permutation tests, can overcome this problem. Permutation tests adjust the selected test statistic based on randomly generated surrogate data.

A more serious problem arises with the increasing number of multiple comparisons. Statistical tests rely on a predefined $p$-value that controls the false positive rate (type-I error). For example, if the desired false positive rate is 0.05, it implies that the null hypothesis might be falsely rejected with a probability of .05. However, as the number of independent statistical tests increases, this probability accumulates. Briefly, if we perform $N$ independent statistical tests with a desired $p$-value $p$, we expect to reject the null hypothesis incorrectly with a probability of $1-(1-p)^N$ as the so-called family-wise error rate (FWER). An aggressive approach by which to adjust the FWER is to use Bonferroni correction, which divides the desired $p$-value by the number of statistical tests (Dunn, 1959).

However, this procedure increase false negative rates (type-II error) and, for a large number of $N$, makes it very unlikely to find significant differences. A reasonable alternative to Bonferroni correction is false discovery rate approach (Benjamini and Yekutieli, 2001). This approach controls type-I and type-II errors based on the expected value of the false discovery rate.

Zalesky et al. proposed a different approach to compare connectivity among groups. They pointed to the fact that, in a network, the differences of interest will propagate through the existing connections (Zalesky et al., 2010). Therefore, the statistical significance of a connection (or a set of connections) cannot be considered as independent of others. Instead, the statistical test should rely on the significance of the differences in the networks as opposed to the connections. Accordingly, they introduced the Network Based Statistics approach for the comparison of whole-brain connectivity among different groups.

### 4.7.1 Clinical Applications of Large-Scale Resting State Connectivity

Apart from the rich properties of RSNs, rs-fMRI provides a very basic experimental procedure that involves no task. Therefore, its applications in clinical populations rely on underlying brain structure and dynamics, which makes the interpretation of the observed alterations in the populations relatively straightforward. The potential use of this new tool was emphasized shortly after its introduction (Greicius, 2008). During the past decade the clinical applications of rs-fMRI have grown exponentially.

A small subset of clinical rs-fMRI studies involves schizophrenia (Bassett et al., 2012), major depression disorder (Wang et al., 2012), bipolar disorder (Vargas et al., 2013), autism (Cerliani et al., 2015; Cherkassky et al., 2006; Lai et al., 2010; Maximo et al., 2013; Nomi and Uddin, 2015; Spisák et al., 2014; Weng et al., 2010), posttraumatic stress disorder (Kennis et al., 2015; Sadeh et al., 2015; Sripada et al., 2012; Yan et al., 2013); multiple sclerosis (Leonardi et al., 2013), Alzheimer's Disease, and other dementias (Brier et al., 2014; Dennis and Thompson, 2014; Filippi and Agosta, 2011). A detailed review of the clinical rs-fMRI studies is beyond the scope of this chapter; however, here we highlight few studies. Bullmore, focusing on complexity, used the Hurst exponent to identify Alzheimer's patients using rs-fMRI (Maxim et al., 2005). Widespread reduction of FC was observed in schizophrenia patients (Liang et al., 2006; Liu et al., 2008; Salomon et al., 2011). Schizophrenia was also found to be associated with increased randomization of functional networks, decreased small-world properties, lower clustering coefficient, and fewer high-degree hubs (Bassett et al., 2012; Liu et al., 2008; Lynall et al., 2010).

Despite the tremendous amount of research in clinical neuroimaging, there is an overlap between the observed significant differences among different clinical populations and inaccurate replications, consequently impeding the differentiation of the diseases (Kapur et al., 2012). Particularly, the DMN was found to be crucial in various mental disorders (Broyd et al., 2009). Similarly, the core region of DMN—the posterior cingulate cortex—was altered in clinical populations (Leech and Sharp, 2014). In a broader context, several researchers have emphasized the overlapping pathogenesis in distinct diagnostic categories of psychiatric disorders (Craddock and Owen, 2010), whereas others also have stressed the heterogeneity across individuals within a single diagnostic category (Cuthbert and Insel, 2013). These concerns have resulted in efforts to develop modern ways to define psychiatric disorders based on behavioral and neurobiological dimensions (Craddock and Owen, 2010; Cuthbert and Insel, 2013). Computational modeling is a valuable approach to link these dimensions in different spatial and temporal scales.

## 4.8  MODELING THE LARGE-SCALE BRAIN ACTIVITY-I: LINKING STRUCTURE AND FUNCTION

As described previously, advances in DWI technology have made it possible to study anatomical connectivity in a noninvasive way. Moreover, the emergent field of graph theory made it possible to investigate and compare the topological features of structural and functional networks (Hagmann et al., 2008; Bullmore and Bassett, 2011). Nevertheless, structure alone cannot account for the emergence of rich dynamical patterns in the functional organization of the brain. Computational modeling, on the other hand, allows researchers to link the structural architecture of the connectivity to large-scale, spontaneous spatiotemporal fluctuations in the brain (Deco and Corbetta, 2011).

Likewise, large-scale simulations that rely on microscopic description of the neurons have been shown to be feasible (Izhikevich and Edelman, 2008). First, however, the complexity of these models makes the interpretation of underlying mechanisms difficult. Secondly, current imaging techniques lack sufficient power to provide empirical evidence that covers micro- and macrospatiotemporal dynamics. An alternative approach is to focus on the mesoscopic models that reduce the complexity of parametric space as well as the spatial resolution, and to explore the link between these models and empirical observations in the current neuroimaging modalities (Deco et al., 2009). Mesoscopic neural-mass models, reinforced by the advancements in the anatomical tracking of connectivity, have successfully replicated the resting-state fluctuations observed in fMRI (Deco et al., 2009; Cabral et al., 2011; Deco and Jirsa, 2012; Ghosh et al., 2008; Honey et al., 2009), magnetoencephalography (Cabral et al., 2014; Nakagawa et al., 2014), and EEG (Hindriks et al., 2014).

Briefly, these models define each brain region as a node and the anatomical link between two brain regions extracted using DWI techniques as an edge in the model. Both quantities are matched to the template that parcellates the brain areas according to a standardized set of regions for the corresponding neuroimaging modality. Then, the simulated rs-fMRI BOLD time series were generated according to a proposed computational model. This allows investigating the optimal parameter set that explains the correspondence between empirical and simulated FC measures (Fig. 4.1).

The very first models studied the spontaneous neuronal fluctuations using conductance-based biophysical models (Breakspear et al., 2003; Honey et al., 2007) and the FitzHugh-Nagumo model (Ghosh et al., 2008) based on the anatomical connectivity of the cortex in macaques (Kötter, 2004). These models were then adapted to simulate rs-FC in healthy human subjects (Honey et al., 2009) and the impact of lesions in human brain (Alstott et al., 2009). Modeling the spontaneous activity in neural-masses in macaque brain using the FitzHugh-Nagumo model (Fitzhugh, 1961; Nagumo et al., 1962), the researchers investigated the role of time-delayed interactions among brain regions. The system lost stability for a critical global coupling value, and illustrated the emergence of DMN for transmission velocities ranging from 5 to 10 m/s. The model also showed noise-driven oscillatory dynamics (10 Hz) at the edge of instability. Later, a similar approach was adapted to the neuronal population firing-rate model (Wilson and Cowan, 1972). This model showed how functional RSNs are dynamically organized near the bifurcation point (Deco et al., 2009). In this model, in the absence of any external input, the internal noise of the system sustained the fluctuations, where the system showed empirically observed properties at the edge of the transition between stable and unstable states. They set the spontaneous background activity and the efficiency of recurrent connections at the edge of a Hopf bifurcation and then investigated the role of global coupling, time delays and noise level. The results showed that the spontaneous gamma (40 Hz) oscillations in the modules that were organized into two competing functional networks. They also showed anticorrelated networks fluctuating in a low-frequency band (0.1 Hz) consistent with the empirical findings (Fox et al., 2005).

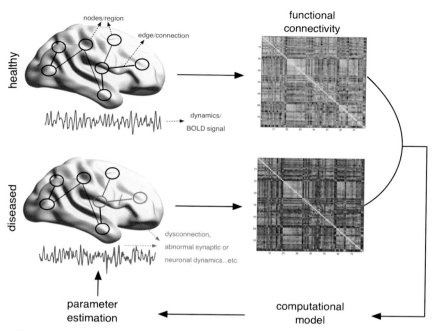

FIGURE 4.1    Large-scale computational modeling framework. Upper panel: parcellated brain regions are represented by nodes and the coupling among brain regions are represented by edges. The neural/blood-oxygen-level dependent (BOLD) dynamics of each node is governed by system of differential equations. The model parameters are estimated by fitting the observed functional connectivity (FC) to the simulated FC. Lower panel: hypothetical alterations in clinical populations. Due to propagation of activity through the coupled dynamical system, the dysconnectivity observed FC may not reflect the underlying pathology. Large-scale computational modeling aims to reveal the underlying mechanisms via parameter estimation.

Following the developments in DWI techniques, conductance-based biophysical models were applied to simulate spontaneous fluctuations in the rs-fMRI BOLD signals of healthy human subjects (Honey et al., 2009). Later, using an attractor network of spiking neurons, Deco and Jirsa (2012) showed that the simulated neuronal dynamics converged with the empirical observations in rs-FC at the edge of the instability in human subjects. The dynamic mean-field approximation of the same attractor model showed the organization of the spontaneous fluctuations constrained due to anatomical connectivity structure (Deco et al., 2013). Another approach to simulate the spontaneous was the attractor model of spiking neurons (Deco and Jirsa, 2012). Based on previously proposed models (Brunel and Wang, 2001), they used a leaky integrate and fire dynamics. The mean-field approximation of spiking neural networks showed that the brain operates at the edge of instability, thus maximizing the dynamic repertoire of the possible states (Deco and Jirsa, 2012).

Simulation of biophysical spiking neural network models is computationally expensive, which limits their scope. To overcome this problem, Deco et al. (2014a) adapted the dynamic mean-field approximation that was previously proposed to model working memory (Wong and Wang, 2006). The model was based on the previously mentioned spiking neural network model. However, the dynamic mean-field provided computationally efficient way to model large-scale brain networks. Later, the dynamic mean-field approach was further extended to show the role of local feedback inhibition in whole-brain spontaneous fluctuations (Deco et al., 2014b). This model filled the gap between the observations of uncorrelated low firing-rate activity and intraarea correlations that were induced by long-range connections. By adding GABAergic inhibitory feedback loops into the mean-field approximation, they showed how large-scale synchronous networks emerged from local asynchronous spontaneous fluctuations (Deco et al., 2014b). Furthermore, the existence of local feedback inhibition not only diminished the critical point but also enhanced the similarity between empirical and simulated FC.

Another novel approach that was proposed by this mean-field study was the analytical approximation of covariance between neural populations using the moments method. Where $\mathbf{J}$ is the Jacobian of the system of equations describing the neural activity in each node, and $\mathbf{Q}$ is input-noise covariance matrix, the covariance between neural populations, $\mathbf{P}$, was described using the equation:

$$\frac{d\mathbf{P}}{dt} = \mathbf{J}\mathbf{P} + \mathbf{P}\mathbf{J}^T + \mathbf{Q}_n$$

Briefly, the moments method allowed the approximation of simulated FC without the need to simulate large-scale simulations. This provided a unified framework to explain rs-FC emerged from diffusion of noise through the cortex via anatomical links.

Another direction in the modeling approach followed the idea of nonlinear interactions between coupled phase-oscillators (Cabral et al., 2011). Using the Kuramoto network model, this approach showed that the time delays of the couplings between nodes shapes the spontaneous synchronization patterns among regions along with the coupling strength between nodes and the level of noise (Cabral et al., 2011). In this model the activity of each region was simulated using the Kuramoto model, which simulates the nonlinear interactions between coupled oscillators (Kuramoto, 1986). They showed that the optimal similarity between simulated and empirical FC was observed at a point between fully synchronized and asynchronous regimes. Furthermore, their model showed that the nonlinear interactions between noisy oscillators led to the emergence of nonstationary connectivity patterns.

The Kuramoto network model captured the alterations in global graph metrics in schizophrenia patients due to the impaired anatomical connectivity structure (Cabral et al., 2012a). Furthermore, the model illustrated the clinical implications of the large-scale simulations of resting-state activity by showing the relationship between alterations in topological properties in schizophrenia and the anatomical decoupling between brain regions (Cabral et al., 2012a,b). Later, a similar approach was adapted to directly simulate the slow fluctuation in BOLD signals (Ponce-Alvarez et al., 2015). In this model, the instantaneous phases of band-pass filtered (0.04–0.07 Hz) rs-fMRI BOLD signals were acquired using the Hilbert transform. This model successfully also demonstrated the emergence of networks in dynamic FC.

One study compared the performance of the proposed models of resting-state activity and it illustrated that given the optimal internal noise of each nose, a linear simple autoregressive model (SAR) might outperform more complex models in predicting FC using SC, but not in showing the nonstationary properties (Messé et al., 2014, 2015).

## 4.9  MODELING THE LARGE-SCALE BRAIN ACTIVITY-II: ADDING DYNAMICS INTO THE EQUATION

Several models have addressed the possible mechanisms underlying dynamic FC. For instance, multivariate modeling of rs-fMRI BOLD signals showed two major states of activation representing high- and low-activity in DMN (Ferguson and Anderson, 2012). In another study, researchers used SC (obtained by DTI) to determine which regions played a role in orchestrating dynamic FC (Lv et al., 2013). They found that the so-called hub routers were located in DMN, and they provided a clinical application of altered routers in posttraumatic stress disorder patients. As an attempt to propose mechanistic model approaches in clinical populations, one study used the Kuramoto model to study the abnormal synchronization patterns in various frequency bands in epilepsy patients using EEG (Schmidt et al., 2014). They found potential regions that might drive the seizures in those patients, and they showed the high predictive power of their analysis. Another study used the FitzHugh–Nagumo model to investigate dynamics of resting-state signals (Vuksanović and Hövel, 2015). They showed that the observed resting-state signals emerged from the maximization of synchrony and variability in synchrony.

Two principal mechanisms were proposed as the means to explain the dynamics underlying rs-FC: noise-driven spontaneous dynamics constrained by anatomical connectivity; and complex interactions between phase-oscillators shaped coupling, delays, and noise. However, the relationship between these two mechanisms is nontrivial and possibly contradictory. The noise-based models suggest that the temporal correlations in spontaneous activity emerge from the propagation of uncorrelated noise through the long-range anatomical connections. In contrast, oscillation-based models explain the rs-FC by means of complex interactions between oscillatory activities in the regions of the brain. Furthermore, the proposed models did not account for the temporal structure of spontaneous fluctuations.

One study to reconcile the discrepancy between these approaches proposed a model based on the normal form of supercritical Hopf bifurcation (Fig. 4.2). This model showed that the brain regions at rest operate close to a critical point near the local supercritical Hopf bifurcation, in which each region manifests noise- and coupling-induced transitions between a stable fixed point (noisy fluctuations), and a limit-cycle attractor (noisy oscillations). This approach not only illustrated the emergence of synchronized networks but also the temporal variations of these synchronized networks. They studied the spatiotemporal structure in dynamic FC, which was quantified as the correlation coefficients between functional states at different time instances (i.e., functional connectivity dynamics, or FCD). They showed that the Kolmogorov–Smirnov distance between empirical and simulated FCD distributions is not only optimal at critical point but is also more sensitive to deviations from the critical point. Furthermore, this model, by means of the local bifurcation parameter, illustrated the role of nodal dynamics in on the large-scale dynamics. Being specific, the model made it possible to study the role of each region's dynamics in orchestrating whole-brain activity patterns.

## 4.10 DISCUSSION

rs-fMRI BOLD signals not only comprise information regarding the functional links between two regions, but also reflects all the dynamical factors related to these links. In the clinical context, the abnormal connectivity reported by resting-state studies recurs in various populations. Particularly, DMN and the regions associated with DMN are commonly found to be altered. This suggests a crucial role of DMN in mental processing. The well-studied research on structural and FC of the DMN showed that it contains highly central, hub regions. For example, one review discussed the role of DMN suppression and its relationship with frontoparietal control network in psychiatric disorders (Anticevic et al., 2012). The evidence suggests that multiple mechanisms may cause this

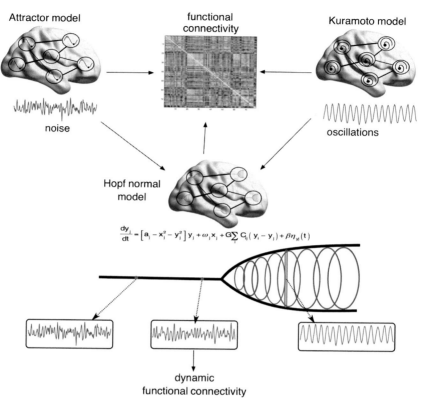

FIGURE 4.2  Overview of Hopf normal model. Upper panel: two common dynamics of computational models comprise noise-driven and oscillation-driven activity. In a noise-driven system, such as an attractor model, each node exhibits asynchronous activity and the functional connectivity (FC) emerges from propagation of noise through synaptic coupling. In an oscillation-driven system, such as Kuramoto model, each node exhibits self-sustained oscillatory activity and the FC emerges from noise, coupling, and time delays in signal transmission. Both dynamics generates similar FC. Lower panel: Hopf normal model is the reduced representation of a nonlinear dynamical system that exhibits Hopf bifurcation. For negative bifurcation parameter values, the model generates dynamics equivalent to a noise-driven system. For positive bifurcation parameter values, the model generates dynamics equivalent to a oscillation-driven system. In both dynamical regimes the system generates similar static FC. When the system is close to its critical point, where it transiently switches between two dynamics, the system explains not only static FC but also dynamic FC.

abnormal connectivity in DMN. These findings are not surprising because any perturbation in the underlying brain dynamics or connectivity would more likely affect the hub regions.

The common interpretation of the increased/decreased FC is that the communication between regions increased or decreases. This interpretation can be misleading. There are several scenarios that might be considered before interpreting the alterations in FC:

- The link between two regions is intact, but the grand-average FC is altered because of the alterations in the dynamics or as a consequence of network effects in at the whole-brain level.
- The link between two regions is impaired, but the alterations in the grand average FC are compensated by the alterations in the dynamics or as a consequence of network effects at the whole-brain level.

Therefore, the intensity of grand average FC is not enough to explain the communication between two regions. It is also crucial to take into account the relationship between static and dynamic FC. It is important to note that this fact will eventually affect our judgments on healthy brain function. The presence of networks with increased FC does not necessarily mean that regions of this network are effectively exchanging information. In fact, a network of increased FC may indicate that the regions involved in this network might not be perturbed enough to manifest a rich dynamic repertoire. Therefore, one must take into account the reasons behind the altered FC to draw conclusions on the functional relevance of this alteration.

Regarding dynamic FC, we believe that carefully designed electrophysiological experiments are needed to understand the interplay between static FC and variability of FC, as well as task-related effects in the whole-brain. To our knowledge there are no studies that directly address these issues. Simultaneous recordings of fMRI with high-temporal resolution techniques such as EEG and local field potential might be useful for the verification of our approaches. Furthermore, noninvasive stimulation techniques such as deep-brain stimulation and transmagnetic stimulation are promising tools to study dynamic interactions among brain regions.

Another underappreciated feature in large-scale computational models is the role of hemodynamic responses in fMRI. Appropriate experimental paradigms may allow theoretical models to address highly controversial problems like nonneuronal noise in rs-fMRI analysis, such as global signal removal. Indeed, the role of nonneuronal signal changes may have significant impact on clinical neuroimaging research.

Computation modeling frameworks are crucial to have a mechanistic understanding of resting-state brain function. Furthermore, using computational modeling it is possible to study how the spontaneous fluctuations in the entire brain respond to different settings such as external stimulation, changes in neurotransmitter levels, excitation—inhibition balance, or cortical heterogeneity. However, computational modeling relies on a priori knowledge about the underlying brain mechanisms or on estimation of the optimal model parameters that corresponds to some observed measures. This requires novel experimental paradigms and empirical measures of large-scale brain function that can be translated into mathematical objects by computational neuroscience community. As discussed in this chapter, the entire set of measures from grand-average FC and dynamic FC to graph theory metric have contributed to the development of computational models of whole-brain network dynamics.

## References

Achard, S., Salvador, R., Whitcher, B., Suckling, J., Bullmore, E., 2006. A resilient, low-frequency, small-world human brain functional network with highly connected association cortical hubs. J. Neurosci. 26 (1), 63–72. http://dx.doi.org/10.1523/JNEUROSCI.3874-05.2006.

Alaerts, K., Geerlings, F., Herremans, L., Swinnen, S.P., Verhoeven, J., Sunaert, S., Wenderoth, N., 2015. Functional organization of the action observation network in autism: a graph theory approach. PLoS One 10 (8), e0137020. http://dx.doi.org/10.1371/journal.pone.0137020.

Albert, N.B., Robertson, E.M., Miall, R.C., 2009. The resting human brain and motor learning. Curr. Biol. 19 (12), 1023–1027. http://dx.doi.org/10.1016/j.cub.2009.04.028.

Allen, E.A., Damaraju, E., Plis, S.M., Erhardt, E.B., Eichele, T., Calhoun, V.D., November 2012. Tracking whole-brain connectivity dynamics in the resting state. Cereb. Cortex bhs352. http://dx.doi.org/10.1093/cercor/bhs352.

Alstott, J., Breakspear, M., Hagmann, P., Cammoun, L., Sporns, O., 2009. Modeling the impact of lesions in the human brain. PLoS Comput. Biol. 5 (6), e1000408. http://dx.doi.org/10.1371/journal.pcbi.1000408.

Anticevic, A., Cole, M.W., Murray, J.D., Corlett, P.R., Wang, X.-J., Krystal, J.H., 2012. The role of default network deactivation in cognition and disease. Trends Cogn. Sci. (Regul. Ed.) 16 (12), 584–592. http://dx.doi.org/10.1016/j.tics.2012.10.008.

Aymé, S., Schmidtke, J., 2007. Networking for rare diseases: a necessity for Europe. Bundesgesundheitsbl 50 (12), 1477–1483. http://dx.doi.org/10.1007/s00103-007-0381-9.

Bandettini, P.A., Jesmanowicz, A., Wong, E.C., Hyde, J.S., 1993. Processing strategies for time-course data sets in functional MRI of the human brain. Magn. Reson. Med. 30 (2), 161–173.

Basser, P.J., Mattiello, J., LeBihan, D., 1994. MR diffusion tensor spectroscopy and imaging. Biophys. J. 66 (1), 259–267. http://dx.doi.org/10.1016/S0006-3495(94)80775-1.

Bassett, D.S., Nelson, B.G., Mueller, B.A., Camchong, J., Lim, K.O., 2012. Altered resting state complexity in schizophrenia. NeuroImage 59 (3), 2196–2207. http://dx.doi.org/10.1016/j.neuroimage.2011.10.002.

Beckmann, C.F., Smith, S.M., 2004. Probabilistic independent component analysis for functional magnetic resonance imaging. IEEE Trans. Med. Imaging 23 (2), 137–152. http://dx.doi.org/10.1109/TMI.2003.822821.

Beckmann, C.F., DeLuca, M., Devlin, J.T., Smith, S.M., 2005. Investigations into resting-state connectivity using independent component analysis. Philos. Trans. R. Soc. B 360 (1457), 1001–1013. http://dx.doi.org/10.1098/rstb.2005.1634.

Benjamini, Y., Yekutieli, D., 2001. The control of the false discovery rate in multiple testing under dependency. Ann. Stat. 29 (4), 1165–1188. http://dx.doi.org/10.1214/aos/1013699998.

Birn, R.M., 2012. The role of physiological noise in resting-state functional connectivity. NeuroImage 62 (2), 864–870. http://dx.doi.org/10.1016/j.neuroimage.2012.01.016.

Biswal, B., Yetkin, F.Z., Haughton, V.M., Hyde, J.S., 1995. Functional connectivity in the motor cortex of resting human brain using echo-planar MRI. Magn. Reson. Med. 34 (4), 537–541.

Boly, M., Garrido, M.I., Gosseries, O., Bruno, M.-A., Boveroux, P., Schnakers, C., Massimini, M., Litvak, V., Laureys, S., Friston, K., 2011. Preserved feedforward but impaired top-down processes in the vegetative state. Science 332 (6031), 858–862. http://dx.doi.org/10.1126/science.1202043.

Breakspear, M., Terry, J.R., Friston, K.J., 2003. Modulation of excitatory synaptic coupling facilitates synchronization and complex dynamics in a biophysical model of neuronal dynamics. Network 14 (4), 703–732.

Brier, M.R., Thomas, J.B., Ances, B.M., 2014. Network dysfunction in Alzheimer's disease: refining the disconnection hypothesis. Brain Connect. 4 (5), 299–311. http://dx.doi.org/10.1089/brain.2014.0236.

Broca, P.P., 1861. Perte de la parole, ramollissement chronique et destruction partielle du lobe antérieur gauche du cerveau. Bull. Soc. Anthropol. 2 (235–238), 301–321.

Brodersen, K.H., Schofield, T.M., Leff, A.P., Ong, C.S., Lomakina, E.I., Buhmann, J.M., Stephan, K.E., 2011. Generative embedding for model-based classification of fMRI data. PLoS Comput. Biol. 7 (6), e1002079. http://dx.doi.org/10.1371/journal.pcbi.1002079.

Broyd, S.J., Demanuele, C., Debener, S., Helps, S.K., James, C.J., Sonuga-Barke, E.J.S., 2009. Default-mode brain dysfunction in mental disorders: a systematic review. Neurosci. Biobehav. Rev. 33 (3), 279–296. http://dx.doi.org/10.1016/j.neubiorev.2008.09.002.

Brunel, N., Wang, X.J., 2001. Effects of neuromodulation in a cortical network model of object working memory dominated by recurrent inhibition. J. Comput. Neurosci. 11 (1), 63–85.

Büchel, C., Friston, K.J., 1997. Modulation of connectivity in visual pathways by attention: cortical interactions evaluated with structural equation modelling and fMRI. Cereb. Cortex 7 (8), 768–778. http://dx.doi.org/10.1093/cercor/7.8.768.

Bullmore, E.T., Bassett, D.S., 2011. Brain graphs: graphical models of the human brain connectome. Annu. Rev. Clin. Psychol. 7, 113–140. http://dx.doi.org/10.1146/annurev-clinpsy-040510-143934.

Bush, K., Zhou, S., Cisler, J., Bian, J., Hazaroglu, O., Gillispie, K., Yoshigoe, K., Kilts, C., August 2015. A deconvolution-based approach to identifying large-scale effective connectivity. Magn. Reson. Imaging. http://dx.doi.org/10.1016/j.mri.2015.07.015.

Buxton, R.B., Wong, E.C., Frank, L.R., 1998. Dynamics of blood flow and oxygenation changes during brain activation: the balloon model. Magn. Reson. Med. 39 (6), 855–864. http://dx.doi.org/10.1002/mrm.1910390602.

Cabral, J., Hugues, E., Sporns, O., Deco, G., 2011. Role of local network oscillations in resting-state functional connectivity. NeuroImage 57 (1), 130–139. http://dx.doi.org/10.1016/j.neuroimage.2011.04.010.

Cabral, J., Hugues, E., Kringelbach, M.L., Deco, G., 2012a. Modeling the outcome of structural disconnection on resting-state functional connectivity. NeuroImage 62 (3), 1342–1353. http://dx.doi.org/10.1016/j.neuroimage.2012.06.007.

Cabral, J., Kringelbach, M.L., Deco, G., 2012b. Functional graph alterations in schizophrenia: a result from a global anatomic decoupling? Pharmacopsychiatry 45 (Suppl. 1), 57–64.

Cabral, J., Luckhoo, H., Woolrich, M., Joensson, M., Mohseni, H., Baker, A., Kringelbach, M.L., Deco, G., 2014. Exploring mechanisms of spontaneous MEG functional connectivity: how delayed network interactions lead to structured amplitude envelopes of band-pass filtered oscillations. NeuroImage 90, 423–435.

Cerliani, L., Mennes, M., Thomas, R.M., Di Martino, A., Thioux, M., Keysers, C., 2015. Increased functional connectivity between subcortical and cortical resting-state networks in autism spectrum disorder. JAMA Psychiatry 72 (8), 767–777. http://dx.doi.org/10.1001/jamapsychiatry.2015.0101.

Chang, C., Glover, G.H., 2010. Time–frequency dynamics of resting-state brain connectivity measured with fMRI. NeuroImage 50 (1), 81–98. http://dx.doi.org/10.1016/j.neuroimage.2009.12.011.

Cherkassky, V.L., Kana, R.K., Keller, T.A., Just, M.A., 2006. Functional connectivity in a baseline resting-state network in autism. Neuroreport 17 (16), 1687–1690. http://dx.doi.org/10.1097/01.wnr.0000239956.45448.4c.

Conturo, T.E., Lori, N.F., Cull, T.S., Akbudak, E., Snyder, A.Z., Shimony, J.S., McKinstry, R.C., Burton, H., Raichle, M.E., 1999. Tracking neuronal fiber pathways in the living human brain. Proc. Natl. Acad. Sci. U.S.A. 96 (18), 10422–10427.

Craddock, N., Owen, M.J., 2010. Data and clinical utility should be the drivers of changes to psychiatric classification. Br. J. Psychiatry 197 (2), 158. http://dx.doi.org/10.1192/bjp.197.2.158 (author reply 158–159).

Cuthbert, B.N., Insel, T.R., 2013. Toward the future of psychiatric diagnosis: the seven pillars of RDoC. BMC Med. 11, 126. http://dx.doi.org/10.1186/1741-7015-11-126.

Damaraju, E., Allen, E.A., Belger, A., Ford, J.M., McEwen, S., Mathalon, D.H., Mueller, B.A., Pearlson, G.D., Potkin, S.G., Preda, A., Turner, J.A., Vaidya, J.G., van Erp, T.G., Calhoun, V.D., 2014. Dynamic functional connectivity analysis reveals transient states of dysconnectivity in schizophrenia. NeuroImage Clin. 5, 298–308. http://dx.doi.org/10.1016/j.nicl.2014.07.003.

Damoiseaux, J.S., Greicius, M.D., 2009. Greater than the sum of its parts: a review of studies combining structural connectivity and resting-state functional connectivity. Brain Struct. Funct. 213 (6), 525–533. http://dx.doi.org/10.1007/s00429-009-0208-6.

Daunizeau, J., Friston, K.J., Kiebel, S.J., 2009. Variational Bayesian identification and prediction of stochastic nonlinear dynamic causal models. Phys. D Nonlinear Phenom. 238 (21), 2089–2118. http://dx.doi.org/10.1016/j.physd.2009.08.002.

Dauwels, J., Vialatte, F., Musha, T., Cichocki, A., 2010. A comparative study of synchrony measures for the early diagnosis of Alzheimer's disease based on EEG. NeuroImage 49 (1), 668–693. http://dx.doi.org/10.1016/j.neuroimage.2009.06.056.

David, O., Guillemain, I., Saillet, S., Reyt, S., Deransart, C., Segebarth, C., Depaulis, A., 2008. Identifying neural drivers with functional MRI: an electrophysiological validation. PLoS Biol. 6 (12), e315. http://dx.doi.org/10.1371/journal.pbio.0060315.

De Luca, M., Beckmann, C.F., De Stefano, N., Matthews, P.M., Smith, S.M., 2006. fMRI resting state networks define distinct modes of long-distance interactions in the human brain. NeuroImage 29 (4), 1359–1367. http://dx.doi.org/10.1016/j.neuroimage.2005.08.035.

Deco, G., Corbetta, M., 2011. The dynamical balance of the brain at rest. Neuroscientist 17 (1), 107–123. http://dx.doi.org/10.1177/1073858409354384.

Deco, G., Jirsa, V.K., 2012. Ongoing cortical activity at rest: criticality, multistability, and ghost attractors. J. Neurosci. 32 (10), 3366–3375. http://dx.doi.org/10.1523/JNEUROSCI.2523-11.2012.

Deco, G., Jirsa, V., McIntosh, A.R., Sporns, O., Kötter, R., 2009. Key role of coupling, delay, and noise in resting brain fluctuations. Proc. Natl. Acad. Sci. U.S.A. 106 (25), 10302–10307. http://dx.doi.org/10.1073/pnas.0901831106.

Deco, G., Ponce-Alvarez, A., Mantini, D., Romani, G.L., Hagmann, P., Corbetta, M., 2013. Resting-state functional connectivity emerges from structurally and dynamically shaped slow linear fluctuations. J. Neurosci. 33.

Deco, G., McIntosh, A.R., Shen, K., Hutchison, R.M., Menon, R.S., Everling, S., Hagmann, P., Jirsa, V.K., 2014a. Identification of optimal structural connectivity using functional connectivity and neural modeling. J. Neurosci. 34 (23), 7910–7916. http://dx.doi.org/10.1523/JNEUROSCI.4423-13.2014.

Deco, G., Ponce-Alvarez, A., Hagmann, P., Romani, G.L., Mantini, D., Corbetta, M., 2014b. How local excitation-inhibition ratio impacts the whole brain dynamics. J. Neurosci. 34 (23), 7886–7898. http://dx.doi.org/10.1523/JNEUROSCI.5068-13.2014.

Demirtaş, M., Tornador, C., Falcón, C., López-Solà, M., Hernández-Ribas, R., Pujol, J., Menchón, J.M., Ritter, P., Cardoner, N., Soriano-Mas, C., Deco, G., 2016. Dynamic functional connectivity reveals altered variability in functional connectivity among patients with major depressive disorder. Hum. Brain Mapp. 37 (8), 2918–2930. http://dx.doi.org/10.1002/hbm.23215.

Denis Jordan, R.I., 2013. Simultaneous electroencephalographic and functional magnetic resonance imaging indicate impaired cortical top-down processing in association with anesthetic-induced unconsciousness. Anesthesiology 5, 119. http://dx.doi.org/10.1097/ALN.0b013e3182a7ca92.

Dennis, E.L., Thompson, P.M., 2014. Functional brain connectivity using fMRI in aging and Alzheimer's disease. Neuropsychol. Rev. 24 (1), 49−62. http://dx.doi.org/10.1007/s11065-014-9249-6.

de Schotten, M.T., Dell'Acqua, F., Ratiu, P., Leslie, A., Howells, H., Cabanis, E., Iba-Zizen, M.T., Plaisant, O., Simmons, A., Dronkers, N.F., Corkin, S., Catani, M., August 2015. From Phineas Gage and Monsieur Leborgne to H.M.: revisiting disconnection syndromes. Cereb. Cortex bhv173. http://dx.doi.org/10.1093/cercor/bhv173.

Di, X., Biswal, B.B., 2014. Identifying the default mode network structure using dynamic causal modeling on resting-state functional magnetic resonance imaging. NeuroImage 86, 53−59. http://dx.doi.org/10.1016/j.neuroimage.2013.07.071.

Dunn, O.J., 1959. Estimation of the medians for dependent variables. Ann. Math. Stat. 30 (1), 192−197. http://dx.doi.org/10.1214/aoms/1177706374.

Felleman, D.J., Van Essen, D.C., 1991. Distributed hierarchical processing in the primate cerebral cortex. Cereb. Cortex 1 (1), 1−47.

Ferguson, M.A., Anderson, J.S., 2012. Dynamical stability of intrinsic connectivity networks. NeuroImage 59 (4), 4022−4031. http://dx.doi.org/10.1016/j.neuroimage.2011.10.062.

Filippi, M., Agosta, F., 2011. Structural and functional network connectivity breakdown in Alzheimer's disease studied with magnetic resonance imaging techniques. J. Alzheimers Dis. 24 (3), 455−474. http://dx.doi.org/10.3233/JAD-2011-101854.

Fillard, P., Descoteaux, M., Goh, A., Gouttard, S., Jeurissen, B., Malcolm, J., Ramirez-Manzanares, A., Reisert, M., Sakaie, K., Tensaouti, F., Yo, T., Mangin, J.-F., Poupon, C., 2011. Quantitative evaluation of 10 tractography algorithms on a realistic diffusion MR phantom. NeuroImage 56 (1), 220−234. http://dx.doi.org/10.1016/j.neuroimage.2011.01.032.

Fitzhugh, R., 1961. Impulses and physiological states in theoretical models of nerve membrane. Biophys. J. 1 (6), 445−466.

Fox, M.D., Raichle, M.E., 2007. Spontaneous fluctuations in brain activity observed with functional magnetic resonance imaging. Nat. Rev. Neurosci. 8 (9), 700−711. http://dx.doi.org/10.1038/nrn2201.

Fox, M.D., Snyder, A.Z., Vincent, J.L., Corbetta, M., Essen, D.C.V., Raichle, M.E., 2005. The human brain is intrinsically organized into dynamic, anticorrelated functional networks. Proc. Natl. Acad. Sci. U.S.A. 102 (27), 9673−9678. http://dx.doi.org/10.1073/pnas.0504136102.

Fransson, P., 2005. Spontaneous low-frequency BOLD signal fluctuations: an fMRI investigation of the resting-state default mode of brain function hypothesis. Hum. Brain Mapp. 26 (1), 15−29. http://dx.doi.org/10.1002/hbm.20113.

Friston, K.J., Ungerleider, L.G., Jezzard, P., Turner, R., 1995. Characterising modulatory interactions between V1 and V2 in human cortex with fMRI. Hum. Brain Mapp. 2, 211−224.

Friston, K.J., Buechel, C., Fink, G.R., Morris, J., Rolls, E., Dolan, R.J., 1997. Psychophysiological and modulatory interactions in neuroimaging. NeuroImage 6 (3), 218−229. http://dx.doi.org/10.1006/nimg.1997.0291.

Friston, K.J., Harrison, L., Penny, W., 2003. Dynamic causal modelling. NeuroImage 19 (4), 1273−1302.

Friston, K., Mattout, J., Trujillo-Barreto, N., Ashburner, J., Penny, W., 2007. Variational free energy and the Laplace approximation. NeuroImage 34 (1), 220−234. http://dx.doi.org/10.1016/j.neuroimage.2006.08.035.

Friston, K.J., Kahan, J., Biswal, B., Razi, A., 2014. A DCM for resting state fMRI. NeuroImage 94, 396–407. http://dx.doi.org/10.1016/j.neuroimage.2013.12.009.

Friston, K.J., 2011. Functional and effective connectivity: a review. Brain Connect. 1 (1), 13–36. http://dx.doi.org/10.1089/brain.2011.0008.

Gazzaniga, M.S., Bogen, J.E., Sperry, R.W., 1963. Laterality effects in somesthesis following cerebral commissurotomy in man. Neuropsychologia 1 (3), 209–215.

Geschwind, N., 1965a. Disconnexion syndromes in animals and man I. Brain 88, 237–294.

Geschwind, N., 1965b. Disconnexion syndromes in animals and man II. Brain 88, 585–644.

Ghosh, A., Rho, Y., McIntosh, A.R., Kötter, R., Jirsa, V.K., 2008. Noise during rest enables the exploration of the Brain's dynamic repertoire. PLoS Comput. Biol. 4 (10), e1000196. http://dx.doi.org/10.1371/journal.pcbi.1000196.

Gigandet, X., Hagmann, P., Kurant, M., Cammoun, L., Meuli, R., Thiran, J.-P., 2008. Estimating the confidence level of white matter connections obtained with MRI tractography. PLoS One 3 (12), e4006. http://dx.doi.org/10.1371/journal.pone.0004006.

Gilson, M., Moreno-Bote, R., Ponce-Alvarez, A., Ritter, P., Deco, G., 2016. Estimation of directed effective connectivity from fMRI functional connectivity hints at asymmetries of cortical connectome. PLoS Comput. Biol. 12 (3), e1004762. http://dx.doi.org/10.1371/journal.pcbi.1004762.

Goebel, R., Roebroeck, A., Kim, D.-S., Formisano, E., 2003. Investigating directed cortical interactions in time-resolved fMRI data using vector autoregressive modeling and Granger causality mapping. Magn. Reson. Imaging 21 (10), 1251–1261. http://dx.doi.org/10.1016/j.mri.2003.08.026.

Greicius, M.D., Krasnow, B., Reiss, A.L., Menon, V., 2003. Functional connectivity in the resting brain: a network analysis of the default mode hypothesis. Proc. Natl. Acad. Sci. U.S.A. 100 (1), 253–258. http://dx.doi.org/10.1073/pnas.0135058100.

Greicius, M.D., Supekar, K., Menon, V., Dougherty, R.F., 2009. Resting-state functional connectivity reflects structural connectivity in the default mode network. Cereb. Cortex 19 (1), 72–78. http://dx.doi.org/10.1093/cercor/bhn059.

Greicius, M., 2008. Resting-state functional connectivity in neuropsychiatric disorders. Curr. Opin. Neurol. 21 (4), 424–430. http://dx.doi.org/10.1097/WCO.0b013e328306f2c5.

Gusnard, D.A., Raichle, M.E., 2001. Searching for a baseline: functional imaging and the resting human brain. Nat. Rev. Neurosci. 2 (10), 685–694. http://dx.doi.org/10.1038/35094500.

Hagmann, P., Kurant, M., Gigandet, X., Thiran, P., Wedeen, V.J., Meuli, R., Thiran, J.-P., 2007. Mapping human whole-brain structural networks with diffusion MRI. PLoS One 2 (7), e597. http://dx.doi.org/10.1371/journal.pone.0000597.

Hagmann, P., Cammoun, L., Gigandet, X., Meuli, R., Honey, C.J., Wedeen, V.J., Sporns, O., 2008. Mapping the structural core of human cerebral cortex. PLoS Biol. 6 (7), e159. http://dx.doi.org/10.1371/journal.pbio.0060159.

Harlow, J., 1868. Recovery from the passage of an iron bar through the head. Publ. Mass. Med. Soc. 2, 327–347.

Harrison, L., Penny, W.D., Friston, K., 2003. Multivariate autoregressive modeling of fMRI time series. NeuroImage 19 (4), 1477–1491. http://dx.doi.org/10.1016/S1053-8119(03)00160-5.

Havlicek, M., Roebroeck, A., Friston, K., Gardumi, A., Ivanov, D., Uludag, K., 2015. Physiologically informed dynamic causal modeling of fMRI data. NeuroImage 122, 355–372. http://dx.doi.org/10.1016/j.neuroimage.2015.07.078.

Hindriks, R., van Putten, M.J.A.M., Deco, G., 2014. Intra-cortical propagation of EEG alpha oscillations. NeuroImage 103, 444–453. http://dx.doi.org/10.1016/j.neuroimage.2014.08.027.

Honey, C.J., Kötter, R., Breakspear, M., Sporns, O., 2007. Network structure of cerebral cortex shapes functional connectivity on multiple time scales. Proc. Natl. Acad. Sci. U.S.A. 104 (24), 10240–10245. http://dx.doi.org/10.1073/pnas.0701519104.

Honey, C.J., Sporns, O., Cammoun, L., Gigandet, X., Thiran, J.P., Meuli, R., Hagmann, P., 2009. Predicting human resting-state functional connectivity from structural connectivity. Proc. Natl. Acad. Sci. U.S.A. 106 (6), 2035−2040. http://dx.doi.org/10.1073/pnas.0811168106.

Horwitz, B., 2003. The elusive concept of brain connectivity. NeuroImage 19 (2 Pt 1), 466−470.

Huettel, S.A., McCarthy, G., 2001. Regional differences in the refractory period of the hemodynamic response: an event-related fMRI study. NeuroImage 14 (5), 967−976. http://dx.doi.org/10.1006/nimg.2001.0900.

Huisman, T.A.G.M., 2003. Diffusion-weighted imaging: basic concepts and application in cerebral stroke and head trauma. Eur. Radiol. 13 (10), 2283−2297. http://dx.doi.org/10.1007/s00330-003-1843-6.

Iturria-Medina, Y., Canales-Rodríguez, E.J., Melie-García, L., Valdés-Hernández, P.A., Martínez-Montes, E., Alemán-Gómez, Y., Sánchez-Bornot, J.M., 2007. Characterizing brain anatomical connections using diffusion weighted MRI and graph theory. NeuroImage 36 (3), 645−660. http://dx.doi.org/10.1016/j.neuroimage.2007.02.012.

Izhikevich, E.M., Edelman, G.M., 2008. Large-scale model of mammalian thalamocortical systems. Proc. Natl. Acad. Sci. U.S.A. 105 (9), 3593−3598. http://dx.doi.org/10.1073/pnas.0712231105.

Jones, D.T., Vemuri, P., Murphy, M.C., Gunter, J.L., Senjem, M.L., Machulda, M.M., Przybelski, S.A., Gregg, B.E., Kantarci, K., Knopman, D.S., Boeve, B.F., Petersen, R.C., Jack Jr., C.R., 2012. Non-stationarity in the "resting Brain's" modular architecture. PLoS One 7 (6), e39731. http://dx.doi.org/10.1371/journal.pone.0039731.

Kapur, S., Phillips, A.G., Insel, T.R., 2012. Why has it taken so long for biological psychiatry to develop clinical tests and what to do about it? Mol. Psychiatry 17 (12), 1174−1179. http://dx.doi.org/10.1038/mp.2012.105.

Kennis, M., Rademaker, A.R., van Rooij, S.J.H., Kahn, R.S., Geuze, E., 2015. Resting state functional connectivity of the anterior cingulate cortex in veterans with and without post-traumatic stress disorder. Hum. Brain Mapp. 36 (1), 99−109. http://dx.doi.org/10.1002/hbm.22615.

Kötter, R., 2004. Online retrieval, processing, and visualization of primate connectivity data from the CoCoMac database. Neuroinformatics 2 (2), 127−144. http://dx.doi.org/10.1385/NI:2:2:127.

Kuramoto, Y., 1986. Chemical oscillations, waves, and turbulence. Springer-Verlag, Berlin, Heidelberg, New York, Tokyo 66 (7), 296. http://dx.doi.org/10.1002/zamm.19860660706.

Lai, M.-C., Lombardo, M.V., Chakrabarti, B., Sadek, S.A., Pasco, G., Wheelwright, S.J., Bullmore, E.T., Baron-Cohen, S., MRC AIMS Consortium, Suckling, J., 2010. A shift to randomness of brain oscillations in people with autism. Biol. Psychiatry 68 (12), 1092−1099. http://dx.doi.org/10.1016/j.biopsych.2010.06.027.

Le Bihan, D., Iima, M., 2015. Diffusion magnetic resonance imaging: what water tells us about biological tissues. PLoS Biol. 13 (7), e1002203. http://dx.doi.org/10.1371/journal.pbio.1002203.

Le Bihan, D., Breton, E., Lallemand, D., Grenier, P., Cabanis, E., Laval-Jeantet, M., 1986. MR imaging of intravoxel incoherent motions: application to diffusion and perfusion in neurologic disorders. Radiology 161 (2), 401−407. http://dx.doi.org/10.1148/radiology.161.2.3763909.

Leech, R., Sharp, D.J., 2014. The role of the posterior cingulate cortex in cognition and disease. Brain 137 (Pt. 1), 12−32. http://dx.doi.org/10.1093/brain/awt162.

Leonardi, N., Richiardi, J., Van De Ville, D., 2013. Functional connectivity eigennetworks reveal different brain dynamics in multiple sclerosis patients. In: 2013 IEEE 10th International Symposium on Biomedical Imaging (ISBI), pp. 528−531. http://dx.doi.org/10.1109/ISBI.2013.6556528.

Lewis, C.M., Baldassarre, A., Committeri, G., Romani, G.L., Corbetta, M., 2009. Learning sculpts the spontaneous activity of the resting human brain. Proc. Natl. Acad. Sci. U.S.A. 106 (41), 17558−17563. http://dx.doi.org/10.1073/pnas.0902455106.

Liang, M., Zhou, Y., Jiang, T., Liu, Z., Tian, L., Liu, H., Hao, Y., 2006. Widespread functional disconnectivity in schizophrenia with resting-state functional magnetic resonance imaging. Neuroreport 17 (2), 209–213.

Liu, Y., Liang, M., Zhou, Y., He, Y., Hao, Y., Song, M., Yu, C., Liu, H., Liu, Z., Jiang, T., 2008. Disrupted small-world networks in schizophrenia. Brain 131 (Pt. 4), 945–961. http://dx.doi.org/10.1093/brain/awn018.

Lizier, J.T., Heinzle, J., Horstmann, A., Haynes, J.-D., Prokopenko, M., 2010. Multivariate information-theoretic measures reveal directed information structure and task relevant changes in fMRI connectivity. J. Comput. Neurosci. 30 (1), 85–107. http://dx.doi.org/10.1007/s10827-010-0271-2.

Lv, P., Guo, L., Hu, X., Li, X., Jin, C., Han, J., Li, L., Liu, T., 2013. Modeling dynamic functional information flows on large-scale brain networks. Med. Image Comput. Comput. Assist. Interv. 16 (Pt 2), 698–705.

Lynall, M.-E., Bassett, D.S., Kerwin, R., McKenna, P.J., Kitzbichler, M., Muller, U., Bullmore, E., 2010. Functional connectivity and brain networks in schizophrenia. J. Neurosci. 30 (28), 9477–9487. http://dx.doi.org/10.1523/JNEUROSCI.0333-10.2010.

Ma, S., Calhoun, V.D., Phlypo, R., Adalı, T., 2014. Dynamic changes of spatial functional network connectivity in healthy individuals and schizophrenia patients using independent vector analysis. NeuroImage 90, 196–206. http://dx.doi.org/10.1016/j.neuroimage.2013.12.063.

Majeed, W., Magnuson, M., Keilholz, S.D., 2009. Spatiotemporal dynamics of low frequency fluctuations in BOLD fMRI of the rat. J. Magn. Reson. Imaging 30 (2), 384–393. http://dx.doi.org/10.1002/jmri.21848.

Majeed, W., Magnuson, M., Hasenkamp, W., Schwarb, H., Schumacher, E.H., Barsalou, L., Keilholz, S.D., 2011. Spatiotemporal dynamics of low frequency BOLD fluctuations in rats and humans. NeuroImage 54 (2), 1140–1150. http://dx.doi.org/10.1016/j.neuroimage.2010.08.030.

Manolio, T.A., Collins, F.S., Cox, N.J., Goldstein, D.B., Hindorff, L.A., Hunter, D.J., McCarthy, M.I., Ramos, E.M., Cardon, L.R., Chakravarti, A., Cho, J.H., Guttmacher, A.E., Kong, A., Kruglyak, L., Mardis, E., Rotimi, C.N., Slatkin, M., Valle, D., Whittemore, A.S., Boehnke, M., Clark, A.G., Eichler, E.E., Gibson, G., Haines, J.L., Mackay, T.F.C., McCarroll, S.A., Visscher, P.M., 2009. Finding the missing heritability of complex diseases. Nature 461 (7265), 747–753. http://dx.doi.org/10.1038/nature08494.

Maxim, V., Şendur, L., Fadili, J., Suckling, J., Gould, R., Howard, R., Bullmore, E., 2005. Fractional Gaussian noise, functional MRI and Alzheimer's disease. NeuroImage 25 (1), 141–158. http://dx.doi.org/10.1016/j.neuroimage.2004.10.044.

Maximo, J.O., Keown, C.L., Nair, A., Müller, R.-A., 2013. Approaches to local connectivity in autism using resting state functional connectivity MRI. Front. Hum. Neurosci. 7, 605. http://dx.doi.org/10.3389/fnhum.2013.00605.

McCarthy, M.I., Abecasis, G.R., Cardon, L.R., Goldstein, D.B., Little, J., Ioannidis, J.P.A., Hirschhorn, J.N., 2008. Genome-wide association studies for complex traits: consensus, uncertainty and challenges. Nat. Rev. Genet. 9 (5), 356–369. http://dx.doi.org/10.1038/nrg2344.

Messé, A., Rudrauf, D., Benali, H., Marrelec, G., 2014. Relating structure and function in the human brain: relative contributions of anatomy, stationary dynamics, and non-stationarities. PLoS Comput. Biol. 10 (3), e1003530. http://dx.doi.org/10.1371/journal.pcbi.1003530.

Messé, A., Benali, H., Marrelec, G., 2015. Relating structural and functional connectivity in MRI: a simple model for a complex brain. IEEE Trans. Med. Imaging 34 (1), 27–37. http://dx.doi.org/10.1109/TMI.2014.2341732.

Milgram, S., 1967. The small-world problem. Psychol. Today 1 (1), 61–67.

Moran, R.J., Jung, F., Kumagai, T., Endepols, H., Graf, R., Dolan, R.J., Friston, K.J., Stephan, K.E., Tittgemeyer, M., 2011. Dynamic causal models and physiological inference: a validation study using isoflurane anaesthesia in rodents. PLoS One 6 (8), e22790. http://dx.doi.org/10.1371/journal.pone.0022790.

Motulsky, A.G., 2006. Genetics of complex diseases. J. Zhejiang Univ. Sci. B 7 (2), 167–168. http://dx.doi.org/10.1631/jzus.2006.B0167.

Murphy, K., Birn, R.M., Handwerker, D.A., Jones, T.B., Bandettini, P.A., 2009. The impact of global signal regression on resting state correlations: are anti-correlated networks introduced? NeuroImage 44 (3), 893–905. http://dx.doi.org/10.1016/j.neuroimage.2008.09.036.

Myers, R.E., Sperry, R.W., 1953. Interocular transfer of a visual form discrimination habit in cats after section of the optic chiasm and corpus callosum. Anat. Rec. 115, 351–352.

Nagumo, J., Arimoto, S., Yoshizawa, S., 1962. An active pulse transmission line simulating nerve axon. Proc. IRE 50, 2061–2070.

Nakagawa, T., Woolrich, M., Luckhoo, H., Joensson, M., Mohseni, H., Kringelbach, M.L., Jirsa, V., Deco, G., 2014. How delays matter in an oscillatory whole-brain spiking-neuron network model for MEG alpha-rhythms at rest. NeuroImage 87, 383–394.

Nomi, J.S., Uddin, L.Q., 2015. Developmental changes in large-scale network connectivity in autism. NeuroImage Clin. 7, 732–741. http://dx.doi.org/10.1016/j.nicl.2015.02.024.

Patel, R.S., Bowman, F.D., Rilling, J.K.A., 2006. Bayesian approach to determining connectivity of the human brain. Hum. Brain Mapp. 27 (3), 267–276. http://dx.doi.org/10.1002/hbm.20182.

Penny, W.D., Roberts, S.J., 2002. Bayesian multivariate autoregressive models with structured priors. IEEE Proc. Vis. Image Signal Process. 149 (1), 33–41. http://dx.doi.org/10.1049/ip-vis:20020149.

Penny, W.D., Stephan, K.E., Mechelli, A., Friston, K.J., 2004. Comparing dynamic causal models. NeuroImage 22 (3), 1157–1172. http://dx.doi.org/10.1016/j.neuroimage.2004.03.026.

Ponce-Alvarez, A., Deco, G., Hagmann, P., Romani, G.L., Mantini, D., Corbetta, M., 2015. Resting-state temporal synchronization networks emerge from connectivity topology and heterogeneity. PLoS Comput. Biol. 11 (2), e1004100. http://dx.doi.org/10.1371/journal.pcbi.1004100.

Quian Quiroga, R., Kraskov, A., Kreuz, T., Grassberger, P., 2002. Performance of different synchronization measures in real data: a case study on electroencephalographic signals. Phys. Rev. E 65 (4), 41903. http://dx.doi.org/10.1103/PhysRevE.65.041903.

Raichle, M.E., MacLeod, A.M., Snyder, A.Z., Powers, W.J., Gusnard, D.A., Shulman, G.L., 2001. A default mode of brain function. Proc. Natl. Acad. Sci. U.S.A. 98 (2), 676–682. http://dx.doi.org/10.1073/pnas.98.2.676.

Ramsey, J.D., Hanson, S.J., Hanson, C., Halchenko, Y.O., Poldrack, R.A., Glymour, C., 2010. Six problems for causal inference from fMRI. NeuroImage 49 (2), 1545–1558. http://dx.doi.org/10.1016/j.neuroimage.2009.08.065.

Rashid, B., Damaraju, E., Pearlson, G.D., Calhoun, V.D., 2014. Dynamic connectivity states estimated from resting fMRI Identify differences among schizophrenia, bipolar disorder, and healthy control subjects. Front. Hum. Neurosci. 8 http://dx.doi.org/10.3389/fnhum.2014.00897.

Razi, A., Kahan, J., Rees, G., Friston, K.J., 2015. Construct validation of a DCM for resting state fMRI. NeuroImage 106, 1–14. http://dx.doi.org/10.1016/j.neuroimage.2014.11.027.

Rogers, B.P., Morgan, V.L., Newton, A.T., Gore, J.C., 2007. Assessing functional connectivity in the human brain by fMRI. Magn. Reson. Imaging 25 (10), 1347–1357. http://dx.doi.org/10.1016/j.mri.2007.03.007.

Rogers, B.P., Katwal, S.B., Morgan, V.L., Asplund, C.L., Gore, J.C., 2010. Functional MRI and multivariate autoregressive models. Magn. Reson. Imaging 28 (8), 1058–1065. http://dx.doi.org/10.1016/j.mri.2010.03.002.

Rubinov, M., Sporns, O., 2010. Complex network measures of brain connectivity: uses and interpretations. NeuroImage 52 (3), 1059–1069. http://dx.doi.org/10.1016/j.neuroimage.2009.10.003.

Sadeh, N., Spielberg, J.M., Miller, M.W., Milberg, W.P., Salat, D.H., Amick, M.M., Fortier, C.B., McGlinchey, R.E., 2015. Neurobiological indicators of disinhibition in posttraumatic stress disorder. Hum. Brain Mapp. 36 (8), 3076–3086. http://dx.doi.org/10.1002/hbm.22829.

Salomon, R., Bleich-Cohen, M., Hahamy-Dubossarsky, A., Dinstien, I., Weizman, R., Poyurovsky, M., Kupchik, M., Kotler, M., Hendler, T., Malach, R., 2011. Global functional connectivity deficits in schizophrenia depend on behavioral state. J. Neurosci. 31 (36), 12972–12981. http://dx.doi.org/10.1523/JNEUROSCI.2987-11.2011.

Schmahmann, J.D., Pandya, D.N., Wang, R., Dai, G., D'Arceuil, H.E., de Crespigny, A.J., Wedeen, V.J., 2007. Association fibre pathways of the brain: parallel observations from diffusion spectrum imaging and autoradiography. Brain 130 (3), 630–653. http://dx.doi.org/10.1093/brain/awl359.

Schmidt, H., Petkov, G., Richardson, M.P., Terry, J.R., 2014. Dynamics on networks: the role of local dynamics and global networks on the emergence of hypersynchronous neural activity. PLoS Comput. Biol. 10 (11), e1003947. http://dx.doi.org/10.1371/journal.pcbi.1003947.

Scoville, W.B., Milner, B., 1957. Loss of recent memory after bilateral hippocampal lesions. J. Neurol. Neurosurg. Psychiatry 20 (1), 11–21. http://dx.doi.org/10.1136/jnnp.20.1.11.

Shimizu, S., Hoyer, P.O., Hyvärinen, A., Kerminen, A., 2006. A linear non-gaussian acyclic model for causal discovery. J. Mach. Learn. Res. 7, 2003–2030.

Smith, J.F., Pillai, A., Chen, K., Horwitz, B., 2010. Identification and validation of effective connectivity networks in functional magnetic resonance imaging using switching linear dynamic systems. NeuroImage 52 (3), 1027–1040. http://dx.doi.org/10.1016/j.neuroimage.2009.11.081.

Smith, S.M., Miller, K.L., Salimi-Khorshidi, G., Webster, M., Beckmann, C.F., Nichols, T.E., Ramsey, J.D., Woolrich, M.W., 2011. Network modelling methods for FMRI. NeuroImage 54 (2), 875–891. http://dx.doi.org/10.1016/j.neuroimage.2010.08.063.

Spisák, T., Jakab, A., Kis, S.A., Opposits, G., Aranyi, C., Berényi, E., Emri, M., 2014. Voxel-wise motion artifacts in population-level whole-brain connectivity analysis of resting-state FMRI. PLoS One 9 (9), e104947. http://dx.doi.org/10.1371/journal.pone.0104947.

Sporns, O., Tononi, G., Kötter, R., 2005. The human connectome: a structural description of the human brain. PLoS Comput. Biol. 1 (4), e42. http://dx.doi.org/10.1371/journal.pcbi.0010042.

Sripada, R.K., King, A.P., Garfinkel, S.N., Wang, X., Sripada, C.S., Welsh, R.C., Liberzon, I., 2012. Altered resting-state amygdala functional connectivity in men with posttraumatic stress disorder. J. Psychiatry Neurosci. 37 (4), 241–249. http://dx.doi.org/10.1503/jpn.110069.

Stephan, K.E., Roebroeck, A., 2012. A short history of causal modeling of fMRI data. Neuro-Image 62 (2), 856–863. http://dx.doi.org/10.1016/j.neuroimage.2012.01.034.

Stephan, K.E., Weiskopf, N., Drysdale, P.M., Robinson, P.A., Friston, K.J., 2007. Comparing hemodynamic models with DCM. NeuroImage 38 (3), 387–401. http://dx.doi.org/10.1016/j.neuroimage.2007.07.040.

Thompson, G.J., Merritt, M.D., Pan, W.-J., Magnuson, M.E., Grooms, J.K., Jaeger, D., Keilholz, S.D., 2013. Neural correlates of time-varying functional connectivity in the rat. NeuroImage 83, 826–836. http://dx.doi.org/10.1016/j.neuroimage.2013.07.036.

Travers, J., Milgram, S., 1969. An experimental study of the small world problem. Sociometry 32 (4), 425–443.

van den Heuvel, M.P., Stam, C.J., Kahn, R.S., Hulshoff Pol, H.E., 2009a. Efficiency of functional brain networks and intellectual performance. J. Neurosci. 29 (23), 7619–7624. http://dx.doi.org/10.1523/JNEUROSCI.1443-09.2009.

van den Heuvel, M.P., Mandl, R.C.W., Kahn, R.S., Hulshoff Pol, H.E., 2009b. Functionally linked resting-state networks reflect the underlying structural connectivity architecture of the human brain. Hum. Brain Mapp. 30 (10), 3127–3141. http://dx.doi.org/10.1002/hbm.20737.

Vargas, C., López-Jaramillo, C., Vieta, E., 2013. A systematic literature review of resting state network — functional MRI in bipolar disorder. J. Affect. Disord. 150 (3), 727–735. http://dx.doi.org/10.1016/j.jad.2013.05.083.

Vedel Jensen, E.B., Thorarinsdottir, T.L., 2007. A spatio-temporal model for functional magnetic resonance imaging data — with a view to resting state networks. Scand. J. Stat. 34 (3), 587–614. http://dx.doi.org/10.1111/j.1467-9469.2006.00554.x.

Vuksanović, V., Hövel, P., 2015. Dynamic changes in network synchrony reveal resting-state functional networks. Chaos 25 (2), 23116. http://dx.doi.org/10.1063/1.4913526.

Wang, L., Hermens, D.F., Hickie, I.B., Lagopoulos, J., 2012. A systematic review of resting-state functional-MRI studies in major depression. J. Affect. Disord. 142 (1–3), 6–12. http://dx.doi.org/10.1016/j.jad.2012.04.013.

Watts, D.J., Strogatz, S.H., 1998. Collective dynamics of "small-world" networks. Nature 393 (6684), 440–442. http://dx.doi.org/10.1038/30918.

Wedeen, V.J., Hagmann, P., Tseng, W.-Y.I., Reese, T.G., Weisskoff, R.M., 2005. Mapping complex tissue architecture with diffusion spectrum magnetic resonance imaging. Magn. Reson. Med. 54 (6), 1377–1386. http://dx.doi.org/10.1002/mrm.20642.

Weng, S.-J., Wiggins, J.L., Peltier, S.J., Carrasco, M., Risi, S., Lord, C., Monk, C.S., 2010. Alterations of resting state functional connectivity in the default network in adolescents with autism spectrum disorders. Brain Res. 1313, 202–214. http://dx.doi.org/10.1016/j.brainres.2009.11.057.

Wilson, H.R., Cowan, J.D., 1972. Excitatory and inhibitory interactions in localized populations of model neurons. Biophys. J. 12 (1), 1–24. http://dx.doi.org/10.1016/S0006-3495(72)86068-5.

Wong, K.-F., Wang, X.-J., 2006. A recurrent network mechanism of time integration in perceptual decisions. J. Neurosci. 26 (4), 1314–1328. http://dx.doi.org/10.1523/JNEUROSCI.3733-05.2006.

Yan, X., Brown, A.D., Lazar, M., Cressman, V.L., Henn-Haase, C., Neylan, T.C., Shalev, A., Wolkowitz, O.M., Hamilton, S.P., Yehuda, R., Sodickson, D.K., Weiner, M.W., Marmar, C.R., 2013. Spontaneous brain activity in combat related PTSD. Neurosci. Lett. 547, 1–5. http://dx.doi.org/10.1016/j.neulet.2013.04.032.

Yu, Q., Erhardt, E.B., Sui, J., Du, Y., He, H., Hjelm, D., Cetin, M.S., Rachakonda, S., Miller, R.L., Pearlson, G., Calhoun, V.D., 2015. Assessing dynamic brain graphs of time-varying connectivity in fMRI data: application to healthy controls and patients with schizophrenia. NeuroImage 107, 345–355. http://dx.doi.org/10.1016/j.neuroimage.2014.12.020.

Zalesky, A., Fornito, A., Bullmore, E.T., 2010. Network-based statistic: identifying differences in brain networks. NeuroImage 53 (4), 1197–1207. http://dx.doi.org/10.1016/j.neuroimage.2010.06.041.

Zalesky, A., Fornito, A., Cocchi, L., Gollo, L.L., Breakspear, M., 2014. Time-resolved resting-state brain networks. Proc. Natl. Acad. Sci. U.S.A. 111 (28), 10341–10346. http://dx.doi.org/10.1073/pnas.1400181111.

# 5

# Dynamic Causal Modeling and Its Application to Psychiatric Disorders

*Jakob Heinzle[1], Klaas E. Stephan[1, 2]*

[1] University of Zurich and Swiss Federal Institute of Technology (ETH), Zurich, Switzerland; [2] University College London, London, United Kingdom

*Computational Psychiatry*
http://dx.doi.org/10.1016/B978-0-12-809825-7.00005-5

**117**

Psychiatric diseases are linked to a variety of pathophysiological mechanisms (Krystal and State, 2014; Stephan et al., 2016). To further our understanding of the biological underpinnings and, even more importantly, to introduce clinical tests (Kapur et al., 2012) for psychiatry it is key to measure the "brain at work." Functional magnetic resonance imaging (fMRI) and electroencephalography (EEG) and magnetoencephalography (MEG) are the most widely used brain imaging methods to measure human brain activity noninvasively. Traditional data analysis methods have focused on regional activations and their differences across conditions. However, the richness of the data allows for more detailed analysis of interactions among different brain regions. Brain connectivity is of particular interest for psychiatry, because many psychiatric disorders are characterized by aberrant synaptic transmission and plasticity (e.g., Stephan et al., 2009a) or neuromodulation (e.g., Howes and Kapur, 2009). As a consequence, the last years have seen a surge in the development and application of methods for characterizing brain connectivity in psychiatry (Anticevic et al., 2015; Stephan et al., 2016).

In this chapter, we focus on one such method called dynamic causal modeling (DCM; Friston et al., 2003) that uses a Bayesian approach to infer the connectivity (and other parameters) of a neural system of interest from brain imaging data. In the first section of this chapter, we briefly introduce the basics of DCM and refer the reader to a series of reviews and original papers for a more detailed account of the mathematical and physiological foundations of DCM. In a second part, we discuss studies that have applied DCM in the context of psychiatry with a focus on schizophrenia. This section discusses three potential applications of DCM in psychiatry. Finally, we will give an outlook on possible translational applications of DCM in psychiatry with a focus on challenges that need to be addressed for this endeavor.

## 5.1 INTRODUCTION TO DYNAMIC CAUSAL MODELING

DCMs (Friston et al., 2003) are a class of generative (Bayesian) models that can be used to infer on parameters and model structure. The general idea is to describe neuronal population dynamics in a system of interest by a set of differential equations, and how these dynamics give rise to the measured data. DCMs are most commonly used for fMRI (Friston et al., 2003) and EEG/MEG (David et al., 2006; Kiebel et al., 2006; Moran et al., 2011a). Both measurement techniques highlight different aspects of cortical activity: fMRI data have a relatively low temporal resolution on the order of seconds but good spatial resolution of about 1 mm; MEG and EEG data can capture very fast temporal dynamics (milliseconds) but are limited in their ability to precisely localize activity within the brain.

A DCM is a generative model that tries to explain the observed data (Fig. 5.1A). This is achieved by using two key ingredients: First, a dynamical system describes the neuronal dynamics:

$$\dot{x} = f(x(t), \theta_x, u(t)) \tag{5.1}$$

Here, $\dot{x}$ is the temporal derivative of the neural states $x(t)$, and $f(x, \theta_x, u)$ is a function that describes how the rate of change of the states of the system depends on a set of parameters $\theta_x$ and external input $u$. The parameters define the dynamics of the system and can sometimes be directly related to physiological quantities, e.g., the conductance strength of a particular ion channel (Moran et al., 2011b).

Second, a forward model $g$ describes how the dynamic neuronal states are mapped to the measured data, e.g., by defining the blood oxygen level dependent (BOLD) signal equation in fMRI (Stephan et al., 2007) or the forward mapping of neural states to the EEG/MEG signal through a lead field (Kiebel et al., 2006):

$$y(t) = g(x(t), \theta_y) + \varepsilon. \tag{5.2}$$

In Eq. (5.2), $y(t)$ is the measured data and $\varepsilon$ is a measurement noise term, which is usually assumed to be Gaussian noise with zero mean and variance $\Sigma : \mathcal{N}(0, \Sigma)$. The function $g$ depends on a different set of parameters $\theta_y$. Together with the noise model these two equations define the likelihood, i.e., the probability of the data

$$p(y|\theta, m) = \mathcal{N}(g(x, \theta), \Sigma) \tag{5.3}$$

given parameters $\theta = \{\theta_x, \theta_y\}$ and a particular model $m$. In other words, it is assumed that the observed data are distributed around the trace predicted by the dynamical system mapped through the forward model. See Stephan and Friston (2010) for a review.

Together with a prior distribution on the parameters, which defines the a priori expected range of parameters, Eqs. (5.1)–(5.3) define a full generative model of the data. To invert this model, i.e., to estimate the posterior probability of the parameters given the data, one can apply Bayes' theorem:

$$p(\theta|y, m) = \frac{p(y|\theta, m)p(\theta|m)}{p(y|m)}. \tag{5.4}$$

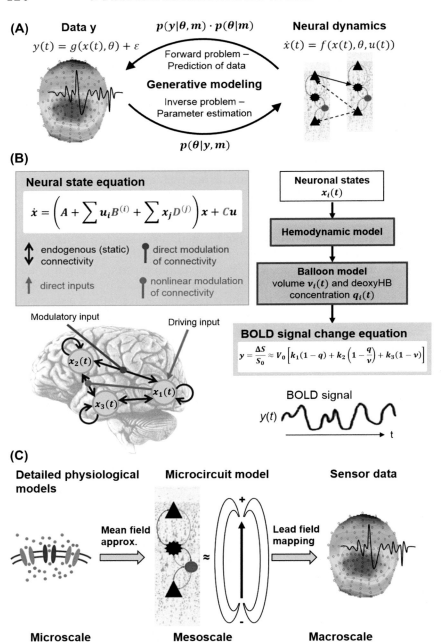

**(A)**     **Data y**     $p(y|\theta, m) \cdot p(\theta|m)$     **Neural dynamics**

$y(t) = g(x(t), \theta) + \varepsilon$             $\dot{x}(t) = f(x(t), \theta, u(t))$

Forward problem –
Prediction of data

**Generative modeling**

Inverse problem –
Parameter estimation

$p(\theta|y, m)$

**(B)**

**Neural state equation**

$$\dot{x} = \left( A + \sum u_i B^{(i)} + \sum x_j D^{(j)} \right) x + Cu$$

↕ endogenous (static) connectivity

↕ direct modulation of connectivity

↑ direct inputs

● nonlinear modulation of connectivity

**Neuronal states** $x_i(t)$

**Hemodynamic model**

**Balloon model**
volume $v_i(t)$ and deoxyHB
concentration $q_i(t)$

Modulatory input       Driving input

$x_2(t)$

$x_1(t)$

$x_3(t)$

**BOLD signal change equation**

$$y = \frac{\Delta S}{S_0} \approx V_0 \left[ k_1(1-q) + k_2 \left( 1 - \frac{q}{v} \right) + k_3(1-v) \right]$$

BOLD signal

$y(t)$           t

**(C)**

**Detailed physiological models**     **Microcircuit model**     **Sensor data**

Mean field approx.         Lead field mapping

+

≈

–

**Microscale**        **Mesoscale**        **Macroscale**

FIGURE 5.1   **Dynamic causal modeling for functional magnetic resonance imaging (fMRI) and electro- and magnetoencephalography (EEG and MEG).** (A) *Illustration of the generative modeling approach of dynamic causal modeling (DCM).* Experimental data, e.g., EEG sensor time series, are explained by a generative model. This model approximates basic

A graphical illustration of the relation of the above quantities to the data is provided in Fig. 5.1A. In DCM, solving the equation for the posterior is not possible analytically, but only by approximate Bayesian methods. We will not discuss the technicalities of Bayesian model inversion here, and refer the reader to the literature. The most commonly used approach is variational Bayes (Friston et al., 2007). For an introduction to variational Bayes see, for example, Bishop (2007). Other methods to invert DCMs for fMRI include the use of Gaussian Processes (Lomakina et al., 2015) and Markov Chain Monte Carlo (MCMC) sampling (Aponte et al., 2016).

## 5.1.1 Dynamic Causal Models for Functional Magnetic Resonance Imaging

The original application domain of DCM was fMRI (Friston et al., 2003). Because of the low temporal resolution of fMRI due the low-pass filtering properties of the BOLD signal, the neural model of DCM for fMRI is kept relatively simple. The data that are modeled are time series from a preselected network of cortical regions (nodes). These neural nodes are modeled as exponentially decaying in activity and receive linear external as well as input from other nodes. In addition to these driving inputs there are also modulatory effects that modulate the strength of connections, either directly via external input (in bilinear DCM) or by the activity in a node of the network (in nonlinear DCM,

---

physiological processes by a dynamical system that describes the evolution of the neural states through a differential equation $\dot{x} = f(x, \theta_x, u)$. The likelihood function is defined by the observation or forward function $y = g(x, \theta_y) + \varepsilon$, where $\varepsilon$ represents measurement noise. Together with the prior the likelihood defines the generative model. During inversion, the parameters are updated so that the model fits the data best (under the constraints given by the prior). (B) *State equations for DCM for fMRI.* The hemodynamic model is not described in detail. The blood oxygen level dependent (BOLD) signal equation has three parameters which depend on scanner settings and magnetic field strength: $k_1 = 4.3\vartheta_0 E_0 TE$, $k_2 = \varepsilon r_0 E_0 TE$ and $k_3 = 1 - \varepsilon$. (C) *Illustration of DCM for EEG/MEG.* For EEG/MEG the neural model describes the mean field activity of a cortical microcircuit (middle). The dynamics model neural signals at a mesoscale and are derived from basic neuronal and synaptic dynamics at the microscale through a mean field approximation. The neural activity, e.g., of the pyramidal cells in deep layers, describes the fluctuations of an electric or current dipole. The forward model in EEG/ MEG maps the activity of this dipole (i.e., the neural activity) onto the sensor data via a so-called lead field, thereby, providing a model of the macroscale data, the EEG or MEG signal. *(B) Adapted with permission from Stephan, K.E., Iglesias, S., Heinzle, J., Diaconescu, A.O., 2015. Translational perspectives for computational neuroimaging. Neuron 87, 716–732 (License obtained from RightsLink), figure of microcircuit taken from Moran, R., Pinotsis, D.A., Friston, K., 2013. Neural masses and fields in dynamic causal modeling. Front. Comput. Neurosci. 7, 57 under license (Attribution 3.0 Unported, CC BY 3.0; https://creativecommons.org/licenses/by/3.0/).*

Stephan et al., 2008). Hence, the full dynamics of the neural network are given by

$$\dot{x}(t) = \left( A + \sum_i u_i(t)B^{(i)} + \sum_j x_j(t)D^{(j)} \right) x(t) + Cu(t) \tag{5.5}$$

Here, bold letters $x$ and $u$ are temporally evolving vectors of the neuronal states and inputs, respectively. The matrix $C$ controls the input weights. The rate of change in the states depends deterministically on all other states through a dynamic connectivity matrix given in the large parenthesis on the right hand side of the equation. As described above, the connectivity between neural nodes has a static component $A$, components $B^{(i)}$ that are directly modulated by external input $u_i(t)$ and a nonlinear component $D^{(j)}$ modulated by activity in the neuronal nodes $x_j(t)$. Fig. 5.1B summarizes this basic form of DCM for fMRI. The modulatory and nonlinear effects are of particular interest for clinical applications as they can represent, for example, neuromodulatory effects.

To map the neuronal states onto the predicted fMRI signals, DCM incorporates a hemodynamic model. This model includes the dynamics of neurovascular coupling and the Balloon–Windkessel model, which describes changes in blood volume and oxygenation level of the venous blood, both directly related to relative signal changes in fMRI (Buxton et al., 1998). The details of the hemodynamic model for DCM are discussed in several papers (Friston et al., 2000; Stephan et al., 2007; Havlicek et al., 2015; Heinzle et al., 2016b) and will not be presented here.

## 5.1.2 Dynamic Causal Models for Electrophysiological Data

Electrophysiological data from EEG and MEG have completely different characteristics than fMRI data: The measured signals have a very high temporal, but lower spatial resolution. Hence, the neuronal model describes activity of populations of neurons at a mesoscopic scale and exhibits complex dynamics fast enough to characterize common EEG/MEG features such as event-related responses or oscillations in the range of EEG frequency bands. In other words, the function $f(x(t), \theta_x, u(t))$ exhibits much richer dynamics than in the case of classical DCM for fMRI. It is beyond the scope of this book chapter to provide the mathematical details of all versions of DCM for electrophysiological data, which are summarized in some excellent reviews (Kiebel et al., 2009; Moran et al., 2013).

Usually, the data modeled in EEG/MEG are the sensor time series, while the neuronal model describes the dynamics of a local canonical circuit at the mesoscopic scale (Fig. 5.1C). The activity of the mesoscopic circuit model describes the fluctuations of a dipole (electric dipole for EEG and current dipole for MEG) at a particular location in the brain. The mapping from the neuronal activations to the sensor data is obtained by applying a lead field, which describes how the electrical field on the surface of the head is generated by a dipole at any location in the brain. The neuronal dynamics of electrophysiological DCMs describe small cortical circuits consisting of several populations of neurons. The dynamics of these circuits are based on mean field dynamics that describe the average activity of neurons and synapses in populations of neurons. See Fig. 5.1C for an illustration of how the different scales are related in DCM for EEG/MEG. The original version of DCM for EEG was based on convolution models (David et al., 2006; Kiebel et al., 2006, 2008) that describe neuronal masses using the formalism by Jansen and Rit (1995). This also forms the basis for a recent DCM that can be fitted to fMRI and EEG data alike (Friston et al., 2017). Later developments include conductance-based models (Moran et al., 2011a; Pinotsis et al., 2012) and models for canonical microcircuits (Bastos et al., 2012; Moran et al., 2013).

### 5.1.3 Model Comparison

Comparing alternative DCMs of the same data rests on the model evidence $p(y|m)$, a principled index of model goodness that takes into account both model fit, or accuracy, as well as the complexity of the model. The model evidence can be used to directly compare different models, i.e., competing hypotheses, and serve to select the model that best explains how a particular experimental finding was generated (Penny et al., 2004; Stephan et al., 2009b, 2010). Having selected this model, one can proceed to interpreting its parameter estimates (see below).

In the simple case of two model hypotheses, model comparison uses the difference in log model evidence[1] between the models to decide which of the two explains the data better. A difference of 3 log units or more is considered strong evidence in favor of the model with the higher log model evidence (Kass and Raftery, 1995).

For clinical applications, the (log) evidence can be used for model selection as a basis for differential diagnostics in individual patients (Stephan et al., 2017). By contrast, in basic science studies, group-level model comparisons are usually more relevant. At the group level, model comparison can be performed either by summing log evidences across subjects (a "fixed-effects" analysis that assumes the same model underlies the data from all subjects) or by estimating the expected frequency of any given model in the population ("random-effects" analysis) (Stephan et al., 2009b).

---

[1] Note that a difference in log model evidence is equal to the log Bayes Factor between two models.

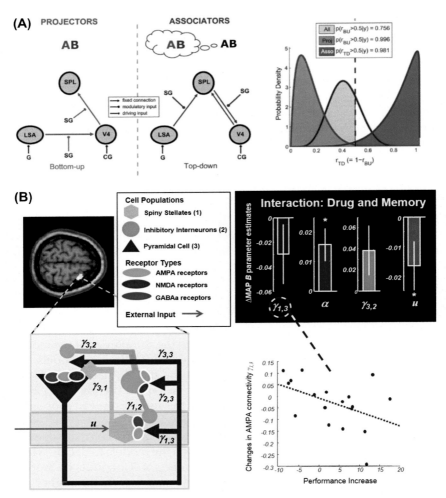

FIGURE 5.2  **Examples of model comparison and parameter estimate.** (A) *Model comparison separates different types of synesthesia.* Left: a three-region dynamic causal modeling (DCM) including a visual letter shape area (LSA), a visual color sensitive region (V4), and a parietal region (superior parietal lobule, SPL) was applied to data from a paradigm where participants saw synesthesia evoking graphemes (SG), graphemes (G) that did not evoke synesthesia, and colored graphemes (CG). Grapheme input always targeted LSA. For colored graphemes there was additional direct input to V4. The SG input was added as a modulation of either bottom-up (projector model) or top-down (associator model) connections in the network. Right: model comparison using the entire group of subjects was not able to distinguish which of the two models had generated the data (gray distribution illustrates the expected distribution of the model in the population). When, however, the participants were split into a projector and associator group based on a questionnaire, model comparison resulted in a clear preference for the bottom-up model in projectors and the top-down model in associators, respectively. Hence, the two groups could be separated based on which model better explained their brain data while viewing graphemes. Importantly, the absence of a finding when considering the whole group illustrates that model comparison correctly inferred that the two models were roughly equally

An illustrative example of how model comparison can help understand differences between groups of participants in terms of effective connectivity was given by van Leeuwen et al. (2011). They compared two groups of participants with grapheme-color synesthesia: so-called "projectors," who report seeing letters in colors, and "associators," who report a vivid association of letters with colors. Model comparison suggested that projectors showed direct effective connectivity between early visual regions for word processing and color, while in associators interactions between these two regions were mediated by a higher region in the parietal cortex (Fig. 5.2A). This example shows, how model comparison can be used to separate underlying physiological mechanisms, i.e., connectivity in the case of fMRI, that differ, for example, among groups with different perception of color during synesthesia.

### 5.1.3.1 Model Family Comparison

Instead of simply comparing individual models, it is also possible to compare families of models (Penny et al., 2010). This is often used to group several models that share the same feature into a joint family and compare them to another family that does not include this feature. By doing so, family comparison averages over all models with a particular feature of interest and, thus, statements about, for example, a specific modulation can be made irrespective of the details of the rest of the model. An example of model family comparison, will be given below (Deserno et al., 2012).

---

likely within the study population. (B) *Estimating synaptic function form MEG signals.* Left: a conductance-based DCM of a single region in dorsolateral prefrontal cortex was used to model neural activity measured with MEG during working memory and no working memory in two drug conditions: L-Dopa versus placebo. The sketch in the bottom left illustrates the modeled circuit. For explanation of symbols and colors, see the legend in the inset. Top right: several parameters of the model showed an interaction effect between drug and working memory. *Asterisks* indicate significant values ($P < .05$). Two of the interactions showed a significant correlation with the change in working memory performance between L-Dopa and placebo: the α-amino-3-hydroxy-5-methyl-4-isoxazolepropionic acid (AMPA) conductance of the connection from the pyramidal cell population to the spiny stellate population ($\gamma_{1,3}$, see bottom right) as well as a term that controls the strength of the N-methyl-d-aspartate (NMDA) conductance (α, not shown). *MAP*, maximum a posteriori. *(A) Adapted under license (Attribution-NonCommercial-ShareAlike 3.0 Unported, CC BY-NC-SA 3.0; https://creativecommons.org/licenses/by-nc-sa/3.0/) from van Leeuwen, T.M., den Ouden, H.E., Hagoort, P., 2011. Effective connectivity determines the nature of subjective experience in grapheme-color synesthesia. J. Neurosci. 31, 9879–9884. (B) Adapted with permission from Stephan, K.E., Iglesias, S., Heinzle, J., Diaconescu, A.O., 2015. Translational perspectives for computational neuroimaging. Neuron 87, 716–732 (Licence obtained from RightsLink) and Moran, R.J., Symmonds, M., Stephan, K.E., Friston, K.J., Dolan, R.J., 2011b. An in vivo assay of synaptic function mediating human cognition. Curr. Biol. 21, 1320–1325. under license (Creative Commons Attribution License, CC BY; https://creativecommons.org/licenses/by/4.0/).*

## 5.1.4 Model Parameter Estimates: Physiological and Clinical Interpretations

### 5.1.4.1 Posterior Parameter Estimates

Once a particular model has been selected, one can proceed to examining its posterior distribution $p(\theta|y, m)$. The posterior represents a complete representation of the model's parameter values that govern the dynamics of the neural system, including the uncertainty about them. In single subjects, the posterior can be used to compute the probability that a particular parameter has a value larger than a threshold of interest; at the group level, classical tests can be applied to subject-wise maximum a posteriori (MAP) estimates.

The physiological interpretability of parameters differs across type of DCMs. In classical DCM for fMRI, coupling parameters are of a phenomenological nature and only allow for a coarse distinction between glutamatergic (cortical long-range connections) and γ-aminobutyric acid (GABA)ergic (local inhibition) processes. In electrophysiological DCMs, parameters are often more tightly linked to physiological properties. For example, in conductance-based DCMs, different ionotropic receptor types are represented by distinct parameters. For example, Moran et al. (2011b) estimated the changes of ion channel conductances during working memory from MEG data under the influence of L-Dopa and placebo. Their results are in agreement with previous studies in nonhuman primates and provide a direct link between measures of synaptic physiology and performance during a working memory task. Fig. 5.2B illustrates this finding, but see below for a more detailed description of this study.

### 5.1.4.2 Model Averaging

It is possible to average parameters over different models (Penny et al., 2010). This Bayesian model averaging (BMA) weights the posterior distribution by the posterior probability of the corresponding model and thus removes the dependency of parameter comparison on the selection of a particular model. It also enables the comparison of parameter estimates across groups in which different models were found to best account for the data (for an example in the context of schizophrenia see Schmidt et al., 2013b).

### 5.1.4.3 Generative Embedding

Instead of comparing individual parameters of DCMs, it is also possible to use all model parameter estimates in their entirety and apply machine learning methods for multivariate classification or clustering. This approach was originally proposed by Brodersen et al. (2011b) and introduced for DCM of fMRI data as "generative embedding" (Brodersen et al., 2011a). The key idea is to use a generative model like DCM for hypothesis-driven dimensionality reduction (from very high-dimensional fMRI data to a small number of parameter estimates) and to render any classification results mechanistically interpretable (in terms of the model's parameters).

Brodersen et al. (2011a) used DCM for fMRI to model an auditory circuit in stroke patients with aphasia and healthy control subjects during a simple auditory task. Crucially, none of the regions included in the DCM was affected by the stroke. A DCM was estimated for each subject individually and the parameters were entered into a support vector machine for classification. Using a cross-validation procedure, patients could be separated from controls with near-perfect (98%) accuracy. The parameters that enabled this classification featured the strengths of interhemispheric connections from the right to the dominant left hemisphere.

Instead of supervised methods such as classification, it is also possible to use unsupervised techniques such as clustering techniques. Brodersen et al. (2014) used data from a working memory DCM study (Deserno et al., 2012) and then applied a variational Gaussian mixture model (vGMM) to the parameter estimates. Again, DCM served as a feature extraction device that provided a low-dimensional parameter vector for each participant (41 schizophrenia patients and 42 controls). These parameter vectors were then used to cluster either the entire sample or the schizophrenia patients only into subgroups. Model comparison enabled the unsupervised selection of an optimal number of clusters in the vGMM. First, patients and healthy controls were correctly clustered into two groups, with a balanced purity[2] of 71%. Second, when examining the patients on their own, three clusters of patients were found. Remarkably, subgroup membership was significantly related to a major index for disease severity, the level of negative symptoms. Fig. 5.3 summarizes the analysis pipeline of this generative embedding approach. See below for a more thorough description of the results and a discussion of this application of generative embedding for dissecting spectral disorders.

## 5.1.5 Other Variants of Dynamic Causal Modeling

While fMRI and MEG/EEG data are classical application domains of DCM, in principle, DCMs can be developed for any kind of measured data where the observation function (from hidden states to measurements) is known. For example, eye movements are a promising candidate, as they are impaired in many psychiatric disorders (e.g., Hutton and Ettinger, 2006; Klein and Ettinger, 2008), and thus constitute a prominent target for computational psychiatry (Heinzle et al., 2016a). Adams et al. (2012) have proposed a model for smooth pursuit that is able to explain

---

[2] Purity is a measure to quantify how well a clustering solution overlaps with the known solution. Balanced purity corrects for biases that can occur due to different cluster sizes. For mathematical definitions of these concepts, see Brodersen et al. (2014).

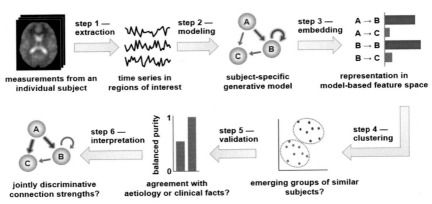

**FIGURE 5.3    Illustration of generative embedding.** A generative embedding analysis starts (step 1) with the extraction of the time series in regions of interest (functional magnetic resonance imaging) or the sensor data for the relevant time window (electroencephalography/magnetoencephalography). These data are then modeled (step 2) with a dynamic causal model whose inversion leads to a generative model with subject specific posterior parameter estimates. From the posterior density a set of features can be extracted, e.g., by taking the maximum a posterior (MAP) estimates (step 3). In step 4, a machine learning algorithm is applied, for example, clustering or classification. The last step of the classification or clustering procedure (step 5) consists of evaluating whether the classification or clustering solution are correct or correspond to some external measure, respectively. Finally, it is possible to investigate which features were most informative for classification or clustering (step 6). This allows for a mechanistic interpretation of the classifier or clustering solution. *Reproduced with under license (Creative Commons Attribution License, CC BY; https://creativecommons.org/licenses/by/4.0/) from Brodersen, K.H., Deserno, L., Schlagenhauf, F., Lin, Z., Penny, W.D., Buhmann, J.M., Stephan, K.E., 2014. Dissecting psychiatric spectrum disorders by generative embedding. NeuroImage Clin. 4, 98–111.*

abnormal smooth pursuit in patients with schizophrenia. This model can be treated as a DCM and applied to eye movement data in a similar fashion as the DCMs for fMRI and EEG data explained above (Adams et al., 2015, 2016). Another physiological quantity that is of relevance for psychiatry, and in particular for anxiety disorders, is skin conductance. A DCM for skin conductance has been proposed by Bach et al. (2010, 2011).

DCM for fMRI in its original form requires an external input that drives the neural system. Hence, it is not suited to model "resting-state" fMRI data, which are widely used in psychiatry (for reviews see, e.g., Greicius, 2008; Karbasforoushan and Woodward, 2012). However, there are variants of DCM that can be used to model resting-state data. The simplest approach is to use the classical DCM but stimulate the network nodes with inputs that represent fluctuations up to certain frequencies (Di and Biswal, 2014). A more principled, but computationally expensive approach models noise in neuronal dynamics explicitly (Li et al., 2011), so-called stochastic DCM. Finally, spectral DCM (Friston et al., 2014) also includes noise at the neuronal level, but is fitted to cross-spectral densities, as a measure of functional connectivity in the frequency domain.

Refinements of classical DCM for fMRI are also needed to address questions that have arisen with new technological developments, for example, ultra high-field MRI (van der Zwaag et al., 2016). For example, it is now possible to measure fMRI at a spatial resolution sufficient to investigate hypotheses about how the layered cortical structure relates to predictive coding (Bastos et al., 2012), which has been suggested to be altered in patients with schizophrenia (Fletcher and Frith, 2009; Corlett et al., 2011). However, studying layered fMRI signals require one to take into account blood draining effects across layers, which are usually ignored in fMRI analyses, because they occur at a spatial scale below commonly used voxel sizes. A layered DCM that takes into account these confounding effects was introduced recently, making it possible to separate neuronal and hemodynamic effects when inferring layer-specific connections (Heinzle et al., 2016b).

# 5.2 APPLICATION OF DYNAMIC CAUSAL MODELING IN PSYCHIATRY

Due to its ability to infer on structure and parameters of a not directly observable dynamical system that generated the data, DCM offers a way to potentially uncover physiological mechanisms linked to psychiatric diseases. Of course, not all models and model parameters have a direct physiological interpretation such as the conductance changes induced by a particular neurotransmitter or neuromodulator. However, even more abstract models like the neural model in DCM for fMRI offer a more mechanistic explanation and provide a more physiological interpretation than, for example, average activations within a cortical region. Conceptually, three main branches of DCM applications to psychiatry exist. The first and also most common application to date is to employ DCM for a better understanding of the mechanisms that cause perceptual or cognitive disturbances in psychiatric diseases. A nice example of this approach was given by Dima et al. (2009) who addressed the question why patients with schizophrenia are less susceptible to the hollow mask illusion than healthy controls. DCM allowed for detecting differences in effective connectivity that were associated with different perception of the hollow mask illusion (see also below for more studies following this approach). The second application rests on the ability of DCM to estimate parameters that have a direct physiological meaning, for example, conductances of ionotropic receptors (Moran et al., 2008, 2013), spike frequency adaptation (Moran et al., 2008), or distinct glutamatergic and GABAergic[3] synaptic strengths (Moran et al., 2011c). Finally, the third and most recent application of DCM

---

[3] GABA: γ-aminobutyric acid.

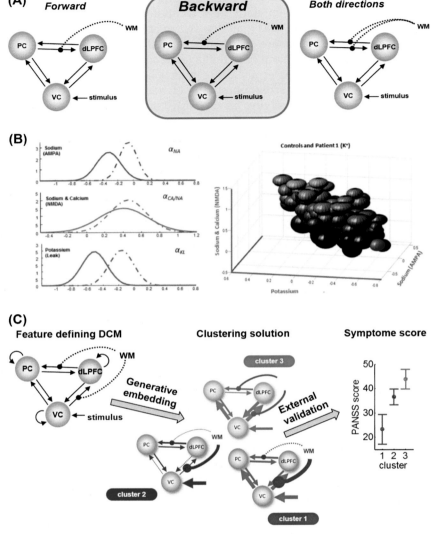

FIGURE 5.4  **Applications of dynamic causal modeling (DCM) in psychiatry.** (A) *Illustration of three model families used to investigate working memory.* The three families differed in the modulation of the connections between parietal cortex (PC) and dorsolateral prefrontal cortex (dlPFC). Each family contained 16 individual models that represented all possible combinations of modulation on connections from and to visual cortex (VC). The model family with backward modulation by working memory (WM) achieved the highest log model evidence (*highlighted*). Subsequent model averaging within this family and comparison between patients and controls showed that the same modulation was significantly different between patients and healthy controls. In addition, the modulation of the connection from VC to PC also differed between the two groups. (B) *Parameter estimates of physiological quantities with conductance-based DCM.* Left: three exemplary comparisons of parameter estimates of a patient with a genetic mutation affecting a potassium current (*solid lines*) and a group of 94 controls

is to use for physiological stratification of heterogeneous clinical groups. For example, Brodersen et al. (2014) have demonstrated that unsupervised clustering of DCM parameters can yield meaningful subgroups within a larger group of patients with schizophrenia.

This section reviews studies that have used DCMs in the context of psychiatry and/or for inferring pathophysiologically relevant parameters such as the conductance of specific ion channels. While this is not an exhaustive review of the literature, we place emphasis on illustrating different ways of applying DCM in psychiatry, which includes all three approaches: model comparison, parameter comparison, and generative embedding. Fig. 5.4 summarizes these conceptual approaches for DCM.

## 5.2.1 Using Dynamic Causal Modeling to Understand Mechanism of Behavioral/Cognitive Dysfunction

The most common application of DCM in general concerns the comparison of alternative mechanisms, in terms of effective connectivity, that could explain observed fMRI or EEG/MEG findings (Stephan et al., 2010). One of the first studies to use DCM in the context of psychiatry investigated differences in processing of a hollow mask between patients

---

*(dashed lines)*. The three plots show (from top to bottom): α-amino-3-hydroxy-5-methyl-4-isoxazolepropionic acid (AMPA) mediated sodium conductance, N-methyl-d-aspartate receptor (NMDA) mediated sodium and calcium conductance, and potassium leak. Note the good separation between patients and controls, in particular for the potassium leak. Right: illustration of the same three parameter estimates for all individual subjects. Ellipsoids indicate Bayesian confidence regions. The patient is shown in *red*. (C) *Dissecting a spectrum disorder.* The winning model of the full model comparison in (A) was used to estimate subject specific DCM parameters for a group of 41 patients with schizophrenia. These parameter vectors were then clustered using a variational Gaussian mixture model. Model comparison on the clustering showed that three clusters provided the best explanation of the data. Representative models of these three clusters are shown in the middle. Finally, the authors performed an external validation by comparing the negative score of the positive and negative syndrome scale (PANSS) for the three clusters (right). *(A) Adapted from Deserno, L., Sterzer, P., Wustenberg, T., Heinz, A., Schlagenhauf, F., 2012. Reduced prefrontal-parietal effective connectivity and working memory deficits in schizophrenia. J. Neurosci. 32, 12–20; Brodersen, K.H., Deserno, L., Schlagenhauf, F., Lin, Z., Penny, W.D., Buhmann, J.M., Stephan, K.E., 2014. Dissecting psychiatric spectrum disorders by generative embedding. NeuroImage Clin. 4, 98–111. under licenses (Attribution-NonCommercial-ShareAlike 3.0 Unported, CC BY-NC-SA 3.0; https://creativecommons.org/ licenses/by-nc-sa/3.0/ and Creative Commons Attribution License, CC BY; https://creativecommons. org/licenses/by/4.0/, respectively). (B) Adapted from Gilbert, J.R., Symmonds, M., Hanna, M.G., Dolan, R.J., Friston, K.J., Moran, R.J., 2016. Profiling neuronal ion channelopathies with non-invasive brain imaging and dynamic causal models: case studies of single gene mutations. NeuroImage 124, 43–53. (C) Adapted from Brodersen, K.H., Deserno, L., Schlagenhauf, F., Lin, Z., Penny, W.D., Buhmann, J.M., Stephan, K.E., 2014. Dissecting psychiatric spectrum disorders by generative embedding. NeuroImage Clin. 4, 98–111 - both under (Creative Commons Attribution License, CC BY; https://creativecommons.org/licenses/by/4.0/).*

---

with schizophrenia and healthy controls (Dima et al., 2009). As expected, patients were less susceptible to the hollow face illusion; this was paralleled by differences in activation of visual areas. The DCM analysis in this study compared two competing models for explaining this finding; these assumed differential modulation of top-down versus bottom-connections when viewing the hollow mask stimulus. While healthy controls showed strong evidence for a model with top-down modulation, patients tended to favor a model with bottom-up modulation. In a follow-up study (Dima et al., 2010), the same authors reproduced their findings using EEG measurements and DCM for event-related potentials. This DCM predicted the EEG measurements based on five equivalent current dipoles placed at the locations of previous fMRI activations. Model comparison gave the same finding as in the previous fMRI study, i.e., in healthy controls a model with modulation of top-down connections was favored, while in patients a model with modulation of bottom-up connections was superior. In summary, these two studies suggested that the mechanism underlying the patients' reduced susceptibility to the hollow mask illusion was a reduced top-down modulation—a physiological difference that is consistent with predictive coding related theories postulating ineffective (low precision) prior expectations during perception in schizophrenia (Adams et al., 2013).

Predictive coding is currently perhaps the most prominent computational framework for understanding abnormalities of perception in schizophrenia (Stephan et al., 2009a; Corlett et al., 2011; Adams et al., 2013). Several DCM studies have investigated possible deficits in predictive coding in psychiatric patients (Dima et al., 2012; Fogelson et al., 2014; Ranlund et al., 2016). A classical paradigm to study predictive coding is the auditory oddball paradigm that results in a prominent EEG/MEG signal, the so-called mismatch negativity (MMN). A basic DCM structure for this paradigm is well established and has been extensively studied in healthy controls (Garrido et al., 2007, 2009a,b). The auditory mismatch was studied both in patients with schizophrenia (Dima et al., 2012) and in a mixed population of patients with psychosis and their first-degree unaffected relatives (Ranlund et al., 2016). In a preliminary study, Dima et al. (2012) compared schizophrenia patients with healthy controls. They employed a classical DCM for auditory MMN (Garrido et al., 2009b) to model MEG data from an auditory mismatch experiment and included three different scenarios of modulation of connections by oddball (deviant) events: forward modulation, backward modulation, and modulation in both directions. The results suggest that modulation in both directions is likely causing the MMN in healthy controls, while patients showed a tendency for modulation of forward connections only. Concerning parameter estimates, the authors reported significant modulatory effects on gain modulation in the right primary auditory cortex and on the top-down connection from the inferior frontal gyrus to the secondary auditory area in the superior temporal gyrus. Ranlund et al. (2016) used the same paradigm in an EEG setting and focused on gain modulation (which, for historical reasons, is often referred to as "intrinsic connections" in the DCM literature) at different levels of the hierarchy. The best model included gain modulation in the inferior frontal gyrus and primary auditory cortex. Both patients and relatives of patients showed reduced synaptic gain in the frontal area compared with healthy controls. This illustrates that DCM can also be used to disambiguate effective connectivity between healthy controls and genetically predisposed but healthy participants with an increased risk for psychosis, i.e., relatives of patients.

Fogelson et al. (2014) used a different MMN paradigm in which a particular sequence of visual stimuli was predictive of the occurrence of a target. EEG was measured in a group of patients with schizophrenia and age-matched healthy controls. A canonical microcircuit DCM was used to model source activity in a set of regions (visual areas V1 and V5 as well as a parietal and inferotemporal area) defined by a source localization analysis of the EEG data. The study focused on a single DCM with task modulation of top-down connections only, and used model comparison to distinguish different hypotheses about differences between patients and healthy controls. A model allowing for differences in both forward and backward connections between the two groups explained the data best. In addition, model comparison also revealed that predictability of the target was related to a change in synaptic gain in both groups. The effect of schizophrenia was strongest for backward connections, while effects of predictability where much stronger in healthy controls than in patients.

A consistent finding across all these DCM studies using predictive coding paradigms or the hollow mask illusion is a change in gain and/or top-down connections. In all studies that included prefrontal regions this change occurred in connections from frontal regions to other brain areas, suggesting that altered connectivity of prefrontal regions is crucially involved in schizophrenia.

A second line of DCM research in schizophrenia has focused on working memory tasks (Crossley et al., 2009; Deserno et al., 2012; Schmidt et al., 2013b). Crossley et al. (2009) used a 1- and 2-back memory task and investigated the connectivity to a region of interest in the superior temporal gyrus from three prefrontal and one parietal region. The three groups—at risk mental state (ARMS) patients, first episode patients (FEP), and healthy controls—differed mainly in the connectivity from the medial frontal gyrus to the superior temporal gyrus, with connection strengths for ARMS patients that were in-between those of healthy controls and FEP. In addition, the authors also reported two ARMS patients that transitioned to FEP within 2 years after scanning. However, there was no obvious relation between effective connectivity estimated with DCM and the transition to the disease. Deserno et al. (2012) investigated 41 schizophrenia patients and 42 controls on a working memory task (numeric 2-back vs. 0-back).

The authors used a three-region DCM including visual, parietal, and dorsolateral prefrontal cortex (dlPFC) to investigate how working memory modulated the effective connectivity among these regions. Using all possible combinations of modulatory effects on connections, they devised three families of DCMs with either feedforward, feedback, or both connections between parietal and prefrontal cortex (Fig. 5.4A). From these three families, the one with backward modulation best explained the data in both healthy controls and patients, but with less clear results in patients. For comparison of modulation and connection strengths, all models of the backward family were averaged using BMA. Four of the eleven parameters of this reduced model differed significantly between healthy controls and patients: the endogenous connection strengths from visual and parietal cortex to the dlPFC as well as the modulation of the connection from dlPFC to parietal cortex and visual cortex to parietal cortex. Schmidt et al. (2013b) used a letter n-back (2- vs. 1- vs. 0-back) task for fMRI of three different groups (ARMS, FEP, and healthy controls). They modeled fMRI activity in a bilateral frontoparietal network for which they devised 12 models with different modulation of connectivity by task demand. BMA showed that healthy controls and patients differed in the connection from prefrontal to parietal regions. This difference was particularly pronounced between controls and nontreated FEP; remarkably, for treated FEPs this value recovered and was no longer significantly different from controls. In summary, the working memory DCM studies also point toward the importance of top-down connections form prefrontal areas, with a marked reduction in patients compared with controls.

In addition to these task-based studies, stochastic DCM has been used to study effective connectivity within the default mode network in FEP with schizophrenia (Bastos-Leite et al., 2015). In this study, the authors investigated a network with four nodes: bilateral parietal regions, anterior frontal cortex, and posterior cingulate cortex. Within this network, patients showed reduced connectivity to the anterior frontal node and the self-connection of the posterior cingulate was correlated to clinical scores.

The work outlined above is not an exhaustive review of DCM-based connectivity studies of schizophrenia to date (for reviews, see also Stephan et al., 2015; Friston et al., 2016a) but provides a focused discussion of some applications. In conclusion, most DCM studies that investigated cognitive tasks or predictive coding in patients with schizophrenia have pointed toward connectivity changes of prefrontal regions.

## 5.2.2 Using Dynamic Causal Modeling to Investigate Synaptic Dysfunction

While the pathophysiological basis of most psychiatric diseases is still debated, a prominent role of synaptic abnormalities for circuit dysfunction is widely accepted. This is perhaps particularly obvious for schizophrenia. Here, several synaptic accounts of pathophysiology have been proposed, for example, the dysconnection hypothesis (Friston, 1998; Stephan et al., 2009a; Friston et al., 2016a), which focuses on abnormal neuromodulatory regulation N-methyl-D-aspartate (NMDA) receptor function, and the dopamine hypothesis (Howes and Kapur, 2009). In the absence of techniques for directly measuring synaptic function noninvasively in the human brain, predictions from these theories must be assessed indirectly, for example, by observing hypothesized changes in neuroimaging or electrophysiological signals. An alternative is to infer synaptic processes using suitably defined DCMs (Moran et al., 2011a,b, 2013). Because of their high temporal resolution EEG and MEG signals offer the possibility to distinguish effects of ionotropic receptors that operate at different time scales, e.g., NMDA versus α-amino-3-hydroxy-5-methyl-4-isoxazolepropionic acid (AMPA) receptors. Moran and colleagues have demonstrated the feasibility of this approach in both invasive animal recordings (Moran et al., 2011c, 2015) and human MEG/EEG studies (Moran et al., 2011b; Gilbert et al., 2016). The latter two studies are of particular interest for applications of DCM to psychiatry.

The first example (Moran et al., 2011b) illustrates how conductance-based DCM can be used to estimate the effect of systemic administration of L-Dopa changes on NMDA and AMPA receptor conductances in human prefrontal cortex during working memory. Eighteen participants were measured with MEG, while they performed a visuospatial working memory task under two conditions—administration of placebo versus L-Dopa (with subject crossover design). L-Dopa was used because it is known from nonhuman primates studies that dopamine enhances the NMDA/AMPA receptor conductance ratio (Williams and Goldman-Rakic, 1995; Zheng et al., 1999); a necessary condition for the sustained firing of pyramidal cells during the delay period of delayed-match-to-sample working memory tasks (Tegner et al., 2002). Moran et al. (2011b) focused on one single cortical region in the superior frontal gyrus, where they observed a significant interaction between working memory and drug. Within this region they modeled spectral signals by a conductance-based DCM and fitted the experimental data for all four conditions of the design (working memory vs. no working memory and L-Dopa vs. placebo). They found that, as predicted, under L-Dopa NMDA receptor conductances were significantly increased, while AMPA receptor conductances were (nonsignificantly) reduced. In addition, both estimates of NMDA and AMPA conductances correlated significantly (but in opposite directions) with drug-induced change in working memory performance (Fig. 5.2B).

The second example (Gilbert et al., 2016) asked the question whether it is possible to separate individuals with dysfunction of a specific ion channel (channelopathy) from a group of healthy controls based on MEG signals only. Two patients with single gene mutations that affect a specific potassium channel and a presynaptic calcium channel, respectively, were compared to 94 control subjects. MEG was measured during an auditory oddball paradigm. The DCM used in this study was a conductance-based version of the classical DCM for event-related responses (Garrido et al., 2008). The patient with a mutation of the potassium channel gene could be separated nearly perfectly from the controls based on the parameter estimates for the potassium leak conductance. Fig. 5.4B illustrates this separation that resulted in an area under the curve (AUC) of 0.94 in a receiver operating characteristic. The second patient, with a presynaptic calcium channel gene mutation, could be separated from controls with a less compelling AUC of 0.76 based on a proxy estimate for presynaptic calcium concentration. In particular the first example shows that not only was it possible to distinguish the patient from the healthy controls, but this distinction could be made based on exactly the parameter that was known to be affected due to the genetic mutation.

These two studies provide a proof of concept that it is feasible to infer specific synaptic processes and channel conductances from electrophysiological data. Several other pharmacological studies in humans and animals exist that illustrate this capacity of DCMs, using a variety of different drugs, including ketamine (Schmidt et al., 2013a; Moran et al., 2015; Muthukumaraswamy et al., 2015) and isoflurane (Moran et al., 2011c).

In conclusion, DCMs of electrophysiological data include parameters that can be interpreted in terms of synaptic physiology. These models might facilitate the search for potential targets of treatment, help understand the pathophysiology of psychiatric disease, and perhaps most importantly, serve as predictors of treatment response in individual patients (Stephan et al., 2015).

## 5.2.3 Using Dynamic Causal Modeling to Dissect Spectrum Disorders

As many psychiatric disorders, schizophrenia is highly heterogeneous, a so-called "spectrum disease": patients with schizophrenia can have different symptoms, and patients with the same symptoms often respond differently to treatment. A central goal for neuromodeling and computational psychiatry is to split up spectrum diseases into subgroups that are mechanistically homogenous (Stephan and Mathys, 2014). Validated tools for achieving such a physiological or computational stratification could provide a powerful basis for differential diagnosis and treatment prediction.

Brodersen et al. (2014) used generative embedding to investigate whether a group of schizophrenia patients could be split into clinically meaningful subgroups based on directed connectivity estimates obtained by DCM. For this, the authors reanalyzed the data from the working memory study by Deserno et al. (2012) described above (Fig. 5.4A). The parameters of the winning model of that study (Fig. 5.4C, left) were used to cluster the participants (41 schizophrenia patients and 42 controls) into different groups. The first analysis included all participants and the best (in terms of log evidence) clustering solution consisted of two clusters that mapped onto patients and controls with a balanced purity of 71%. While a model-based distinction between DSM-defined patients and controls is practically not very relevant, it is noteworthy that generative embedding as described in Fig. 5.3 achieved this separation by only considering fMRI activity from three regions (or more specifically the few parameters of the DCM explaining this activity). The second analysis focused on the patients only. Any successful clustering within this group would constitute a separation of schizophrenia into subgroups based on physiological parameters. Model comparison in this case suggested that three clusters provided the best solution. Fig. 5.4C illustrates the models (in terms of average parameter estimates) of the three groups. The authors then tested whether clinical scores would differ across the connectivity-defined subgroups. They found that the three subgroups differed significantly in negative symptoms (positive and negative syndrome scale, Fig. 5.4C, right). Hence, the model parameter estimates allowed for unsupervised detection of unknown subgroups, based on fMRI data alone, that related to independent clinical scores. This example illustrates how it might be possible to dissect spectrum diseases based on physiologically meaningful parameters in DCM.

### 5.2.4 Current Dynamic Causal Modeling Limitations

The evolving DCM framework faces several interpretational and methodological challenges that need to be addressed in forthcoming work. First, as noted, the application of DCM to fMRI faces the challenge that hemodynamic signals captured by this technique are sluggish and provide low-pass filtered information about neuronal events. As discussed above, this puts constraints on inferring neurophysiological processes and interregional causal influences. Combining EEG and fMRI could help resolve some of these limitations formally. For example, the recent development of DCMs that can equally be applied to fMRI and EEG data (Friston et al., 2017) allows, in principle, for a sequential inference procedure that rests on Bayesian belief updating. That is, neuronal parameters could be estimated in a first pass from EEG, and the ensuing posterior estimates could serve as priors for a second inference step, now using fMRI (with concomitant estimation of hemodynamic parameters).

Second, as described, DCM faces limitations when applied to "resting state" signals (i.e., task free states). Key issues include computational complexity (run time) and the assumption of stationarity (of neuronal processes) across the measurement. Further methodological optimizations of stochastic and spectral DCMs will be important to exploit the potential utility of "resting-state" fMRI for clinical applications.

Finally, all of the described DCM "computational psychiatry" studies in this chapter face the hurdle of being restricted to a few nodes/areas. This restriction to a few areas is appropriate in specific situations (for example, examining interactions between the inferior frontal gyrus and superior temporal gyrus in the context of MMN in schizophrenia). However, this restriction becomes less appropriate when dealing with more distributed functions that require network-level computations potentially involving numerous areas. The computational challenge here is model inversion that, in the classical DCM setting, can become intractable with a large number of network nodes. One recent development that offers a solution, at least in the context of fMRI, is regression DCM (Frässle et al., 2017). This method rests on a simplified DCM that scales gracefully and can include hundreds of regions; we return to this in the next section.

## 5.3 OUTLOOK

In this last part, we briefly outline potential applications of DCM in a clinical context. The search for clinical tests in psychiatry is ongoing (Kapur et al., 2012). One of the central challenges is to develop methods that allow for (treatment) predictions in individual patients (Stephan et al., 2017). In principle, these methods could come in several forms. The most direct approach would be to use single model parameter estimates of distinct therapeutic targets, such as specific neuromodulatory transmitters or ionotropic receptors, for clinical predictions (Gilbert et al., 2016). A second option is to use multiple parameter estimates from a DCM in a generative embedding approach to learn, in a supervised manner, which is the most effective treatment for a given patient. A third approach is to use generative embedding for splitting up a spectrum disease in an unsupervised manner and then establish relations to clinically relevant outcome measures. All of these approaches require large prospective patient studies that allow for assessing the predictive power of model parameter estimates to clinical outcome measures.

Irrespective of which approach will be chosen, it is critically important that parameter estimation is as robust as possible at the single subject level. While variational Bayes has been the main workhorse for model inversion in DCM so far (Friston et al., 2007), more robust estimation may be achieved by Gaussian process optimization (Lomakina et al., 2015) or MCMC techniques that, however, require a lot of computational resources. Using graphical processing units (GPUs) it is now feasible to invert DCMs for fMRI with MCMC efficiently (Aponte et al., 2016). An additional way to make estimates of parameters more robust is to apply hierarchical Bayesian models (Friston et al., 2016b; Raman et al., 2016). Including a second (group) layer of inference makes these models more robust by regularizing the parameters at the first (subject) level. Hierarchical models can be used for classical comparison between parameters (Friston et al., 2016b), but also for unsupervised clustering (Raman et al., 2016). However, hierarchical models tend to require considerable computational resources. This is an area of ongoing research.

A slightly different strategy to enhance the potential of DCM for clinical applications is to widen the spatial scope of modeling. As outlined above, computational restrictions mean that DCMs for fMRI focus on a few selected regions, as specified by a particular question. This is useful to test specific hypotheses about brain function, but could be affected by missing nodes/inputs of unknown importance. A novel variant of DCM—regression DCM (Frässle et al., 2017)—rests on a simplified linear form of DCM and converts the challenge of model inversion into a Bayesian linear regression problem in the frequency domain. This is computationally sufficiently efficient that treatment of whole-brain networks (with tens to hundreds of nodes) becomes feasible. Having said this, this technique is presently still in its infancy and several major methodological developments will still be necessary to unlock its potential.

In this book chapter, we have reviewed, in all brevity, the foundations of DCM for both fMRI and electrophysiological data (MEG and EEG) and existing applications to problems in psychiatry. In summary, DCM represents an important tool for physiological inference in computational psychiatry and has considerable potential for implementing clinically useful tests. This will depend on further methodological improvements, for example, with respect to the robustness of numerical procedures for model inversion, but also successful prospective validation studies in patients with clinically relevant outcome measures. If these conditions are met, generative models such as DCM could enable clinically relevant tests for treatment prediction that help reduce the burden and suffering of patients.

## Acknowledgments

K.E.S. is supported by the University of Zurich and the René and Susanne Braginsky Foundation.

# References

Adams, R.A., Perrinet, L.U., Friston, K., 2012. Smooth pursuit and visual occlusion: active inference and oculomotor control in schizophrenia. PLoS One 7, e47502.

Adams, R.A., Aponte, E., Marshall, L., Friston, K.J., 2015. Active inference and oculomotor pursuit: the dynamic causal modelling of eye movements. J. Neurosci. Methods 242, 1–14.

Adams, R.A., Bauer, M., Pinotsis, D., Friston, K.J., 2016. Dynamic causal modelling of eye movements during pursuit: confirming precision-encoding in V1 using MEG. NeuroImage 132, 175–189.

Adams, R.A., Stephan, K.E., Brown, H.R., Frith, C.D., Friston, K.J., 2013. The computational anatomy of psychosis. Front. Psychiatry 4, 47.

Anticevic, A., Hu, X., Xiao, Y., Hu, J., Li, F., Bi, F., Cole, M.W., Savic, A., Yang, G.J., Repovs, G., Murray, J.D., Wang, X.J., Huang, X., Lui, S., Krystal, J.H., Gong, Q., 2015. Early-course unmedicated schizophrenia patients exhibit elevated prefrontal connectivity associated with longitudinal change. J. Neurosci. 35, 267–286.

Aponte, E.A., Raman, S., Sengupta, B., Penny, W.D., Stephan, K.E., Heinzle, J., 2016. mpdcm: A toolbox for massively parallel dynamic causal modeling. J. Neurosci. Methods 257, 7–16.

Bach, D.R., Daunizeau, J., Friston, K.J., Dolan, R.J., 2010. Dynamic causal modelling of anticipatory skin conductance responses. Biol. Psychol. 85, 163–170.

Bach, D.R., Daunizeau, J., Kuelzow, N., Friston, K.J., Dolan, R.J., 2011. Dynamic causal modeling of spontaneous fluctuations in skin conductance. Psychophysiology 48, 252–257.

Bastos-Leite, A.J., Ridgway, G.R., Silveira, C., Norton, A., Reis, S., Friston, K.J., 2015. Dysconnectivity within the default mode in first-episode schizophrenia: a stochastic dynamic causal modeling study with functional magnetic resonance imaging. Schizophr. Bull. 41, 144–153.

Bastos, A.M., Usrey, W.M., Adams, R.A., Mangun, G.R., Fries, P., Friston, K.J., 2012. Canonical microcircuits for predictive coding. Neuron 76, 695–711.

Bishop, C.M., 2007. Pattern Recognition and Machine Learning (Information Science and Statistics). Springer-Verlag New York Inc.

Brodersen, K.H., Schofield, T.M., Leff, A.P., Ong, C.S., Lomakina, E.I., Buhmann, J.M., Stephan, K.E., 2011a. Generative embedding for model-based classification of fMRI data. PLoS Comput. Biol. 7, e1002079.

Brodersen, K.H., Deserno, L., Schlagenhauf, F., Lin, Z., Penny, W.D., Buhmann, J.M., Stephan, K.E., 2014. Dissecting psychiatric spectrum disorders by generative embedding. NeuroImage Clin. 4, 98–111.

Brodersen, K.H., Haiss, F., Ong, C.S., Jung, F., Tittgemeyer, M., Buhmann, J.M., Weber, B., Stephan, K.E., 2011b. Model-based feature construction for multivariate decoding. NeuroImage 56, 601–615.

Buxton, R.B., Wong, E.C., Frank, L.R., 1998. Dynamics of blood flow and oxygenation changes during brain activation: the balloon model. Magn. Reson. Med. 39, 855–864.

Corlett, P.R., Honey, G.D., Krystal, J.H., Fletcher, P.C., 2011. Glutamatergic model psychoses: prediction error, learning, and inference. Neuropsychopharmacology 36, 294–315.

Crossley, N.A., Mechelli, A., Fusar-Poli, P., Broome, M.R., Matthiasson, P., Johns, L.C., Bramon, E., Valmaggia, L., Williams, S.C., McGuire, P.K., 2009. Superior temporal lobe dysfunction and frontotemporal dysconnectivity in subjects at risk of psychosis and in first-episode psychosis. Hum. Brain Mapp. 30, 4129–4137.

David, O., Kiebel, S.J., Harrison, L.M., Mattout, J., Kilner, J.M., Friston, K.J., 2006. Dynamic causal modeling of evoked responses in EEG and MEG. NeuroImage 30, 1255–1272.

Deserno, L., Sterzer, P., Wustenberg, T., Heinz, A., Schlagenhauf, F., 2012. Reduced prefrontal-parietal effective connectivity and working memory deficits in schizophrenia. J. Neurosci. 32, 12−20.

Di, X., Biswal, B.B., 2014. Identifying the default mode network structure using dynamic causal modeling on resting-state functional magnetic resonance imaging. NeuroImage 86, 53−59.

Dima, D., Dietrich, D.E., Dillo, W., Emrich, H.M., 2010. Impaired top-down processes in schizophrenia: a DCM study of ERPs. NeuroImage 52, 824−832.

Dima, D., Frangou, S., Burge, L., Braeutigam, S., James, A.C., 2012. Abnormal intrinsic and extrinsic connectivity within the magnetic mismatch negativity brain network in schizophrenia: a preliminary study. Schizophr. Res. 135, 23−27.

Dima, D., Roiser, J.P., Dietrich, D.E., Bonnemann, C., Lanfermann, H., Emrich, H.M., Dillo, W., 2009. Understanding why patients with schizophrenia do not perceive the hollow-mask illusion using dynamic causal modelling. NeuroImage 46, 1180−1186.

Fletcher, P.C., Frith, C.D., 2009. Perceiving is believing: a Bayesian approach to explaining the positive symptoms of schizophrenia. Nat. Rev. Neurosci. 10, 48−58.

Fogelson, N., Litvak, V., Peled, A., Fernandez-del-Olmo, M., Friston, K., 2014. The functional anatomy of schizophrenia: a dynamic causal modeling study of predictive coding. Schizophr. Res. 158, 204−212.

Frässle, S., Lomakina, E.I., Razi, A., Friston, K.J., Buhmann, J.M., Stephan, K.E., 2017. Regression DCM for fMRI. NeuroImage 155, 406−421.

Friston, K., Brown, H.R., Siemerkus, J., Stephan, K.E., 2016a. The dysconnection hypothesis (2016). Schizophr. Res. 176, 83−94.

Friston, K., Mattout, J., Trujillo-Barreto, N., Ashburner, J., Penny, W., 2007. Variational free energy and the Laplace approximation. NeuroImage 34, 220−234.

Friston, K.J., 1998. The disconnection hypothesis. Schizophr. Res. 30, 115−125.

Friston, K.J., Harrison, L., Penny, W., 2003. Dynamic causal modelling. NeuroImage 19, 1273−1302.

Friston, K.J., Mechelli, A., Turner, R., Price, C.J., 2000. Nonlinear responses in fMRI: the Balloon model, Volterra kernels, and other hemodynamics. NeuroImage 12, 466−477.

Friston, K.J., Kahan, J., Biswal, B., Razi, A., 2014. A DCM for resting state fMRI. NeuroImage 94, 396−407.

Friston, K.J., Preller, K.H., Mathys, C., Cagnan, H., Heinzle, J., Razi, A., Zeidman, P., 2017. Dynamic causal modelling revisited. NeuroImage. http://dx.doi.org/10.1016/j.neuroimage.2017.02.045.

Friston, K.J., Litvak, V., Oswal, A., Razi, A., Stephan, K.E., van Wijk, B.C.M., Ziegler, G., Zeidman, P., 2016b. Bayesian model reduction and empirical Bayes for group (DCM) studies. NeuroImage 128, 413−431.

Garrido, M.I., Kilner, J.M., Kiebel, S.J., Friston, K.J., 2009a. Dynamic causal modeling of the response to frequency deviants. J. Neurophysiol. 101, 2620−2631.

Garrido, M.I., Kilner, J.M., Stephan, K.E., Friston, K.J., 2009b. The mismatch negativity: a review of underlying mechanisms. Clin. Neurophysiol. 120, 453−463.

Garrido, M.I., Kilner, J.M., Kiebel, S.J., Stephan, K.E., Friston, K.J., 2007. Dynamic causal modelling of evoked potentials: a reproducibility study. NeuroImage 36, 571−580.

Garrido, M.I., Friston, K.J., Kiebel, S.J., Stephan, K.E., Baldeweg, T., Kilner, J.M., 2008. The functional anatomy of the MMN: a DCM study of the roving paradigm. NeuroImage 42, 936−944.

Gilbert, J.R., Symmonds, M., Hanna, M.G., Dolan, R.J., Friston, K.J., Moran, R.J., 2016. Profiling neuronal ion channelopathies with non-invasive brain imaging and dynamic causal models: case studies of single gene mutations. NeuroImage 124, 43−53.

Greicius, M., 2008. Resting-state functional connectivity in neuropsychiatric disorders. Curr. Opin. Neurol. 21, 424−430.

Havlicek, M., Roebroeck, A., Friston, K., Gardumi, A., Ivanov, D., Uludag, K., 2015. Physiologically informed dynamic causal modeling of fMRI data. NeuroImage 122, 355–372.

Heinzle, J., Aponte, E.A., Stephan, K.E., 2016a. Computational models of eye movements and their application to schizophrenia. Curr. Opin. Behav. Sci. 11, 21–29.

Heinzle, J., Koopmans, P.J., den Ouden, H.E., Raman, S., Stephan, K.E., 2016b. A hemodynamic model for layered BOLD signals. NeuroImage 125, 556–570.

Howes, O.D., Kapur, S., 2009. The dopamine hypothesis of schizophrenia: version III—the final common pathway. Schizophr. Bull. 35, 549–562.

Hutton, S.B., Ettinger, U., 2006. The antisaccade task as a research tool in psychopathology: a critical review. Psychophysiology 43, 302–313.

Jansen, B.H., Rit, V.G., 1995. Electroencephalogram and visual evoked potential generation in a mathematical model of coupled cortical columns. Biol. Cybern. 73, 357–366.

Kapur, S., Phillips, A.G., Insel, T.R., 2012. Why has it taken so long for biological psychiatry to develop clinical tests and what to do about it? Mol. Psychiatry 17, 1174–1179.

Karbasforoushan, H., Woodward, N.D., 2012. Resting-state networks in schizophrenia. Curr. Top. Med. Chem. 12, 2404–2414.

Kass, R.E., Raftery, A.E., 1995. Bayes factors. J. Am. Stat. Assoc. 90, 773–795.

Kiebel, S.J., David, O., Friston, K.J., 2006. Dynamic causal modelling of evoked responses in EEG/MEG with lead field parameterization. NeuroImage 30, 1273–1284.

Kiebel, S.J., Garrido, M.I., Moran, R.J., Friston, K.J., 2008. Dynamic causal modelling for EEG and MEG. Cogn. Neurodynamics 2, 121–136.

Kiebel, S.J., Garrido, M.I., Moran, R., Chen, C.C., Friston, K.J., 2009. Dynamic causal modeling for EEG and MEG. Hum. Brain Mapp. 30, 1866–1876.

Klein, C., Ettinger, U., 2008. A hundred years of eye movement research in psychiatry. Brain Cogn. 68, 215–218.

Krystal, J.H., State, M.W., 2014. Psychiatric disorders: diagnosis to therapy. Cell 157, 201–214.

Li, B., Daunizeau, J., Stephan, K.E., Penny, W., Hu, D., Friston, K., 2011. Generalised filtering and stochastic DCM for fMRI. NeuroImage 58, 442–457.

Lomakina, E.I., Paliwal, S., Diaconescu, A.O., Brodersen, K.H., Aponte, E.A., Buhmann, J.M., Stephan, K.E., 2015. Inversion of hierarchical Bayesian models using Gaussian processes. NeuroImage 118, 133–145.

Moran, R., Pinotsis, D.A., Friston, K., 2013. Neural masses and fields in dynamic causal modeling. Front. Comput. Neurosci. 7, 57.

Moran, R.J., Stephan, K.E., Dolan, R.J., Friston, K.J., 2011a. Consistent spectral predictors for dynamic causal models of steady-state responses. NeuroImage 55, 1694–1708.

Moran, R.J., Symmonds, M., Stephan, K.E., Friston, K.J., Dolan, R.J., 2011b. An in vivo assay of synaptic function mediating human cognition. Curr. Biol. 21, 1320–1325.

Moran, R.J., Jones, M.W., Blockeel, A.J., Adams, R.A., Stephan, K.E., Friston, K.J., 2015. Losing control under ketamine: suppressed cortico-hippocampal drive following acute ketamine in rats. Neuropsychopharmacology 40, 268–277.

Moran, R.J., Stephan, K.E., Kiebel, S.J., Rombach, N., O'Connor, W.T., Murphy, K.J., Reilly, R.B., Friston, K.J., 2008. Bayesian estimation of synaptic physiology from the spectral responses of neural masses. NeuroImage 42, 272–284.

Moran, R.J., Jung, F., Kumagai, T., Endepols, H., Graf, R., Dolan, R.J., Friston, K.J., Stephan, K.E., Tittgemeyer, M., 2011c. Dynamic causal models and physiological inference: a validation study using isoflurane anaesthesia in rodents. PLoS One 6, e22790.

Muthukumaraswamy, S.D., Shaw, A.D., Jackson, L.E., Hall, J., Moran, R., Saxena, N., 2015. Evidence that subanesthetic doses of ketamine cause sustained disruptions of NMDA and AMPA-mediated frontoparietal connectivity in humans. J. Neurosci. 35, 11694–11706.

Penny, W.D., Stephan, K.E., Mechelli, A., Friston, K.J., 2004. Comparing dynamic causal models. NeuroImage 22, 1157–1172.

Penny, W.D., Stephan, K.E., Daunizeau, J., Rosa, M.J., Friston, K.J., Schofield, T.M., Leff, A.P., 2010. Comparing families of dynamic causal models. PLoS Comput. Biol. 6, e1000709.

Pinotsis, D.A., Moran, R.J., Friston, K.J., 2012. Dynamic causal modeling with neural fields. NeuroImage 59, 1261–1274.

Raman, S., Deserno, L., Schlagenhauf, F., Stephan, K.E., 2016. A hierarchical model for integrating unsupervised generative embedding and empirical Bayes. J. Neurosci. Methods 269, 6–20.

Ranlund, S., Adams, R.A., Diez, A., Constante, M., Dutt, A., Hall, M.H., Maestro Carbayo, A., McDonald, C., Petrella, S., Schulze, K., Shaikh, M., Walshe, M., Friston, K., Pinotsis, D., Bramon, E., 2016. Impaired prefrontal synaptic gain in people with psychosis and their relatives during the mismatch negativity. Hum. Brain Mapp. 37, 351–365.

Schmidt, A., Diaconescu, A.O., Kometer, M., Friston, K.J., Stephan, K.E., Vollenweider, F.X., 2013a. Modeling ketamine effects on synaptic plasticity during the mismatch negativity. Cereb. Cortex 23, 2394–2406.

Schmidt, A., Smieskova, R., Aston, J., Simon, A., Allen, P., Fusar-Poli, P., McGuire, P.K., Riecher-Rossler, A., Stephan, K.E., Borgwardt, S., 2013b. Brain connectivity abnormalities predating the onset of psychosis: correlation with the effect of medication. JAMA Psychiatry 70, 903–912.

Stephan, K.E., Friston, K.J., 2010. Analyzing effective connectivity with fMRI. Wiley Interdiscip. Rev. Cogn. Sci. 1, 446–459.

Stephan, K.E., Mathys, C., 2014. Computational approaches to psychiatry. Curr. Opin. Neurobiol. 25C, 85–92.

Stephan, K.E., Friston, K.J., Frith, C.D., 2009a. Dysconnection in schizophrenia: from abnormal synaptic plasticity to failures of self-monitoring. Schizophr. Bull. 35, 509–527.

Stephan, K.E., Iglesias, S., Heinzle, J., Diaconescu, A.O., 2015. Translational perspectives for computational neuroimaging. Neuron 87, 716–732.

Stephan, K.E., Weiskopf, N., Drysdale, P.M., Robinson, P.A., Friston, K.J., 2007. Comparing hemodynamic models with DCM. NeuroImage 38, 387–401.

Stephan, K.E., Penny, W.D., Daunizeau, J., Moran, R.J., Friston, K.J., 2009b. Bayesian model selection for group studies. NeuroImage 46, 1004–1017.

Stephan, K.E., Penny, W.D., Moran, R.J., den Ouden, H.E., Daunizeau, J., Friston, K.J., 2010. Ten simple rules for dynamic causal modeling. NeuroImage 49, 3099–3109.

Stephan, K.E., Kasper, L., Harrison, L.M., Daunizeau, J., den Ouden, H.E., Breakspear, M., Friston, K.J., 2008. Nonlinear dynamic causal models for fMRI. NeuroImage 42, 649–662.

Stephan, K.E., Schlagenhauf, F., Huys, Q.J.M., Raman, S., Aponte, E.A., Brodersen, K.H., Rigoux, L., Moran, R.J., Daunizeau, J., Dolan, R.J., Friston, K.J., Heinz, A., 2017. Computational neuroimaging strategies for single patient predictions. NeuroImage 145 (Pt. B), 180–199.

Stephan, K.E., Bach, D.R., Fletcher, P.C., Flint, J., Frank, M.J., Friston, K.J., Heinz, A., Huys, Q.J., Owen, M.J., Binder, E.B., Dayan, P., Johnstone, E.C., Meyer-Lindenberg, A., Montague, P.R., Schnyder, U., Wang, X.J., Breakspear, M., 2016. Charting the landscape of priority problems in psychiatry, part 1: classification and diagnosis. Lancet Psychiatry 3, 77–83.

Tegner, J., Compte, A., Wang, X.J., 2002. The dynamical stability of reverberatory neural circuits. Biol. Cybern. 87, 471–481.

van der Zwaag, W., Schafer, A., Marques, J.P., Turner, R., Trampel, R., 2016. Recent applications of UHF-MRI in the study of human brain function and structure: a review. NMR Biomed. 29, 1274–1288.

van Leeuwen, T.M., den Ouden, H.E., Hagoort, P., 2011. Effective connectivity determines the nature of subjective experience in grapheme-color synesthesia. J. Neurosci. 31, 9879–9884.

Williams, G.V., Goldman-Rakic, P.S., 1995. Modulation of memory fields by dopamine D1 receptors in prefrontal cortex. Nature 376, 572–575.

Zheng, P., Zhang, X.X., Bunney, B.S., Shi, W.X., 1999. Opposite modulation of cortical N-methyl-D-aspartate receptor-mediated responses by low and high concentrations of dopamine. Neuroscience 91, 527–535.

# CHAPTER

# 6

# Systems Level Modeling of Cognitive Control in Psychiatric Disorders: A Focus on Schizophrenia

*Deanna M. Barch, Adam Culbreth, Julia Sheffield*

**Washington University in St. Louis, St. Louis, MO, United States**

## OUTLINE

*Computational Psychiatry*
http://dx.doi.org/10.1016/B978-0-12-809825-7.00006-7

**145**

# 6.1 INTRODUCTION

Researchers who work in the field of psychopathology have increasingly come to recognize the transdiagnostic nature of many of the core features of mental illness. By transdiagnostic we mean the fact that many putatively different psychiatric disorders appear to share what on the surface appear to be similar symptoms, albeit to differing degrees of severity. There are many such symptoms, but one that clearly cuts across diagnostic boundaries is a deficit in cognitive control. By cognitive control we mean the set of cognitive functions that are employed to encode and maintain task or goal representations so as to regulate one's thoughts and actions (Botvinick and Braver, 2015). These functions are accomplished through the recruitment of neural systems that are also involved in supporting memory, perception, attention, action selection, and inhibition, among other functions (Miller and Cohen, 2001; Botvinick and Braver, 2015), making cognitive control a central feature of human behavior and a key pathway through which we regulate the decision-making processes that are necessary in our daily lives.

Given the centrality of cognitive control impairments to psychopathology (Bora et al., 2010; Snyder et al., 2015), and the relationships of such impairments to function in everyday life, formal modeling approaches that attempt to identify and understand both the psychological and neural mechanisms of cognitive controls have long been of interest to the field. Such models can be generated at a number of different levels of analysis (Huys et al., 2016; Maia and Frank, 2017). For example, one level of analysis is biophysical, identifying synaptic, and microcircuit mechanisms that generate complex cortical oscillations (Wang, 2010). This work has been extended to include information about how different receptor contributions, at different time scales, can generate persistent and recurrent neuronal firing that supports computations that contribute to higher-level cognitive processes such as working memory and or cognitive control (Murray et al., 2014; Murray and Anticevic, 2017). At a different, more abstracted level, are system-level computational modeling approaches, often in the parallel distributed processing (PDP) framework, that have generated and tested hypotheses about the role of particular neurotransmitters and brain systems that may be critical to cognitive control, including dopamine (DA), norepinephrine (NE), dorsal frontal–parietal systems, and dorsal anterior cingulate (Cohen and Servan-Schreiber, 1992; Durstewitz et al., 1999; Durstewitz and Seamans, 2002). At yet another level of analysis, often bridging between more biophysical versus systems levels analyses are mathematical formalisms of cognition and behavior. These models often focus on processes such as reinforcement learning or reward processing (Maia and Frank, 2011), but have also been expanded to include mechanisms of behavioral control and prediction (Frank et al., 2001; O'Reilly and Frank, 2006; Otto et al., 2013a,b, 2015; Doll et al., 2015; Gillan et al., 2015).

Models at each of these levels of analyses have been fruitfully used to understand processes and mechanisms relevant to cognitive control, with perhaps the largest body of work focusing on psychosis, most specifically on schizophrenia. Here we provide a selective review of formal modeling approaches to understanding cognitive control in psychopathology. We focus on schizophrenia as an example domain of psychopathology, as it has received the most attention in this literature and is the form of psychopathology with arguably the strongest documentation for evidence of cognitive control impairments (Barch and Sheffield, 2014, 2017). Regarding computational modeling approaches, we focus primarily on system-level approaches, as those, again arguably, have most explicitly attempted to model the mechanisms of cognitive control. However, mathematical formalisms have offered much in this domain and thus we bring such models into this review when they are relevant to the cognitive control deficit literature in schizophrenia. First, we outline a general model of the core mechanisms of cognitive control and their impairment in schizophrenia. Second, we provide an overview of modeling approaches that have worked to capture these mechanisms, and what hypotheses about the mechanisms of cognitive control impairment in schizophrenia they have helped to generate.

# 6.2 MECHANISMS OF CONTROL: PROACTIVE AND REACTIVE

Much empirical and theoretical work has been devoted to understanding the mechanisms contributing to cognitive control in human cognition (e.g., Braver, 2012). One such model, termed the dual mechanisms of control theory (DMC, Braver, 2012), makes a distinction between proactive and reactive modes of cognitive control. The concept of proactive control builds on earlier ideas (Cohen and Servan-Schreiber, 1992) to argue for flexible mechanisms of cognitive control that enables humans to deal with the diversity of challenges that we face in everyday life. A proactive control mechanism can be thought of as a form of "early selection," in which goal-relevant information is actively maintained in a sustained or anticipatory manner, before the occurrence of cognitively demanding events. Such proactive control allows for optimal biasing of attention, perception, and action systems in a goal-driven manner. In this framework, goals refer to the information one needs to accomplish a particular task or the intended outcome of a series of actions or mental operations. In real life, such goal information may include the main points one wishes to communicate in a conversation, the need to organize a shopping trip so that one can make sure to get everything that is needed, or even the desire to preserve a friendship by not yelling when upset. In contrast, in the reactive mode, control is recruited as a "late correction" mechanism that is mobilized only when needed, such as after a high-interference event is detected (e.g., there is a need to retrieve the topic of conversation, an unexpected distracting stimulus is encountered, or a friend says something that upsets you). Thus, proactive control relies on the *anticipation* and prevention of interference before it occurs, whereas reactive control relies on the *detection* and resolution of interference after its onset.

Like many other models of cognitive control, the DMC theory postulates that proactive control depends on actively representing information in lateral prefrontal cortex (PFC), and that the updating and maintenance of such information relies on inputs from a range of neurotransmitter systems, including DA,γ-Aminobutyric acid (GABA), and glutamate (Wang and Arnsten, 2015). As outlined in detail in Braver (2012) proactive and reactive control functions are not mutually exclusive, and some balance between the two modes is necessary to successfully meet most ongoing cognitive demands. However, Braver (2012) has argued that the two control modes can be distinguished based on their temporal characteristics (e.g., *when* they are engaged in the course of cognitive processing) and the requirement that control representations are actively maintained over time for proactive control. Furthermore, Braver has suggested that there may be biases to favor one processing mode over the other, which may be dependent on task demands (e.g., high conflict situations may push toward a proactive control mode), as well as on individual differences in factors such as working memory capacity, fluid intelligence, and even personality traits such as reward sensitivity. Proactive control may also be particularly vulnerable to disruption, given that it is resource demanding and dependent on temporally precise active maintenance mechanisms (e.g., Lesh et al., 2013).

### 6.2.1 Proactive Versus Reactive Control Deficits in Schizophrenia

We and others have argued that a central feature of cognitive control impairment in schizophrenia is a deficit in the ability to use proactive control mechanisms to actively represent goal information needed to guide behavior in working memory. Furthermore, we argue that this deficit reflects impairments in the function of the dorsolateral prefrontal cortex (DLPFC), its interactions with other brain regions such as the dorsal anterior cingulate, the thalamus, and the striatum, and the even impairments in of neurotransmitter systems such as DA, GABA, and glutamate (Barch et al., 2009; Edwards et al., 2010; Lesh et al., 2011). Consistent with this hypothesis, numerous studies have provided evidence consistent with impaired proactive control in schizophrenia (for a review, see Barch and Braver, 2007), as well as evidence for impairments in individuals at risk for schizophrenia (e.g., Snitz et al., 2006), suggesting that such deficits may be associated with liability to schizophrenia as well as manifest illness. Furthermore, there is ample evidence for an association between impairments in DLPFC activity and deficits of proactive control in schizophrenia (e.g., Lesh et al., 2013) as well as those at risk for the development of schizophrenia (e.g., MacDonald et al., 2009), as will be reviewed in more detail below.

### 6.2.2 Computational Models of Proactive and Reactive Control

A key class of models that have been used to investigate the mechanisms of cognitive control, including both proactive and reactive control, are those coming from the connectionist or PDP framework, sometimes also referred to as "guided activation theories" (McClelland and Rumelhart, 1986; Rumelhart and McClelland, 1986; Botvinick and Cohen, 2014). These models originated from a tradition that prioritized understanding the mechanisms driving cognition from a psychological perspective, but using principles that were thought to capture computations as they might be carried out in the brain. At a simplistic level, in such models, units are summing devices that collect inputs from other units in the model and change their outputs when the inputs change. These models represent information as graded patterns of activity over populations of units where processing occurs as the flow of activity from one set of units to another. Learning occurs through the modification of the connection strengths between these units. In some models, this learning is meant to be analogous to the type of Hebbian learning known to drive changes in connection strength between neurons. It can be accomplished using algorithms referred to as "back-propagation" (Rumelhart et al., 1986). In back-propagation, the model is presented with an input pattern, and the output layer is allowed to settle into some response. The difference between the response generated by the model and the expected response (the target) is computed as an error signal. These error signals are propagated back through the network toward the input layer to adjust the weights or connections between units. In one sense, this is not a biologically plausible learning mechanism, given that we do not typically have access to the "correct" response in a way that can shape learning. As such, connectionist models may help us to understand key principles of neural computation, but may not always map onto specific neurobiologically realistic mechanisms. However, alternative learning mechanisms that are putatively more biologically realistic have also been incorporated into such guided activation models (O'Reilly et al., 2007, 2013; Aisa et al., 2008).

In such connectionist models, the units do not correspond one-to-one with individual neurons, but instead represent the processing accomplished by groups of neurons that are thought to operate in ways that capture the general principles of neuronal information processing. Such models are often simplified and capture brain-*style* computation. They do not necessarily commit to the details of any particular neural system or subsystem. At the same time, such connectionist models help to build bridges between our understanding of the low-level properties of neural systems, and their participation in higher-level (system) behavior. They are able to capture a range of complex behaviors and neural system interactions that currently cannot be as straightforwardly modeled using more biophysically based models (McClelland and Elman, 1986; McClelland et al., 1989; Plaut and Farah, 1990; Servan-Schreiber et al., 1990; McClelland, 1991; Cohen and Servan-Schreiber, 1992; Plaut and Shallice, 1993; Plaut, 1996; Plaut et al., 1996), though with the sacrifice of biological complexity and realism as a more detailed level (Murray et al., 2014; Murray and Anticevic, 2017). This is not meant to argue that one type of model is better than another, but rather to point out that they have different strengths and weakness. There will be situations where each type of model is best suited. A useful quote capturing the utility of connectionist models is provided by Hoffman (p. 1683) (Hoffman, 1997):

> Computer models are central to scientific disciplines ranging from meteorology to physical chemistry. Their usefulness lies in simulating complex, interactive systems. A good model does not recreate 'reality' in its entirety—if that were the case, the best model would be the real-life system itself. Instead, model construction proceeds by incorporating certain properties of the system in a much simplified form which, when simulated by computer, exhibits characteristic properties or behaviors that have been previously unexplained.

### 6.2.2.1 Connectionist Modeling of Proactive Control in Schizophrenia: Guided Activation Framework

An important set of guided activation models developed by Cohen and Servan-Schreiber (Botvinick and Cohen, 2014) instantiated the hypothesis that proactive control arises from interactions between the DA neuro-transmitter system and the PFC. These models suggested that goal-related information, or context information in the original terminology, was actively maintained in PFC (see Fig. 6.1) and used to bias stimulus response mappings represented in posterior cortex, serving as a source of top-down support for controlling behavior. In these models, the "context" or task goal module was specifically associated with the functions of the DLPFC. Active maintenance in the absence of external inputs was assumed to occur via recurrent excitation (Fig. 6.1), as suggested by

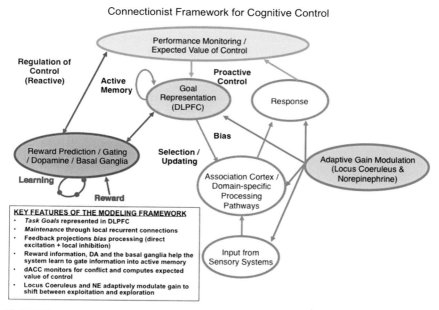

FIGURE 6.1  **Computational model of cognitive control.** Modifications and elaborations to the original model developed by Cohen and Servan-Schreiber (model components in gray) that include mechanisms of both proactive and reactive control. The model includes roles for: (1) dopamine (DA) and/or the basal ganglia in gating information into active memory (model components in purple; Braver and Cohen, 1999; Frank et al., 2001), and in potentially providing feedback about prediction errors (Holroyd et al., 2005); (2) dorsal anterior cingulate cortex (dACC) in monitoring for conflict, computing the expected value of control, and dynamically regulating cognitive control (model components in orange; Botvinick et al., 2001; Shenhav et al., 2013); (3) Locus Coeruleus and Norepinephrine (NE) in adaptive gain modulation to shift between exploitation and exploration (model components in green; Aston-Jones and Cohen, 2005). *DLPFC*, dorsolateral prefrontal cortex. *From Anticevic, A., Murray, J.D., Barch, D.M., 2015. Bridging levels of understanding in schizophrenia through computational modeling. Clin. Psychol. Sci. 3 (3), 433−459. http://dx.doi.org/10.1177/2167702614562041.*

neurophysiological data (Funahashi et al., 1989). This model assumed that projections from the goal representation module biased processing in other systems via direct excitation, but that competition between representations within a processing stream occurred via lateral inhibition (Cohen et al., 1990, 1992a,b). The effects of DA were assumed to be modulatory, such that changes in DA activity could either increase or decrease activity depending on the nature of the inputs that DA was modulating. In other words, increases in DA were assumed to increase the signal-to-noise ratio of a unit's activation value in relation to its input. Thus, the effects of DA were simulated as a change in the gain parameter of the model, which is the parameter that relates a unit's activity to its input. With excitatory input, higher gain means that the same level of input leads to higher activation. In contrast, with inhibitory input, higher gain leads to more negative values and lower activation.

To validate this framework, the same basic model was used to capture the patterns of behavioral data shown by healthy individuals in many different tasks that were all thought to depend on the ability to represent and maintain context/goal information in DLPFC, as has been reviewed in detail elsewhere (Cohen et al., 1999; Anticevic et al., 2015). More importantly for this review, these models were also used to test the hypothesis that reductions in DA input into DLPFC lead to impairments in proactive control, or a deficit in the ability to represent and maintain context information in DLPFC, which could in turn lead to deficits in a range of cognitive tasks among individuals with schizophrenia (Cohen et al., 1992a,b; Cohen and Servan-Schreiber, 1993; Servan-Schreiber et al., 1996). Testing this hypothesis in simulation was instantiated by implementing a reduction in DA inputs into DLPFC via a reduction in the gain parameter that modulates the signal-to-noise ratio of units in the context/DLPFC module. This same manipulation was used to capture reduced DA across models of several different tasks, which all showed that the disrupted model captured the key aspects of performance changes among individuals with schizophrenia. These predictions about the nature of proactive control deficits in schizophrenia have been supported in numerous studies using the types of tasks modeled in this guided activation framework, with evidence that such proactive control impairments are present in individuals with schizophrenia who have manifest illness (Cohen et al., 1999; Barch and Braver, 2007; Zandbelt et al., 2011; Henderson et al., 2012; Fassbender et al., 2014), those with more subsyndromal forms of psychosis (Barch et al., 2004; McClure et al., 2007, 2008; Niendam et al., 2014), and those genetically at risk for psychosis (Delawalla et al., 2008). Furthermore, such models have been used to motivate approaches designed to enhance proactive control in schizophrenia and other disorders (Paxton et al., 2008; Edwards et al., 2010).

#### 6.2.2.1.1 Using Proactive Control Models to Make Predictions About Dorsolateral Prefrontal Cortex Activity

Such connectionist computational models can also be used to make and test predictions about neural mechanisms, though as noted above these predictions are at a somewhat abstracted level. For example, one of the hypotheses embodied in the Cohen and Servan-Schreiber guided activation models was that the DLPFC is responsible for the representation and maintenance of tasks goals and context, and thus central to implementing proactive control. The models predicted that individuals with schizophrenia should show reduced activation of the DLPFC during tasks that require proactive control through the anticipatory representation and maintenance of goal information. The idea that individuals with schizophrenia should show reduced DLPFC function is not a particularly novel or a unique hypothesis and would fit with many models. For example, many studies using a variety of paradigms have shown reduced DLPFC activity in schizophrenia (Minzenberg et al., 2009). However, the advantage of committing hypotheses to formal models is that they can help generate more specificity about the exact conditions under which DLPFC deficits should be apparent, including the timing and duration of such deficits. For example, the Cohen and Servan-Schreiber models made specific predictions that individuals with schizophrenia should show reduced DLPFC activity in conditions with a strong demand for the representation and maintenance of context information.

Consistent with these predictions, individuals with schizophrenia show reduced DLPFC activity during task conditions where proactive control must be maintained (Perlstein et al., 2003; Holmes et al., 2005; MacDonald et al., 2005; Yoon et al., 2008; Edwards et al., 2010), and reduced activity in response to cues that predict the need to use proactive control to overcome a prepotent response (Snitz et al., 2005; Lesh et al., 2013; Niendam et al., 2014; Poppe et al., 2016). Furthermore, the models make predictions about the timing of when DLPFC deficits should occur in schizophrenia. Empirical evidence demonstrates that individuals with schizophrenia show the predictive failure in the early onset of DLPFC activity, in response to cues predicting the need to engage cognitive control (Braver et al., 2009). Importantly, other work has shown that focused training on the use of practice control does lead to increased DLPFC activity to cued tasks designed to specifically tap proactive control in schizophrenia (Edwards et al., 2010).

### 6.2.2.2 Reactive Control—When to Engage or Upregulate Control

The computational framework described above focuses on the "proactive" elements of cognitive control, using the maintenance of task or goal representations to prepare prospectively for cognitive challenges. However, this framework has been extended to include processes that help us understand both reactive control—the engagement of control in response to experienced conflict—and the mechanisms by which future use of proactive control may be recruited or modulated. There are several ways in which this computational framework has been extended that address these questions. One way is by focusing on the role of errors and feedback about performance. This approach brought to bear a large body of literature on reinforcement learning and postulated that error-monitoring systems generate both negative and positive prediction errors that might feedback to modulate control (Sutton and Barto, 1981; Nasser et al., 2017). This hypothesis about error-monitoring systems was instantiated in a combined framework that involved both an expansion of the guided activation model of Cohen et al. (Holroyd et al., 2005; Holroyd and Umemoto, 2016) and components of reinforcement learning models. In the computational instantiation of this model, there is a monitor module that detects these error signals, and can do so in "real-time" to signal feedback for the need for more attentional control, or "reactive" control in the DMC framework. In identifying neurobiological mechanisms, the Holroyd et al. extension postulated critical roles for DA, the striatum and the dorsal anterior cingulate cortex (dACC) in this feedback process (Fig. 6.1) (Holroyd et al., 2005; Holroyd and Umemoto, 2016).

Holroyd's hypotheses about the role of the dACC was based in part on studies of the error-related negativity (ERN), an evoked response potential component that occurs with errors in a variety of experimental paradigms that involve speeded responses (Gehring et al., 1990; Falkenstein et al., 1995). The ERN has been consistently localized to the dACC (Dahaene et al., 1994). In earlier work, Coles et al. suggested that the ERN might index the function of a comparator process that compares the actual output with the "correct" or intended output and then signals a mismatch when an error occurs. An update on this idea explicitly incorporates information about the dopaminergic inputs to dACC to hypothesize that

> when human participants commit errors in reaction time tasks, the mesencephalic dopamine system conveys a negative reinforcement learning signal to the frontal cortex, where it generates the ERN by disinhibiting the apical dendrites of motor neurons in the anterior cingulate cortex. Furthermore, we suggest that the error signals are used to train the anterior cingulate cortex, ensuring that control over the motor system will be released to a motor controller that is best suited for the task at hand. (Holroyd and Coles 2002, p. 679)

Although this error-monitoring hypothesis about reactive control or control regulation captures important components of changes in cognitive control in response to poor performance, it did not address the fact that modulation of control can occur without explicit errors. As such, Botvinick et al. postulated an alternative mechanism and an extension to the guided activation model. In this model (Fig. 6.1), conflict in ongoing processing was used as a feedback signal to upregulate actively maintained task representations, which served to upregulate cognitive control in a dynamic fashion (Botvinick et al., 2001). Conflict was operationalized as the simultaneous activation of incompatible representations, and measured computationally by using Hopfield energy. Simulations with this added "conflict-detection" module showed that it could capture the pattern of ERNs and dACC activation in a variety of tasks (Botvinick et al., 2001). A further extension that incorporated a feedback loop to the task representation model was able to capture a number of behavioral phenomena relevant to understanding modulations of cognitive control in response to errors and conflict (Botvinick et al., 2001).

In more recent work, Brown et al. provided a further modification combining both connectionist and reinforcement learning formalisms to support a predictive coding model of PFC and goal-directed behavior and control. This model is referred to as the Hierarchical Error Representation (HER) model (Alexander and Brown, 2015, 2016). The HER model builds on the working suggesting that the PFC is organized along a gradient from rostral to caudal regions, with more rostral regions representing more abstract rules and task sets, and more caudal regions representing more specific stimulus—response associations (Koechlin et al., 2003; Badre, 2008; Badre and D'Esposito, 2009). The HER model suggests a parallel organization for subregions of the dACC/medial PFC (Alexander and Brown, 2015, 2016), with error information about different levels of rule abstraction computed in different regions of dACC, along a similar rostral–caudal gradient as PFC. Furthermore, the HER model suggests that the "prediction errors" generation in the dACC serve to train the DLPFC to represent information in working memory that serves to predict, and thus presumably reduce, further prediction errors. This HER model represents another type of computationally instantiated conflict-to-control loop that provides a mechanism for the modulation of cognitive control during ongoing processing.

A crucial aspect of the Holroyd, Botvinick, and Brown hypotheses, implemented in their respective computational frameworks, is that the output of dACC's response to error and/or conflict computations serves as an indication to engage immediate cognitive control or to upregulate control over the longer term (see below for a more recent extension in terms of the expected value of control). In the Cohen framework and in the DMC theory, the actual implementation of control adjustments following conflict or error detection has been linked to the function of DLPFC. Consistent with these hypotheses, the degree of anterior cingulate cortex (ACC) activity on previous high-conflict trials, presumably signaling the need for more control, predicts the degree of DLPFC activity on the subsequent trial (Kerns et al., 2004; Kerns, 2006; Hendrick et al., 2010; Kim et al., 2014), for a review see Carter and van Veen (2007) and

Mansouri et al. (2017) for some conflicting data. Furthermore, the degree of DLPFC activity predicts the degree of conflict adaptation (Kerns et al., 2004; Egner and Hirsch, 2005; Kerns, 2006; Marco-Pallares et al., 2008), which is thought to be a measure of the recruitment of increased control, though there is evidence that the degree of ACC activity can also predict conflict adaptation in some conditions (Kim et al., 2014).

#### 6.2.2.2.1 Performance Monitoring and Reactive Control in Schizophrenia

These various theories differ in the specific computational functions that they attribute to the dACC. Nonetheless, they all would predict that dACC disturbances would have a detrimental effect on a range of cognitive functions, in that detecting and responding to errors/conflict may be a critical means by which control is engaged and regulated. At the behavioral level, there is very mixed evidence in schizophrenia regarding impairments on indices thought to reflect the ability to detect or respond to conflict or errors. For example, some studies have found that individuals with schizophrenia detect and/or correct errors as frequently as healthy controls (Kopp and Rist, 1994), while other studies have found reduced rates of error correction among individuals with schizophrenia (Turken et al., 2003). In addition, some studies have found that individuals with schizophrenia show an intact "Rabbit" effect, which is the slowing of responses on trials following errors (Alain et al., 2002; Mathalon et al., 2002; Laurens et al., 2003; Morris et al., 2006; Polli et al., 2006), while other studies have found that this effect is reduced in patients with schizophrenia (Carter et al., 2001; Alain et al., 2002; Kerns et al., 2005; Becerril et al., 2011). In fact a recent review of this area concluded that no conclusions could be drawn yet, at least in terms of conflict adaptation (Abrahamse et al., 2016).

In contrast, event related potential, functional magnetic resonance imaging (fMRI), and positron emission tomography studies have provided more consistent evidence for altered conflict/error processing. For example, a number of studies have shown reduced ERN amplitudes or ACC activity in patients with schizophrenia on error trials (Kopp and Rist, 1994, 1999; Carter et al., 2001; Alain et al., 2002; Bates et al., 2002, 2004; Mathalon et al., 2002, 2009; Laurens et al., 2003, 2010; Kerns et al., 2005; Morris et al., 2006, 2008, 2011; Polli et al., 2008; Becerril et al., 2011; Krawitz et al., 2011; Horan et al., 2012; Perez et al., 2012; Simmonite et al., 2012; Minzenberg et al., 2014; de la Asuncion et al., 2015; Foti et al., 2016; Charles et al., 2017), even when these same individuals show intact behavioral indices or error detection (Kopp and Rist, 1994, 1999; Mathalon et al., 2002; Laurens et al., 2003; Morris et al., 2006).

Interestingly, however, individuals with schizophrenia show intact feedback-related negativity (Morris et al., 2008, 2011; Horan et al., 2012; Llerena et al., 2016) in the context of impaired ERN. This pattern of findings has been interpreted as suggesting that individuals with schizophrenia can process error-related information when it is provided externally but not when it is dependent on internal representations of tasks or outcomes. Our own recent work on dACC function in schizophrenia is consistent with this interpretation, as we found intact ACC function and posterror adjustments in a task that provided information about the required response on a trial-to-trial basis (Becerril and Barch, 2013), thus providing external support for representations about correct versus incorrect outcomes.

### 6.2.2.3 Relationships Between Proactive Control and Reactive Control in Schizophrenia

It is important to note that many studies have shown robust activation of dACC among individuals with schizophrenia, and even in some cases *increased* dACC activation, as compared with *decreased* dACC activation. For example, two major metaanalyses of cognitive task-related activation in schizophrenia—one focused on executive function (Minzenberg et al., 2009) and one focused on episodic memory (Ragland et al., 2009)—both showed consistent evidence for *increased* ACC activity in individuals with schizophrenia as compared with controls. Thus, there is evidence for both increased and decreased dACC activity in schizophrenia. Such findings raise interesting questions about the relationship between deficits in dACC and DLPFC function in schizophrenia, in that theories about primary deficits in either area would also predict deficits in the function of the other region as a consequence. For example, if individuals with schizophrenia have impaired DLPFC function and impaired proactive control, they may experience greater conflict or errors in information processing. According to the conflict monitoring or HER models, increased conflict and errors would, in turn, predict *increased* dACC activity in individuals with schizophrenia. Alternatively, impaired DLPFC function and proactive control might contribute to degraded representations of the predictive information needed to drive error correcting DA signals. If so, the Holroyd or HER error-detection models might predict reduced dACC activity under such conditions, if dACC reflects the presence of prediction errors. It is possible that each of these situations might arise either under different task conditions (e.g., easier vs. harder tasks that are more or less dependent on proactive control) or at different points at time in the course of schizophrenia (e.g., early vs. late in the disease process). Interestingly, there is also mixed evidence as to whether individuals with schizophrenia show deficits in prediction error-related activity in striatum. A recent metaanalysis provided modest evidence of reduced prediction error activity (Radua et al., 2015), but two recent studies showed intact prediction error activity in the striatum among medicated individuals (Culbreth et al., 2016c; Dowd et al., 2016). Clearly more research is needed that focuses on the relationship between conflict/ error detection and the engagement of control processing in individuals with schizophrenia to help provide a better understanding of the dynamic processes that give rise to deficits in cognitive control in schizophrenia and across the spectrum of psychopathology.

# 6.3 UPDATING CONTROL REPRESENTATIONS—DOPAMINE, THE STRIATUM AND A GATING MECHANISM

The modeling framework described above is able to capture the performance of healthy individuals on a range of tasks, to capture a number of key aspects of both proactive and reactive control, to make predictions about the role of specific neural mechanisms, and to make testable predictions about impairments in schizophrenia. However, one issue not captured by these models was how to balance the tradeoff between appropriate updating of task or goal information in the service of proactive control (with such representations putatively maintained in the DLPFC), versus protecting already stored information from irrelevant information that could disrupt ongoing maintenance. Thus, the computational framework described above was expanded to include a gating mechanism postulating that the DA projection to DLPFC serves a possible "gating" function. This is accomplished by regulating access of context representations into active memory, whereby the system learns what information should be gated into DLPFC via the reinforcement learning functions (see Fig. 6.1). There are actually two variations on this gating theme. One was put forth by Braver et al., and suggested that phasic changes in DA activity mediate both gating and learning effects in PFC through similar neuro-modulatory mechanisms, though possibly through different DA receptor subtypes (Braver et al., 1999; Braver and Cohen, 1999). Similar to the idea that DA potentiates the gain of signals into DLPFC, the gating effect in the Braver model occurs through the transient potentiation of both excitatory afferent and local inhibitory input. In contrast, the learning effect occurs through Hebbian-type modulation of synaptic weights, and is driven by errors between predicted and received rewards. The coincidence of the gating and learning signals produces cortical associations between the information being gated and a triggering of the gating signal in the future. Another variation on this gating model was put forth by Frank et al., who also suggested DA was important but more for learning about the timing of a gating signal that was actually implemented by the basal ganglia (Frank et al., 2001).

## 6.3.1 Dopamine and Gating in Schizophrenia

The original Cohen and Servan-Schreiber guided activation model suggested that a reduction in DA inputs into DLPFC was one source of impaired proactive control. Interestingly, the Braver gating model embodied a somewhat different hypothesis about the nature of impaired DA function in schizophrenia, arguing that schizophrenia may reflect a noisy DA gating signal, which results in increased tonic and decreased phasic DA function (Braver and Cohen, 1999). The model with the noisy gating function was able to capture patterns of impaired performance shown by individuals with schizophrenia and suggested that individuals with schizophrenia might both fail to update task representations when needed, and sometimes update with inappropriate information (Braver and Cohen, 1999). A recent neuroimaging study provided data indirectly consistent with these hypotheses (Ceaser and Barch, 2015). This study found that during trials when patients should protect working memory from interference, patients were more likely to inappropriately encode distracters. Furthermore, in patients, both prefrontal and striatal activity was significantly greater when they inappropriately identified the distracter as correct compared to activity during distracter rejection. In contrast, on trials when patients should have updated working memory, prefrontal and striatal activity was significantly lower for incorrect than correct updating trials.

While these findings are intriguing, it is not clear whether this hypothesis about the role of DA in schizophrenia matches the existing data about DA in schizophrenia. Recent metaanalyses of DA in schizophrenia suggests evidence a significant *increase* in DA availability in schizophrenia (Fusar-Poli and Meyer-Lindenberg, 2013a,b; Howes et al., 2012). There is also evidence for increased occupancy of $D_2$ receptors by DA in schizophrenia (Abi-Dargham et al., 2000). These types of data are consistent with the hypothesis that *tonic* DA levels are increased in schizophrenia, consistent with the proposed schizophrenia-related deficits in the Braver gating model. However, there is also strong evidence that individuals with schizophrenia show an *increase* in DA release in response to pharmacological challenge (Laruelle et al., 1996; Abi-Dargham et al., 1998, 2009; Abi-Dargham, 2004). This increased DA release response is more consistent with an *increased* phasic release of DA, rather than a *reduction* in phasic release. However, more work is needed with methods that can capture faster time scale release of DA to understand how the alterations in DA function present in schizophrenia influence the type of fast phase-related DA release simulated in many computational models. Furthermore, such models will need to more fully incorporate the role of different DA receptor subtypes, their contributions to PFC function, and their alterations in schizophrenia, a goal which might be facilitated by incorporating more biophysically realistic components to the modeling framework.

# 6.4 COGNITIVE CONTROL, VALUE, AND EFFORT ALLOCATION

An intriguing recent extension of the Cohen et al. cognitive control computational framework is the idea that there may be mechanisms in the brain that compute the expected value of control. More specifically, the idea is that the use of cognitive control mechanisms comes at a price of computational effort, and that there may be a need to generate cost—benefit analyses that guide when it may be most valuable to increase control exertion and to determine which type of increase is most likely to result in valued outcomes (Shenhav et al., 2013, 2016). Shenhav et al. hypothesized that the dACC may help to compute the expected value of control, and that this role may help to account for why the dACC seems to be important for both cognitive control and decision-making, helping to integrate these domains. A detailed description of this model and the mechanisms is beyond the scope of this chapter, but Shenhav et al. provide a detailed account of the computational mechanism by which this may occur and how it may be implemented neurobiologically (Shenhav et al., 2013, 2016).

Importantly, disturbances in the mechanisms that compute the expected value of control might lead individuals to either overestimate or underestimate the cost of control, and fail to adjust cognitive control accordingly. For example, priori research has shown that individuals with schizophrenia show greater discounting as a function of cognitive effort, which was strongly related to motivation and pleasure negative symptoms (Culbreth et al., 2016a). It is possible that such discounting reflects either an objectively greater cost of cognitive control for individuals with schizophrenia (though the effect remains even when controlling for task performance levels) or dysfunction in the systems that compute the expected value of control. Incorporating modeling of motivational symptoms into models typically focused on cognitive function may help to broaden the utility of these models for understanding key symptom domains of psychopathology.

## 6.4.1 Cognitive Control, Utility, and Exploitation Versus Exploration

Another extension to the original Cohen and Servan-Schreiber guided activity theory involved including a role for utility computation as support by the orbital frontal cortex (OFC) (Wallis, 2007), and the role of NE as what has been referred to as an "adaptive" gain signal (Fig. 6.1). This extension to the model was designed to capture the need for humans and animals to alternate between modes of exploitation (staying with the current response pattern to achieve as much reward as possible) and exploration (seeking new response options when reward is no longer available). Cohen and Aston-Jones hypothesized that if the system is achieving regular reward feedback and little conflict information, then value estimates for the available response options are high and cost/benefit analyses are favorable. When this occurs, the person (or animal) should continue to choose the same response options, as these choices are continuing to pay off and an optimal system should try to maximize available reward. In contrast, when reward feedback reduces and/or conflict increases, the value associated with various response options will reduce, and in turn cost/benefit computations will be less favorable. When this occurs, the system should shift to an exploration mode and try alternative responses that may be associated with greater reward. Cohen and Aston-Jones hypothesized that the locus coeruleus—norepinephrine system (LC—NE) helps to modulate this shift, with a phasic bursting LC—NE mode associated with exploitation and a more tonic firing mode associated with exploration. Furthermore, the shift between these tonic and phasic modes it thought to be governed by inputs from the ACC and the OFC. This set of mechanisms relate to the ideas outlined above under the expected cost of control; Shenhav et al. have suggested that value representations in OFC may be one of the sources of information for the computational model about expected value made by the dACC.

These hypotheses have been instantiated in extensions of the guided activation model that included roles for both DA and LC—NE. This model captures both behavior (Usher et al., 1995; McClure et al., 2005) and the firing of LC neurons as recorded from primates (Usher et al., 1999), and there is recent work using pupil dilation and pharmacological manipulation consistent with the adaptive gain control hypothesis (Warren et al., 2016). Some intriguing initial data suggest that schizophrenia may be associated with impaired in uncertainty-driven exploration (Kasanova et al., 2011; Strauss et al., 2011), a construct similar to that put forth by Cohen and Aston-Jones. However, as of yet, the Cohen and Aston-Jones model has not been used directly to help understand the mechanisms of cognitive control in schizophrenia or other forms of psychopathology.

### 6.4.2 Model-Based Learning and a Decision-Making as a Form of Cognitive Control

The review above focuses on a guided activation framework with a number of extensions. This framework has been used extensively over many years to understand the mechanisms of cognitive control. Recently, there is additional framework that has been popular in computational neuroscience that also provides specific computational definitions to aspects of decision-making and learning that are relevant to cognitive control (Doya et al., 2002; Daw et al., 2005). This framework posits that decision-making arises from two competing systems (Doya et al., 2002; Daw et al., 2005). The first system, a "model-free" learning system is thought to learn the value of actions retrospectively by repeating previously rewarded actions. Such model-free learning algorithms recreate Thorndike's Law of Effect and closely resemble the definition of habitual actions, learning the value of actions by increasing the value of previously rewarded actions and decreasing the value of previously punished or unrewarded actions. Learning in this system is computationally straightforward, but relatively inflexible to rapid changes in the environment or in outcome value. A second "model-based" learning system is thought to implement more goal-directed behavior. It learns the value of actions prospectively by leveraging a cognitive "map" of actions and their future consequences, implementing a form of proactive control over decision-making. Because such cognitive maps or "decision trees" are constructed "on the fly," model-based learning allows for rapid updating of action values. However, simulating such future states requires high levels of representational space and maintenance, making model-based learning computationally expensive (Otto et al., 2013a,b, 2015).

Model-free learning is putatively driven by reward prediction error signaling, which as noted above, refers to mismatches between the expected and observed outcomes of an action (Daw et al., 2005, 2011). Primate studies have suggested that reward prediction error signaling is coded in the phasic firing of dopaminergic neurons in the ventral tegmental area and the substania nigra pars compacta (Schultz et al., 1997; Schultz, 1998). Using fMRI, human neuroimaging studies have shown that blood oxygenation level dependent activation patterns in the ventral striatum show consistencies with the DA response patterns demonstrated in nonhuman primate studies, varying as a function of prediction error (McClure et al., 2003; O'Doherty, 2004; Haruno and Kawato, 2006; Glascher et al., 2010). Thus, DA innervation of the striatum is a likely neurobiological candidate for model-free updating. Furthermore, recent evidence has suggested that model-free learning correlates with activation patterns in the dorsal striatum (Lee et al., 2014) and with the functional connectivity between dorsal striatum and supplementary motor areas (Morris et al., 2016).

In contrast to model-free learning, model-based learning is putatively driven by state prediction error signaling, which refers to mismatches between expected and encountered states in one's environment. While the neural underpinnings of state prediction error signaling are less well characterized in comparison to reward prediction error signaling, recent reports have localized this function to the lateral PFC (Glascher et al., 2010; Doll et al., 2015), providing a link to the other models of cognitive control reviewed above. Further evidence for the lateral PFS's critical role in state prediction errors comes from a study which used theta-burst transcranial magnetic stimulation to effectively make the DLPFC "inactive." This study showed reduced model-based learning, consistent with the hypothesis that lateral PFC activation is causally related to model-based learning (Smittenaar et al., 2013). However, recent work also suggest that DA modulation may be important for model-based learning as well as model-free learning, both of which seem to be influenced by prediction errors generated in the striatum (Daw et al., 2011; Deserno et al., 2015; Sharp et al., 2016).

Interestingly, recent work has shown that individuals with schizophrenia demonstrate a deficit in a model-based learning (Culbreth et al., 2016b), which is consistent with other evidence of impairments in cognitive control. At the group level, model-free learning was intact, although higher levels of hedonic impairment were associated with reduced model-free learning in individuals with schizophrenia. However, one concern with the task used in this study (Daw's Two-Stage Sequential-Learning Task) is that the model-based learning might be inherently more difficult and more likely to elicit poor performance in any population. For example, several other disorders including binge eating, drug use, and obsession—compulsive disorder show a bias toward model-free learning (Voon et al., 2015). While this may reflect the ubiquitous role of cognitive control deficits in psychopathology, future work needs to rule out a differential difficulty explanation, as has been done for other paradigms assessing proactive control (Henderson et al., 2012).

## 6.4.3 Cognitive Control Impairments as a Core Feature of Psychopathology

The review above focused on cognitive control deficits in schizophrenia and potential computational mechanisms that might help us understand the source of these deficits at either the psychological or neural level. Importantly, there is growing work identifying the neural systems that support cognitive control as a potentially transdiagnostic set of neurobiological impairments that may track with cognitive control impairments across diagnostic boundaries. Specifically, two recent metaanalyses identified deficits in the dACC and bilateral insula that were present across many forms of psychopathology, including both psychotic and nonpsychotic disorders. In the first metaanalysis, Goodkind et al. (2015) identified a set of regions that showed gray matter volume reductions across both psychotic and nonpsychotic disorders. These regions included the dACC and bilateral anterior insula, critical nodes within a network referred to as both the "multiple cognitive demand network" and the cingulo-opercular network. Gray matter volume in these regions was positively associated with executive functioning ability, as measured through a composite of task switching, inhibition, and working memory

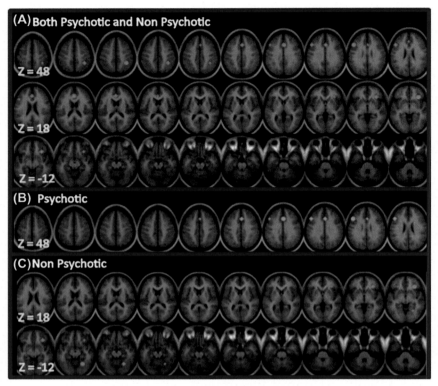

FIGURE 6.2 **Cognitive control—related activation deficits across diagnostic boundaries.** Illustration of the regions found in the McTeague et al. study of altered task-related activation during cognitive control tasks transdiagnostically. (A) Results from combining across psychotic and nonpsychotic disorders; (B) Results in individuals with psychotic disorders; and (C) Results in individuals with nonpsychotic disorders.

performance, providing further evidence of these regions' role in broad executive function deficits. McTeague et al. conducted a metaanalysis of activation during cognitive controls tasks, and again found evidence for transdiagnostic alterations in the dACC and the right insula that overlapped with the regions found in their gray matter analysis (Fig. 6.2A). These results support the hypothesis that cognitive control impairments are a shared feature of many forms of psychopathology, though they do not reveal whether the same behavioral or computational mechanisms are disrupted across psychopathology. At the same time, these results also suggest that there may be diagnostically specific neural mechanisms that also contribute to cognitive control impairment. Specifically, McTeague et al. found that individuals with psychotic disorders showed greater evidence of reduced activation in left DLPFC, consistent with the evidence reviewed above about the role of DLPFC is proactive control deficits in psychosis (Fig. 6.2B). An important next step would be to examine the nature of cognitive control deficits transdiagnostically using computationally informed paradigms that could help determine whether the same mechanisms were leading to cognitive control deficits across disorders.

## 6.5 SUMMARY AND FUTURE DIRECTIONS

As this selective review highlights, individuals with schizophrenia show significant deficits on a wide variety of tasks that index various aspects of cognitive control. We have suggested that there is a common mechanism contributing to these deficits—an impairment in proactive control that can influence performance in a wide variety of cognitive domains. Furthermore, we have suggested that at the neural level, a common denominator to such deficits may be impaired function of DLPFC and the dACC, and their relationships to each other. We have reviewed computational formalisms that help us to understand how these mechanisms may arise out of biologically plausible, albeit not always detailed, mechanisms that support a range of processes relevant to cognitive control. Furthermore, we have reviewed the intriguing, but as of yet still limited, evidence that such cognitive control deficits may be a transdiagnostic feature of psychopathology. There are several important next steps in this line of work. First, we need more behavioral and neuroimaging studies with computationally informed paradigms that include many forms of psychopathology, so that we can begin to understanding whether cognitive control deficits are the results of a shared set of computational, psychological, and neural mechanisms transdiagnostically, or the equifinal outcome of diverging distal pathways to impairment. Second, we need to move toward incorporating modeling approaches at differing levels of analyses so as to benefit from the combined strengths of differing approach. Our selective review highlighted several cases where merging guided activation and reinforcement learning modeling approaches yielded fruitful advances. Going forward, additional work that also integrates more biophysical components to these models may help us make further advances in delineating more detailed aspects of the neurobiology and their role in cognitive control. The modeling frameworks reviewed here primarily focused on connectionist and reinforcement learning type models. These have many strengths in terms of capturing key mechanisms, but lack to detail of biophysical models that may help target treatment development and application.

# References

Abi-Dargham, A., 2004. Do we still believe in the dopamine hypothesis? New data bring new evidence. Int. J. Neuropsychopharmacol. 7 (Suppl. 1), S1—S5.

Abi-Dargham, A., Gil, R., et al., 1998. Increased striatal dopamine transmission in schizophrenia: confirmation in a second cohort. Am. J. Psychiatry 155 (6), 761—767.

Abi-Dargham, A., Rodenhiser, J., et al., 2000. Increased baseline occupancy of D2 receptors by dopamine in schizophrenia. Proc. Natl. Acad. Sci. U.S.A. 97 (14), 8104—8109.

Abi-Dargham, A., van de Giessen, E., et al., 2009. Baseline and amphetamine-stimulated dopamine activity are related in drug-naive schizophrenic subjects. Biol. Psychiatry 65 (12), 1091—1093.

Abrahamse, E., Ruitenberg, M., et al., 2016. Conflict adaptation in schizophrenia: reviewing past and previewing future efforts. Cogn. Neuropsychiatry 21 (3), 197—212.

Aisa, B., Mingus, B., et al., 2008. The emergent neural modeling system. Neural Netw. 21 (8), 1146—1152.

Alain, C., McNeely, H.E., et al., 2002. Neurophysiological evidence of error monitoring deficits in patients with schizophrenia. Cereb. Cortex 12, 840—846.

Alexander, W.H., Brown, J.W., 2015. Hierarchical error representation: a computational model of anterior cingulate and dorsolateral prefrontal cortex. Neural Comput. 27 (11), 2354—2410.

Alexander, W.H., Brown, J.W., 2016. Frontal cortex function derives from hierarchical predictive coding. BioRxiv.

Anticevic, A., Murray, J.D., et al., 2015. Bridging levels of understanding in schizophrenia through computational modeling. Clin. Psychol. Sci. 3 (3), 433—459.

Aston-Jones, G., Cohen, J.D, 2005. An integrative theory of locus coeruleus-norepinephrine function: adaptive gain and optimal performance. Annu. Rev. Neurosci. 28, 403—450. http://dx.doi.org/10.1146/annurev.neuro.28.061604.135709.

Badre, D., 2008. Cognitive control, hierarchy, and the rostro-caudal organization of the frontal lobes. Trends Cogn. Sci. 12 (5), 193—200.

Badre, D., D'Esposito, M., 2009. Is the rostro-caudal axis of the frontal lobe hierarchical? Nat. Rev. Neurosci. 10 (9), 659—669.

Barch, D.M., Braver, T.S., 2007. Cognitive control in schizophrenia: psychological and neural mechanisms. In: Engle, R.W., Sedek, G., von Hecker, U., McIntosh, A.M. (Eds.), Cognitive Limitations in Aging and Psychopathology. Cambridge University Press, Cambridge, pp. 122—159.

Barch, D.M., Braver, T.S., et al., 2009. CNTRICS final task selection: executive control. Schizophr. Bull. 35 (1), 115—135.

Barch, D.M., Mitropoulou, V., et al., 2004. Context-processing deficits in schizotypal personality disorder. J. Abnorm. Psychol. 113 (4), 556—568.

Barch, D.M., Sheffield, J.M., 2014. Cognitive impairments in psychotic disorders: common mechanisms and measurement. World Psychiatry 13 (3), 224—232.

Barch, D.M., Sheffield, J., 2017. Cognitive control in schizophrenia. In: Egner, T. (Ed.), Handbook of Cognitive Control. Wiley Press, pp. 556—580.

Bates, A.T., Kiehl, K.A., et al., 2002. Error-related negativity and correct response negativity in schizophrenia. Clin. Neurophysiol. 113 (9), 1454—1463.

Bates, A.T., Liddle, P.F., et al., 2004. State dependent changes in error monitoring in schizophrenia. J. Psychiatr. Res. 38 (3), 347—356.

Becerril, K.E., Barch, D.M., 2013. Conflict and error processing in an extended cingulo-opercular and cerebellar network in schizophrenia. Neuroimage Clin. 3, 470—480.

Becerril, K.E., Repovs, G., et al., 2011. Error processing network dynamics in schizophrenia. NeuroImage 54 (2), 1495—1505.

Bora, E., Yucel, M., et al., 2010. Cognitive impairment in schizophrenia and affective psychoses: implications for DSM-V criteria and beyond. Schizophr. Bull. 36 (1), 36—42.

Botvinick, M., Braver, T., 2015. Motivation and cognitive control: from behavior to neural mechanism. Annu. Rev. Psychol. 66, 83–113.

Botvinick, M.M., Braver, T.S., et al., 2001. Conflict monitoring and cognitive control. Psychol. Rev. 108, 624–652.

Botvinick, M.M., Cohen, J.D., 2014. The computational and neural basis of cognitive control: charted territory and new frontiers. Cogn. Sci. 38 (6), 1249–1285.

Braver, T.S., 2012. The variable nature of cognitive control: a dual mechanisms framework. Trends Cogn. Sci. 16 (2), 106–113.

Braver, T.S., Barch, D.M., et al., 1999. Cognition and control in schizophrenia: a computational model of dopamine and prefrontal function. Biol. Psychiatry 46, 312–328.

Braver, T.S., Cohen, J.D., 1999. Dopamine, cognitive control, and schizophrenia: the gating model. Prog. Brain Res. 121, 327–349.

Braver, T.S., Paxton, J.L., et al., 2009. Flexible neural mechanisms of cognitive control within human prefrontal cortex. Proc. Natl. Acad. Sci. U.S.A. 106 (18), 7351–7356.

Carter, C.S., MacDonald III, A.W., et al., 2001. Anterior cingulate cortex activity and impaired self-monitoring of performance in patients with schizophrenia: an event-related fMRI study. Am. J. Psychiatry 158, 1423–1428.

Carter, C.S., van Veen, V., 2007. Anterior cingulate cortex and conflict detection: an update of theory and data. Cogn. Affect. Behav. Neurosci. 7 (4), 367–379.

Ceaser, A.E., Barch, D.M., 2015. Striatal activity is associated with deficits of cognitive control and aberrant salience for patients with schizophrenia. Front. Hum. Neurosci. 9, 687.

Charles, L., Gaillard, R., et al., 2017. Conscious and unconscious performance monitoring: evidence from patients with schizophrenia. NeuroImage 144 (Pt A), 153–163.

Cohen, J.D., Barch, D.M., et al., 1999. Context-processing deficits in schizophrenia: converging evidence from three theoretically motivated cognitive tasks. J. Abnorm. Psychol. 108, 120–133.

Cohen, J.D., Dunbar, K., et al., 1990. On the control of automatic processes: a parallel distributed processing account of the Stroop effect. Psychol. Rev. 97 (3), 332–361.

Cohen, J.D., Servan-Schreiber, D., 1992. Context, cortex and dopamine: a connectionist approach to behavior and biology in schizophrenia. Psychol. Rev. 99 (1), 45–77.

Cohen, J.D., Servan-Schreiber, D., 1993. A theory of dopamine function and cognitive deficits in schizophrenia. Schizophr. Bull. 19 (1), 85–104.

Cohen, J.D., Servan-Schreiber, D., et al., 1992a. A parallel distributed processing approach to automaticity. Am. J. Psychol. 105, 239–269.

Cohen, J.D., Targ, E., et al., 1992b. The fabric of thought disorder: a cognitive neuroscience approach to disturbances in the processing of context in schizophrenia. In: Stein, D.J., Young, J.E. (Eds.), Cognitive Science and Clinical Disorders. Academic Press, NY, pp. 101–127.

Culbreth, A., Westbrook, A., et al., 2016a. Negative symptoms are associated with an increased subjective cost of cognitive effort. J. Abnorm. Psychol. 125 (4), 528–536.

Culbreth, A.J., Westbrook, A., et al., 2016b. Reduced model-based decision-making in schizophrenia. J. Abnorm. Psychol. 125 (6), 777–787.

Culbreth, A.J., Westbrook, A., et al., 2016c. Intact ventral striatal prediction error signaling in medicated schizophrenia patients. Biol. Psychiatry Cogn. Neurosci. Neuroimaging 1 (5), 474–483.

Dahaene, S., Posner, M.I., et al., 1994. Localization of a neural system for error detection and compensation. Psychol. Sci. 5, 303–305.

Daw, N.D., Gershman, S.J., et al., 2011. Model-based influences on humans' choices and striatal prediction errors. Neuron 69 (6), 1204–1215.

Daw, N.D., Niv, Y., et al., 2005. Uncertainty-based competition between prefrontal and dorsolateral striatal systems for behavioral control. Nat. Neurosci. 8 (12), 1704–1711.

de la Asuncion, J., Docx, L., et al., 2015. Neurophysiological evidence for diminished monitoring of own, but intact monitoring of other's errors in schizophrenia. Psychiatry Res. 230 (2), 220–226.

Delawalla, Z., Csernansky, J.G., Barch, D.M., 2008. Prefrontal cortex function in nonpsychotic siblings of individuals with schizophrenia. Biol. Psychiatry 63 (5), 490–497. http://dx.doi.org/10.1016/j.biopsych.2007.05.007.

Deserno, L., Huys, Q.J., et al., 2015. Ventral striatal dopamine reflects behavioral and neural signatures of model-based control during sequential decision making. Proc. Natl. Acad. Sci. U.S.A. 112 (5), 1595–1600.

Doll, B.B., Duncan, K.D., et al., 2015. Model-based choices involve prospective neural activity. Nat. Neurosci. 18 (5), 767–772.

Dowd, E.C., Frank, M.J., et al., 2016. Probabilistic reinforcement learning in patients with schizophrenia: relationships to anhedonia and avolition. Biol. Psychiatry Cogn. Neurosci. Neuroimaging 1 (5), 460–473.

Doya, K., Samejima, K., et al., 2002. Multiple model-based reinforcement learning. Neural Comput. 14 (6), 1347–1369.

Durstewitz, D., Kelc, M., et al., 1999. A neurocomputational theory of the dopaminergic modulation of working memory functions. J. Neurosci. 19 (7), 2807–2822.

Durstewitz, D., Seamans, J.K., 2002. The computational role of dopamine D1 receptors in working memory. Neural Netw. 15 (4–6), 561–572.

Edwards, B.G., Barch, D.M., et al., 2010. Improving prefrontal cortex function in schizophrenia through focused training of cognitive control. Front. Hum. Neurosci. 4, 32.

Egner, T., Hirsch, J., 2005. The neural correlates and functional integration of cognitive control in a Stroop task. NeuroImage 24 (2), 539–547.

Falkenstein, M., Hohnsbein, J., et al., 1995. Event related potential correlates of errors in reaction tasks. In: Karmos, G., Molnar, M., Csepe, V., Czigler, I., Desmedt, J.E. (Eds.), Perspectives of Event-Related Potentials Research. Elsevier Science B. V., Amsterdam, pp. 287–296.

Fassbender, C., Scangos, K., et al., 2014. RT distributional analysis of cognitive-control-related brain activity in first-episode schizophrenia. Cogn. Affect. Behav. Neurosci. 14 (1), 175–188.

Foti, D., Perlman, G., et al., 2016. Impaired error processing in late-phase psychosis: four-year stability and relationships with negative symptoms. Schizophr. Res. 176 (2–3), 520–526.

Frank, M.J., Loughry, B., et al., 2001. Interactions between frontal cortex and basal ganglia in working memory: a computational model. Cogn. Affect. Behav. Neurosci. 1 (2), 137–160.

Funahashi, S., Bruce, C.J., et al., 1989. Mnemonic coding of visual space in the monkey's dorsolateral prefrontal cortex. J. Neurophysiol. 61 (2), 331–349.

Fusar-Poli, P., Meyer-Lindenberg, A., 2013a. Striatal presynaptic dopamine in schizophrenia, part I: meta-analysis of dopamine active transporter (DAT) density. Schizophr. Bull. 39 (1), 22–32. http://dx.doi.org/10.1093/schbul/sbr111.

Fusar-Poli, P., Meyer-Lindenberg, A., 2013b. Striatal presynaptic dopamine in schizophrenia, part II: meta-analysis of [(18)F/(11)C]-DOPA PET studies. Schizophr. Bull. 39 (1), 33–42.

Gehring, W.J., Coles, M.G.H., et al., 1990. The error-related negativity: an event-related potential accompanying errors. Psychophysiology 27, S34.

Gillan, C.M., Otto, A.R., et al., 2015. Model-based learning protects against forming habits. Cogn. Affect. Behav. Neurosci. 15 (3), 523–536.

Glascher, J., Daw, N., et al., 2010. States versus rewards: dissociable neural prediction error signals underlying model-based and model-free reinforcement learning. Neuron 66 (4), 585–595.

Goodkind, M., Eickhoff, S.B., et al., 2015. Identification of a common neurobiological substrate for mental illness. JAMA Psychiatry 72 (4), 305–315.

Haruno, M., Kawato, M., 2006. Different neural correlates of reward expectation and reward expectation error in the putamen and caudate nucleus during stimulus-action-reward association learning. J. Neurophysiol. 95 (2), 948–959.

Henderson, D., Poppe, A.B., et al., 2012. Optimization of a goal maintenance task for use in clinical applications. Schizophr. Bull. 38 (1), 104–113.

Hendrick, O.M., Ide, J.S., et al., 2010. Dissociable processes of cognitive control during error and non-error conflicts: a study of the stop signal task. PLoS One 5 (10), e13155.

Hoffman, R.E., 1997. Neural network simulations, cortical connectivity, and schizophrenic psychosis. M.D. Comput. 14 (3), 200−208.

Holmes, A.J., MacDonald 3rd, A., et al., 2005. Prefrontal functioning during context processing in schizophrenia and major depression: an event-related fMRI study. Schizophr. Res. 76 (2−3), 199−206.

Holroyd, C.B., Coles, M.G., 2002. The neural basis of human error processing: reinforcement learning, dopamine, and the error-related negativity. Psychol. Rev. 109 (4), 679−709.

Holroyd, C.B., Umemoto, A., 2016. The research domain criteria framework: the case for anterior cingulate cortex. Neurosci. Biobehav. Rev. 71, 418−443.

Holroyd, C.B., Yeung, N., et al., 2005. A mechanism for error detection in speeded response time tasks. J. Exp. Psychol. Gen. 134 (2), 163−191.

Horan, W.P., Foti, D., et al., 2012. Impaired neural response to internal but not external feedback in schizophrenia. Psychol. Med. 42 (8), 1637−1647.

Howes, O.D., Kambeitz, J., Kim, E., Stahl, D., Slifstein, M., Abi-Dargham, A., Kapur, S., 2012. The nature of dopamine dysfunction in schizophrenia and what this means for treatment. Arch. Gen. Psychiatry 69 (8), 776−786. http://dx.doi.org/10.1001/archgenpsychiatry.2012.169.

Huys, Q.J., Maia, T.V., et al., 2016. Computational psychiatry as a bridge from neuroscience to clinical applications. Nat. Neurosci. 19 (3), 404−413.

Kasanova, Z., Waltz, J.A., et al., 2011. Optimizing vs. matching: response strategy in a probabilistic learning task is associated with negative symptoms of schizophrenia. Schizophr. Res. 127 (1−3), 215−222.

Kerns, J.G., 2006. Anterior cingulate and prefrontal cortex activity in an FMRI study of trial-to-trial adjustments on the Simon task. NeuroImage 33 (1), 399−405.

Kerns, J.G., Cohen, J.D., et al., 2004. Anterior cingulate conflict monitoring and adjustments in control. Science 303 (5660), 1023−1026.

Kerns, J.G., Cohen, J.D., et al., 2005. Decreased conflict- and error-related activity in the anterior cingulate cortex in subjects with schizophrenia. Am. J. Psychiatry 162 (10), 1833−1839.

Kim, C., Johnson, N.F., et al., 2014. Conflict adaptation in prefrontal cortex: now you see it, now you don't. Cortex 50, 76−85.

Koechlin, E., Ody, C., et al., 2003. The architecture of cognitive control in the human prefrontal cortex. Science 302 (5648), 1181−1185.

Kopp, B., Rist, F., 1994. Error-correcting behavior in schizophrenic patients. Schizophr. Res. 13 (1), 11−22.

Kopp, B., Rist, F., 1999. An event-related brain potential substrate of disturbed response monitoring in paranoid schizophrenic patients. J. Abnorm. Psychol. 108 (2), 337−346.

Krawitz, A., Braver, T.S., et al., 2011. Impaired error-likelihood prediction in medial prefrontal cortex in schizophrenia. NeuroImage 54 (2), 1506−1517.

Laruelle, M., D'Souza, G.D., et al., 1996. SPECT measurement of dopamine synaptic concentration in the resting state. J. Nucl. Med. 37 (5), 32.

Laurens, K.R., Hodgins, S., et al., 2010. Error-related processing dysfunction in children aged 9 to 12 years presenting putative antecedents of schizophrenia. Biol. Psychiatry 67 (3), 238−245.

Laurens, K.R., Ngan, E.T., et al., 2003. Rostral anterior cingulate cortex dysfunction during error processing in schizophrenia. Brain 126, 610−622.

Lee, S.W., Shimojo, S., et al., 2014. Neural computations underlying arbitration between model-based and model-free learning. Neuron 81 (3), 687−699.

Lesh, T.A., Niendam, T.A., et al., 2011. Cognitive control deficits in schizophrenia: mechanisms and meaning. Neuropsychopharmacology 36 (1), 316−338.

Lesh, T.A., Westphal, A.J., et al., 2013. Proactive and reactive cognitive control and dorsolateral prefrontal cortex dysfunction in first episode schizophrenia. Neuroimage Clin. 2, 590–599.

Llerena, K., Wynn, J.K., et al., 2016. Patterns and reliability of EEG during error monitoring for internal versus external feedback in schizophrenia. Int. J. Psychophysiol. 105, 39–46.

MacDonald, A., Carter, C.S., et al., 2005. Specificity of prefrontal dysfunction and context processing deficits to schizophrenia in a never medicated first-episode psychotic sample. Am. J. Psychiatry 162, 475–484.

MacDonald 3rd, A.W., Thermenos, H.W., et al., 2009. Imaging genetic liability to schizophrenia: systematic review of FMRI studies of patients' nonpsychotic relatives. Schizophr. Bull. 35 (6), 1142–1162.

Maia, T.V., Frank, M.J., 2011. From reinforcement learning models to psychiatric and neurological disorders. Nat. Neurosci. 14 (2), 154–162.

Maia, T.V., Frank, M.J., 2017. An integrative perspective on the role of dopamine in schizophrenia. Biol. Psychiatry 81 (1), 52–66.

Mansouri, F.A., Egner, T., et al., 2017. Monitoring demands for executive control: shared functions between human and nonhuman primates. Trends Neurosci. 40 (1), 15–27.

Marco-Pallares, J., Camara, E., et al., 2008. Neural mechanisms underlying adaptive actions after slips. J. Cogn. Neurosci. 20 (9), 1595–1610.

Mathalon, D.H., Fedor, M., et al., 2002. Response-monitoring dysfunction in schizophrenia: an event-related brain potential study. J. Abnorm. Psychol. 111 (1), 22–41.

Mathalon, D.H., Jorgensen, K.W., et al., 2009. Error detection failures in schizophrenia: ERPs and FMRI. Int. J. Psychophysiol. 73 (2), 109–117.

McClelland, J., Rumelhart, D.E., 1986. Parallel distributed processing explorations in the microstructure of cognition. In: Psychological and Biological Models, vol. 2. MIT Press, Cambridge, MA.

McClelland, J.L., 1991. Stochastic interactive processes and the effect of context on perception. Cogn. Psychol. 23, 1–44.

McClelland, J.L., Elman, J.L., 1986. The TRACE model of speech perception. Cogn. Psychol. 18, 1–86.

McClelland, J.L., St. John, M., et al., 1989. Sentence comprehension: a parallel distributed processing approach. Lang. Cogn. Process. 4 (3\4), 287–335.

McClure, M.M., Barch, D.M., et al., 2008. Context processing in schizotypal personality disorder: evidence of specificity of impairment to the schizophrenia spectrum. J. Abnorm. Psychol. 117 (2), 342–354.

McClure, M.M., Barch, D.M., et al., 2007. The effects of guanfacine on context processing abnormalities in schizotypal personality disorder. Biol. Psychiatry 61 (10), 1157–1160.

McClure, S.M., Berns, G.S., et al., 2003. Temporal prediction errors in a passive learning task activate human striatum. Neuron 38 (2), 339–346.

McClure, S.M., Gilzenrat, M., et al., 2005. An Exploration-Exploitation Model Based on Norepinephrine and Dopamine Activity. MIT Press, Cambridge, MA.

Miller, E.K., Cohen, J.D., 2001. An integrative theory of prefrontal cortex function. Annu. Rev. Neurosci. 21, 167–202.

Minzenberg, M.J., Gomes, G.C., et al., 2014. Disrupted action monitoring in recent-onset psychosis patients with schizophrenia and bipolar disorder. Psychiatry Res. 221 (1), 114–121.

Minzenberg, M.J., Laird, A.R., et al., 2009. Meta-analysis of 41 functional neuroimaging studies of executive function in schizophrenia. Arch. Gen. Psychiatry 66 (8), 811–822.

Morris, L.S., Kundu, P., et al., 2016. Fronto-striatal organization: defining functional and microstructural substrates of behavioural flexibility. Cortex 74, 118–133.

Morris, S.E., Heerey, E.A., et al., 2008. Learning-related changes in brain activity following errors and performance feedback in schizophrenia. Schizophr. Res. 99 (1–3), 274–285.

Morris, S.E., Holroyd, C.B., et al., 2011. Dissociation of response and feedback negativity in schizophrenia: electrophysiological and computational evidence for a deficit in the representation of value. Front. Hum. Neurosci. 5, 123.

Morris, S.E., Yee, C.M., et al., 2006. Electrophysiological analysis of error monitoring in schizophrenia. J. Abnorm. Psychol. 115 (2), 239—250.

Murray, J.D., Anticevic, A., 2017. Toward understanding thalamocortical dysfunction in schizophrenia through computational models of neural circuit dynamics. Schizophr. Res. 180, 70—77.

Murray, J.D., Anticevic, A., et al., 2014. Linking microcircuit dysfunction to cognitive impairment: effects of disinhibition associated with schizophrenia in a cortical working memory model. Cereb. Cortex 24 (4), 859—872.

Nasser, H.M., Calu, D.J., et al., 2017. The dopamine prediction error: contributions to associative models of reward learning. Front. Psychol. 8, 244.

Niendam, T.A., Lesh, T.A., et al., 2014. Impaired context processing as a potential marker of psychosis risk state. Psychiatry Res. 221 (1), 13—20.

O'Doherty, J.P., 2004. Reward representations and reward-related learning in the human brain: insights from neuroimaging. Curr. Opin. Neurobiol. 14 (6), 769—776.

O'Reilly, R.C., Frank, M.J., 2006. Making working memory work: a computational model of learning in the prefrontal cortex and basal ganglia. Neural Comput. 18 (2), 283—328.

O'Reilly, R.C., Frank, M.J., et al., 2007. PVLV: the primary value and learned value Pavlovian learning algorithm. Behav. Neurosci. 121 (1), 31—49.

O'Reilly, R.C., Wyatte, D., et al., 2013. Recurrent processing during object recognition. Front. Psychol. 4, 124.

Otto, A.R., Gershman, S.J., et al., 2013a. The curse of planning: dissecting multiple reinforcement-learning systems by taxing the central executive. Psychol. Sci. 24 (5), 751—761.

Otto, A.R., Raio, C.M., et al., 2013b. Working-memory capacity protects model-based learning from stress. Proc. Natl. Acad. Sci. U.S.A. 110 (52), 20941—20946.

Otto, A.R., Skatova, A., et al., 2015. Cognitive control predicts use of model-based reinforcement learning. J. Cogn. Neurosci. 27 (2), 319—333.

Paxton, J.L., Barch, D.M., et al., 2008. Cognitive control, goal maintenance, and prefrontal function in healthy aging. Cereb. Cortex 18 (5), 1010—1028.

Perez, V.B., Ford, J.M., et al., 2012. Error monitoring dysfunction across the illness course of schizophrenia. J. Abnorm. Psychol. 121 (2), 372—387.

Perlstein, W.M., Dixit, N.K., et al., 2003. Prefrontal cortex dysfunction mediates deficits in working memory and prepotent responding in schizophrenia. Biol. Psychiatry 53, 25—38.

Plaut, D.C., 1996. Relearning after damage in connectionist networks: toward a theory of rehabilitation. Brain Lang. 52, 25—82.

Plaut, D.C., Farah, M.J., 1990. Visual object representation: interpreting neurophysiological data within a computational framework. J. Cogn. Neurosci. 2 (4), 320—343.

Plaut, D.C., McClelland, J.L., et al., 1996. Understanding normal and impaired word reading: computational principles in quasi-regular domains. Psychol. Rev. 103, 56—115.

Plaut, D.C., Shallice, T., 1993. Deep dyslexia: a case study of connectionist neuropsychology. Cogn. Neuropsychol. 10 (5), 377—500.

Polli, F.E., Barton, J.J., et al., 2008. Reduced error-related activation in two anterior cingulate circuits is related to impaired performance in schizophrenia. Brain 131 (Pt 4), 971—986.

Polli, F.E., Barton, J.J., et al., 2006. Schizophrenia patients show intact immediate error-related performance adjustments on an antisaccade task. Schizophr. Res. 82 (2—3), 191—201.

Poppe, A.B., Barch, D.M., et al., 2016. Reduced frontoparietal activity in schizophrenia is linked to a specific deficit in goal maintenance: a multisite functional imaging study. Schizophr. Bull. 42 (5), 1149—1157.

Radua, J., Schmidt, A., et al., 2015. Ventral striatal activation during reward processing in psychosis: a neurofunctional meta-analysis. JAMA Psychiatry 72 (12), 1243–1251.

Ragland, J.D., Laird, A.R., et al., 2009. Prefrontal activation deficits during episodic memory in schizophrenia. Am. J. Psychiatry 166 (8), 863–874.

Rumelhart, D.E., Hinton, G.E., et al., 1986. Learning internal representations by backpropogating errors. Nature 323, 533–536.

Rumelhart, D.E., McClelland, J.L., 1986. Parallel Distributed Processing: Explorations in the Microstructure of Cognition. MIT Press, Cambridge, MA.

Schultz, W., 1998. Predictive reward signal of dopamine neurons. J. Neurophysiol. 80 (1), 1–27.

Schultz, W., Dayan, P., et al., 1997. A neural substrate of prediction and reward. Science 275, 1593–1599.

Servan-Schreiber, D., Cohen, J.D., et al., December 1996. Schizophrenic deficits in the processing of context: a test of a theoretical model. Arch. Gen. Psychiatry 53, 1105–1113.

Servan-Schreiber, D., Printz, H., et al., 1990. A network model of catecholamine effects: gain, signal-to-noise ratio, and behavior. Science 249, 892–895.

Sharp, M.E., Foerde, K., et al., 2016. Dopamine selectively remediates 'model-based' reward learning: a computational approach. Brain 139 (Pt 2), 355–364.

Shenhav, A., Botvinick, M.M., et al., 2013. The expected value of control: an integrative theory of anterior cingulate cortex function. Neuron 79 (2), 217–240.

Shenhav, A., Cohen, J.D., et al., 2016. Dorsal anterior cingulate cortex and the value of control. Nat. Neurosci. 19 (10), 1286–1291.

Simmonite, M., Bates, A.T., et al., 2012. Error processing-associated event-related potentials in schizophrenia and unaffected siblings. Int. J. Psychophysiol. 84 (1), 74–79.

Smittenaar, P., FitzGerald, T.H., et al., 2013. Disruption of dorsolateral prefrontal cortex decreases model-based in favor of model-free control in humans. Neuron 80 (4), 914–919.

Snitz, B.E., MacDonald 3rd, A., et al., 2005. Lateral and medial hypofrontality in first-episode schizophrenia: functional activity in a medication-naive state and effects of short-term atypical antipsychotic treatment. Am. J. Psychiatry 162 (12), 2322–2329.

Snitz, B.E., Macdonald 3rd, A.W., et al., 2006. Cognitive deficits in unaffected first-degree relatives of schizophrenia patients: a meta-analytic review of putative endophenotypes. Schizophr. Bull. 32 (1), 179–194.

Snyder, H.R., Miyake, A., et al., 2015. Advancing understanding of executive function impairments and psychopathology: bridging the gap between clinical and cognitive approaches. Front. Psychol. 6, 328.

Strauss, G.P., Frank, M.J., et al., 2011. Deficits in positive reinforcement learning and uncertainty-driven exploration are associated with distinct aspects of negative symptoms in schizophrenia. Biol. Psychiatry 69 (5), 424–431.

Sutton, R.S., Barto, A.G., 1981. Toward a modern theory of adaptive networks: expectation and prediction. Psychol. Rev. 88 (2), 135–170.

Turken, A.U., Vuilleumier, P., et al., 2003. Are impairments of action monitoring and executive control true dissociative dysfunctions in patients with schizophrenia? Am. J. Psychiatry 160 (10), 1881–1883.

Usher, M., Cohen, J.D., et al., 1999. The role of locus coeruleus in the regulation of cognitive performance. Science 283, 549–554.

Usher, M., Cohen, J.D., et al., 1995. A Computational Model of Locus Coeruleus Function and Its Influence on Cognitive Performance. Department of Psychology, Carnegie Mellon University.

Voon, V., Derbyshire, K., et al., 2015. Disorders of compulsivity: a common bias towards learning habits. Mol. Psychiatry 20 (3), 345–352.

Wallis, J.D., 2007. Orbitofrontal cortex and its contribution to decision-making. Annu. Rev. Neurosci. 30, 31–56.

Wang, M., Arnsten, A.F., 2015. Contribution of NMDA receptors to dorsolateral prefrontal cortical networks in primates. Neurosci. Bull. 31 (2), 191—197.

Wang, X.-J., 2010. Neurophysiological and computational principles of cortical rhythms in cognition. Physiol. Rev. 90 (3), 1195—1268.

Warren, C.M., Eldar, E., et al., 2016. Catecholamine-mediated increases in gain enhance the precision of cortical representations. J. Neurosci. 36 (21), 5699—5708.

Yoon, J.H., Minzenberg, M.J., et al., 2008. Association of dorsolateral prefrontal cortex dysfunction with disrupted coordinated brain activity in schizophrenia: relationship with impaired cognition, behavioral disorganization, and global function. Am. J. Psychiatry 165 (8), 1006—1014.

Zandbelt, B.B., van Buuren, M., et al., 2011. Reduced proactive inhibition in schizophrenia is related to corticostriatal dysfunction and poor working memory. Biol. Psychiatry 70 (12), 1151—1158.

# 7

# Bayesian Inference, Predictive Coding, and Computational Models of Psychosis

*Rick A. Adams*

University College London, London, United Kingdom

## OUTLINE

## 7.1 HIERARCHICAL MODELS AND PREDICTIVE CODING

The Bayesian brain hypothesis (Yuille and Kersten, 2006; Knill and Pouget, 2004; Friston, 2005) proposes that the central nervous system uses the experience garnered over a lifetime—in the form of "prior beliefs"—to infer the causes of its sensory data collected from moment to moment. Neither its sensory data nor its prior beliefs are completely reliable, and so the brain must use both sources of information—taking into account their uncertainty—to perform its task. The optimal combination of uncertain information is given by Bayes' theorem, in which a "prior" (the initial expectation of the state of the environment) is combined with a "likelihood" (the probability of the sensory input, given that expectation) to compute a "posterior" (an updated estimation of the state of the environment). The contributions of the prior and likelihood to the posterior are weighted by their certainty or precision (inverse variance). For simplicity, these probability distributions are often assumed to be of a kind that can be represented by a few "sufficient statistics." For instance, the mean and precision of a Normal distribution, in this case both prior and likelihood, can be conveniently weighted by their (scalar) precision.

As well as being somewhat uncertain, the statistics of natural sensory stimulation are also extremely complex. Nevertheless, because the environment itself contains hierarchical structure—such as the phonemes, words, sentences, and stories in speech—they contain patterns. These patterns are easiest to interpret if the brain's prior beliefs recapitulate the hierarchical structure in its sensory data—i.e., if they take the form of a hierarchical model (also discussed elsewhere, Adams et al., 2016). Hierarchical models explain complex patterns of low-level data features in terms of more abstract causes, e.g., the shape that describes a collection of pixels, or the climate that describes annual variation in weather. Hierarchical models are particularly important in the face of complex situations, which can emerge both from behavioral and sensory information. Such models allow for highly efficient decompositions that greatly support planning and simplify optimal decision-making (Botvinick et al., 2009; Friston, 2008; Huys et al., 2015).

Hierarchical generative models can use predictive coding (or other message passing schemes, Jardri and Denève, 2013) to predict low-level data using their high-level descriptions, e.g., reconstructing the missing part of an image (Mathys et al., 2011; Rao and Ballard, 1999). In predictive coding, a unit at a given level in the hierarchy sends messages to one or more units at lower levels, which predict their activity; discrepancies between these predictions and the actual input are then passed back up the hierarchy in the form of prediction errors. These prediction errors revise the higher-level predictions, and this hierarchical message passing continues in an iterative fashion.

Exactly which predictions ought to be changed to explain away a given prediction error is a crucial question for hierarchical models. An approximately Bayesian solution to this problem is to make the biggest updates to the level whose uncertainty is greatest relative to the incoming data at the level below. That is, if you are very uncertain about your beliefs, but the source of your information is very reliable, you ought to update your beliefs a lot.

Say, for example, an individual is walking at dusk, and the individual perceives the movement of a bush in their peripheral vision. One might explain this at various hierarchical levels as follows: (1) the bush didn't actually move; it was a trick of the light; (2) the wind was moving the bush; (3) an animal was moving the bush, (4) a man hiding in the bush, intending to rob the person, moved it. The conclusion that one draws will be determined by how precise their beliefs (at each level) are that: (1) they in fact saw movement; (2) the wind probably didn't cause it; (3) there are probably no animals in the vicinity; (4) there is probably no mugger in the vicinity. The most uncertain (least precise) of these beliefs will have to change (assuming, for the sake of argument, that their likelihoods are equivalent), with very distinct consequences for subsequent behavior. Put more formally, the uncertainty (inverse precision) at each level helps determine the learning rate at that level, i.e., the size of the adjustments that are made to explain new data (Mathys et al., 2011).

A classic psychology experiment illustrates uncertainty at different levels (see Fig. 7.1). Imagine you are shown two jars of beads, one containing 85% green and 15% red beads, the other 85% red and 15% green. The jars are then hidden and a sequence of beads is drawn (with replacement)—GGRGG. You are asked to guess the color of the next bead. Even if you are quite certain of the identity of the jar (say, green), you will still be only 85% certain that the next bead will be green: This is "outcome uncertainty" or risk. Imagine you see more beads—the total sequence is GGRGGRR. Now you are very uncertain about the identity of the jar: This is "state uncertainty" or ambiguity. Imagine you see a much longer sequence—GGRGGRRRRRGGGRGGGGGGGRGGGG. From this it seems that the jar changes from green (5 draws), to red (5 draws), to green (remaining sequence). Such temporal changes in hidden "causes" give rise to "volatility" (Yu and Dayan, 2005; Payzan-LeNestour and Bossaerts, 2011); e.g., someone surreptitiously switching the jar during the experiment.

Now suppose that although the real proportions are 85% and 15%, a malicious experimenter misleadingly told you that they are 99.9% and 0.1%. From the 25 draw sequence above, you might reasonably conclude that the jars had actually changed eight times—whenever the color changed. This is what happens when the precision at the bottom of a hierarchical model (input) is too high relative to the precision at the top (prior). Following a sensory prediction error, the model concludes there must have been a change in the environment (in this simplistic example, the jar), rather than "putting it down to chance."

This precision imbalance might contribute to the well-known "jumping to conclusions" (JTC) reasoning bias in individuals with schizophrenia (Fine et al., 2007) and the formation of delusional beliefs themselves. Such beliefs commonly arise in an atmosphere of vivid sensory experiences and

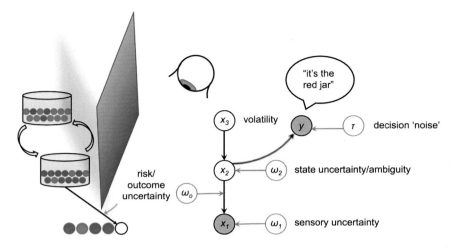

Possible uncertainty changes in schizophrenia:

$\uparrow \omega_2, \tau, \quad \downarrow \omega_1, \omega_0$

FIGURE 7.1   **A hierarchical generative model, illustrated using the "beads" or "urn" task.** On the left, two jars are hidden behind a screen, one containing mostly green and some red balls, the other the converse. A sequence of balls is being drawn from one of these jars, in view of an observer, who is asked to guess from which jar they are coming. We have illustrated a simple hierarchical generative model of this process on the right: the observer is using such a model to make his/her guess. Variables in shaded circles are observed, and variables in unshaded circles are "hidden" (i.e., part of the model only). At the bottom of the model is $x_1$, the color of the currently observed bead. Uncertainty about this quantity (e.g., if the light is low or if the subject is colorblind) is denoted as $\omega_1$. At the next level of the model is $x_2$, the belief about the identity of the current jar, and its associated uncertainty $\omega_2$, known as state uncertainty or ambiguity. Another form of uncertainty, risk or outcome uncertainty ($\omega_0$) governs the relationship between the identity of the jar and the next outcome; even if we are sure of the jar's identity, we cannot be certain of the color of the next bead. At the top of the model is the belief about the probability that the jars could be swapped at any time, known as volatility. We have not shown them here but this could have its own associated uncertainty, and there could be further levels above this. Last, the subject must use his/her belief about the identity of the jar to make a guess; the mapping between this belief and the response $y$ is affected by a degree of stochasticity or decision "noise," $\tau$. In schizophrenia, there may be too much uncertainty (i.e., lower precision) in higher hierarchical areas that encode states or make decisions, and an underestimation of uncertainty in lower (sensory) areas. *Reproduced from Adams, R.A., Huys, Q.J.M., Roiser, J.P., January 2016. Computational psychiatry: towards a mathematically informed understanding of mental illness. J. Neurol. Neurosurg. Psychiatry 87 (1), 53–63.*

strange coincidences (Corlett and Fletcher, 2014)—a subject that the chapter will return to later.

We now turn to a discussion of inference in schizophrenia. Specifically, we explore how various neurobiological abnormalities in schizophrenia might be characterized in computational terms, and how these characterizations might aid our understanding of the disorder. It is well established that schizophrenia is a remarkable heterogeneous syndrome with a complex underlying neurobiology that impacts multiple neurotransmitter pathways that impact synaptic communication in cortical and subcortical circuits. That said, one hallmark symptom noted above is the formation of aberrant and stable beliefs that are not in concert with reality—namely delusions. Here we specifically discuss alterations in "synaptic gain" (broadly defined change in the influence of presynaptic input on post-synaptic responses) in higher-order associative hierarchical areas (e.g., prefrontal cortex (PFC)) and how this might contribute to increased belief updating in these areas. We then discuss how this might relate to the formation of delusions, and whether (and how) different computational mechanisms might contribute to their maintenance.

## 7.2 PSYCHOSIS, SYNAPTIC GAIN, AND PRECISION

What are the main neurobiological abnormalities in schizophrenia and what do they have in common (discussed in detail elsewhere, Adams et al., 2013)? One key abnormality is thought to be hypofunction of the $N$-methyl-$D$-aspartate receptor (NMDA-R) (Abi-Saab et al., 1998; Olney and Farber, 1995)—a glutamate receptor with profound effects on both synaptic gain (due to its prolonged opening time) and synaptic plasticity (via long-term potentiation or depression)—in both PFC and hippo-campus (HC) and likely elsewhere (e.g., higher-order thalamic nuclei). A second well-established neurobiological abnormality is the reduced synthesis of $\gamma$-aminobutyric acid (GABA) by inhibitory interneurons in PFC (Lewis et al., 2012)—perhaps related to the hypofunction of NMDA-Rs on those interneurons themselves. A third abnormality is the hypo-activation of $D_1$ receptors in PFC (Goldman-Rakic et al., 2004) (we shall discuss striatal hyperactivation of $D_2$ receptors later on).

Collectively, these abnormalities could all reduce synaptic gain in PFC or HC, i.e., around the top of the cortical hierarchy. Synaptic gain (or "short term" synaptic plasticity Stephan et al., 2006) refers to a multiplicative change in the influence of presynaptic input on postsynaptic responses. NMDA-R hypofunction and $D_1$ receptor hypoactivity are most easily related to a change in synaptic gain. Similarly, a GABAergic deficit might cause a loss of "synchronous" gain. Sustained oscillations in neuronal populations are facilitated through their rhythmic inhibition by GABAergic interneurons, putatively increasing communication between neurons that oscillate in phase with each other (Fries, 2005).

How can synaptic gain (and its loss) be understood in computational terms? One answer rests on the idea that the brain approximates and simplifies Bayesian inference by using probability distributions that can be encoded by a few "sufficient statistics," e.g., the mean and its precision (or inverse variance). Whilst precision determines the influence one piece of information has over another in Bayesian inference, synaptic gain determines the influence one neural population has over another in neural message passing. The neurobiological substrate of precision could therefore be synaptic gain (Feldman and Friston, 2010), and a loss of synaptic gain in a given area could reduce the precision of information encoded there.

As mentioned above, NMDA-R hypofunction in schizophrenia may have its strongest effects not on pyramidal cells but rather on GABAergic inhibitory interneurons (Lewis et al., 2012). Because GABAergic cells connect with many local pyramidal cells, a pyramidal cell whose activity is sustained through time can inhibit the local spread of this activity via its projections to GABAergic cells. A loss of synaptic gain in these projections would lead to overexcitation of other pyramidal cells—known as an increased excitatory/inhibitory (E/I) balance—and a decrease in the signal-to-noise ratio (SNR) in signaling across areas. In addition to its relevance to belief formation discussed here, this model (Murray et al., 2014) also accounts for cognitive deficits such as the increased susceptibility to local distractors in spatial working memory tasks in schizophrenia (Mayer and Park, 2012)—a hallmark feature of the illness.

A loss of synaptic gain in PFC or HC would diminish their influence over lower-level areas. Indeed, in schizophrenia, higher-order regions of cortex (especially lateral PFC and HC) have diminished connectivity to the thalamus relative to controls, whereas primary sensory areas are coupled more strongly with this region (Anticevic et al., 2014). An increase in E/I balance in higher hierarchical areas might increase their connectivity with lower areas (Anticevic et al., 2015), but decrease the SNR (i.e., precision) of any descending information. In the model, these changes would correspond to a loss of influence of the model's more abstract priors over the more concrete sensory data, making the world look less predictable and more surprising to individuals with this type of neural deficit. This simple computational change can describe a great variety of trait phenomena in schizophrenia (see Fig. 7.2; more references and simulations of some of these phenomena are elsewhere, Adams et al., 2013):

- At a neurophysiological level, responses to predictable stimuli resemble responses to unpredicted stimuli, and vice versa, in both perceptual electrophysiology experiments (e.g., the P50 or P300 responses to tones, Turetsky et al., 2007) and cognitive neuroscience neuroimaging paradigms;
- At a perceptual level, a greater resistance to visual illusions (Silverstein and Keane, 2011) (which exploit the effects of visual priors on ambiguous images, for example, the famous "hollow-mask" illusion, Dima et al., 2009) and a failure to attenuate the

sensory consequences of one's own actions, which could diminish one's sense of agency (Shergill et al., 2005);
- At a behavioral level, impaired smooth visual pursuit of a predictably moving target, but better tracking of a sudden unpredictable change in a target's motion (Hong et al., 2008);

An alternative interpretation of these changes is that pathology in the PFC or HC (e.g., in postsynaptic signaling and neurotrophic pathways, Network and Pathway Analysis Subgroup of the Psychiatric Genomics Consortium, 2015) might impair the formation and representation of prior beliefs more generally, rather than directly and selectively affecting only a separate representation of their precision. Indeed, if this were the case, then reducing the influence of aberrant beliefs on sensory processing might even be computationally adaptive (and physiological rather than pathophysiological), because it would reduce their (possibly misleading) influence on inference.

How do the above ideas relate to the actual clinical symptoms of psychosis? A reasonable hypothesis would be that a loss of high-level precision in the brain's hierarchical model might result in diffuse, generalized cognitive problems (as are routinely found in schizophrenia, Dickinson et al., 2007) and overattention to sensory stimuli (as is found in the "delusional mood"; c.f., the loss of central coherence in autism, Pellicano and Burr, 2012). In addition, it could lead to the formation of more specific unusual beliefs, as the reduced high-level precision permits updates to beliefs that are larger and less constrained. This is because the precision of (low-level) prediction errors is much higher, relative to the (high-level) prior beliefs. However, one might expect that these unusual beliefs should be fleeting—as they themselves would be vulnerable to rapid updating—unlike delusions (we discuss the maintenance of delusions later).

In fact, some recent studies have presented findings seemingly at odds with a simple account in which all prior beliefs have less precision in relation to sensory evidence in the schizophrenia spectrum. Teufel et al. (2015) asked healthy volunteers with a range of schizotypy scores and (unmedicated) subjects with emerging psychosis to identify whether a person was present in an ambiguous two-tone image. The blocks of two-tone images were presented both before and after blocks of test (full color) images from which they were derived, hence improved performance at the second presentation must be due to perceptual prior beliefs garnered from seeing the color image. Subjects with both high schizotypy and those with emerging psychosis performed as well as controls at the first presentation, but *better* than controls at the second. Schmack et al. (2013) used a task in which ambiguously moving dots created bistable percepts moving either clockwise or anticlockwise, and subjects

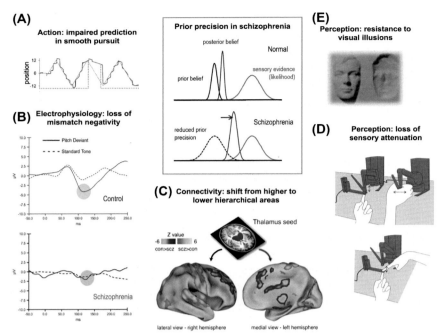

FIGURE 7.2   **Effects of a hierarchical precision imbalance in schizophrenia.** A loss of precision encoding in higher hierarchical areas would bias inference away from prior beliefs and toward sensory evidence (the likelihood), illustrated schematically in the middle panel. This single change could manifest in many ways (moving anticlockwise from left to right). (A) A loss of the ability to smoothly pursue a target moving predictably (in this plot the subject with schizophrenia constantly falls behind the target in his eye tracking, and has to saccade to catch up again); when the target is briefly stabilized on his retina (to reveal the purely predictive element of pursuit), shown as the *red unbroken line*, his/her eye velocity drops very significantly. (B) These graphs illustrate averaged electrophysiological responses in a mismatch negativity paradigm, in which a series of identical tones is followed by a deviant (oddball) tone; in the control subject, the oddball causes a pronounced negative deflection at around 120 ms (*blue circle*), but in a subject with schizophrenia, there is no such deflection (*red circle*); i.e., the brain responses to predictable and unpredictable stimuli are very similar. (C) One physiological change underlying the precision imbalance is a relative decrease in synaptic gain in high-hierarchical areas, and a relative increase in lower-hierarchical areas. This change would also manifest as an alteration in connectivity, shown here as significant whole brain differences in connectivity with a thalamic seed between controls and subjects with schizophrenia; *red/yellow areas* are more strongly coupled in those with schizophrenia, and include primary sensory areas (auditory, visual, motor, and somatosensory); *blue areas* are more weakly coupled, and include higher-hierarchical areas (medial and lateral prefrontal cortex, cingulate cortex, and hippocampus); and the striatum. (D) An imbalance in hierarchical precision may lead to a failure to attenuate the sensory consequences of one's own actions (Shergill et al., 2005), here illustrated by the force-matching paradigm used to measure this effect. In this paradigm, the subject must match a target force by either pressing on a bar with their finger (below) or using a mechanical transducer (top); control subjects tend to exert more force than necessary in the former condition, but

indicated perceptual switches with a button press; subjects were also led to believe that glasses they wore made the dots more likely to move in a particular direction. They showed that (healthy) subjects with greater delusional conviction scores exhibited lower perceptual stability but a *greater* influence of belief-induced bias on perception, and also greater belief-dependent functional connectivity between orbitofrontal cortex (OFC) and V5.

These studies seem to imply that some prior beliefs have greater influence over sensory data in individuals who are higher on the psychosis spectrum: why this should be is unclear. A loss of precision at higher-hierarchical levels is consistent with greater belief updating and perhaps (transiently) greater precision of recently learned prior beliefs; it would be interesting to test whether the more precise priors in these experiments also decayed faster than those in the less delusional subjects.

Not all recently learned visual priors are of higher precision in schizophrenia; however, Valton, Seriès et al. (personal communication) have found that medicated individuals diagnosed with schizophrenia performing a dot motion discrimination task (in which motion directions are sampled from a distribution centered around $0 \pm 30$ degrees) form prior beliefs of similar mean and precision as do healthy controls. However, patients were *less* likely than controls to hallucinate movement when none is present. One might expect that medication could explain this, but the latter effect correlated with worse positive symptoms, but not with medication dose. This may suggest that lower higher-order precision is associated with lower likelihood of imposing structure when there is none.

These studies show that simple accounts of schizophrenia as uniform reductions in the precision of all prior beliefs and/or increases in the precision of sensory evidence are unlikely to explain the range of experimental observations. Furthermore, they do not seem to fully account for

◄─────────────────────────────────────────────

schizophrenic subjects do not. (E) A loss of the precision of prior beliefs can cause a resistance to visual illusions that rely on those prior beliefs for their perceptual effects. Control subjects perceive the face on the right as a convex face lit from below, due to a powerful prior belief that faces are convex, whereas subjects with schizophrenia tend to perceive it veridically as a concave (hollow) face lit from above. *(A) Adapted from Hong, L.E., Turano, K.A., O'Neill, H., et al., 2008. Refining the predictive pursuit endophenotype in schizophrenia. Biol. Psychiatry 63 (5), 458–464; (B) Adapted from Turetsky, B.I., Calkins, M.E., Light, G.A., Olincy, A., Radant, A.D., Swerdlow, N.R., 2007. Neurophysiological endophenotypes of schizophrenia: the viability of selected candidate measures. Schizophr. Bull. 33 (1), 69–94; (C) Adapted from Anticevic, A., Cole, M.W., Repovs, G., et al., 2014. Characterizing thalamo-cortical disturbances in schizophrenia and bipolar illness. Cereb. Cortex 24 (12), 3116–3130; (D) Adapted from Pareés, I., Brown, H., Nuruki, A., et al., 2014. Loss of sensory attenuation in patients with functional (psychogenic) movement disorders. Brain J. Neurol. 137 (Pt 11), 2916–2921; (E) Reproduced from Adams, R.A., Huys, Q.J.M., Roiser, J.P., January 2016. Computational psychiatry: towards a mathematically informed understanding of mental illness. J. Neurol. Neurosurg. Psychiatry 87 (1), 53–63.*

changes induced by disease stage, medication, and subtleties in the behavior. Attempts to model processes underlying the formation and maintenance of delusions have also revealed similar complexity, which we turn to next.

## 7.3 COMPUTATIONALLY MODELING THE FORMATION OF DELUSIONS

Some attempts to understand the genesis of delusions have investigated whether individuals with schizophrenia are impaired at logical inference compared with controls. In fact, patients are just as good as syllogistic reasoning as controls, with any decrements in performance explained purely by loss of IQ rather than psychosis per se (Mirian et al., 2011). Interestingly, when "common sense" and logic conflict, subjects with schizophrenia are more likely to use logic than are control subjects (Owen et al., 2007)! This stands in contrast to the parsimonious cognitive hypothesis that pure logical inference is impaired in schizophrenia.

For this reason, researchers have focused on probabilistic rather than logical inference across the schizophrenia spectrum. Groups have used a commonly employed paradigm in this field—namely the beads or urn task (Garety et al., 1991) (described above). The two commonest variants of this task are the "draws to decision" (DTD) and "probability estimates" (PE); in the former, the subject is asked to stop the sequence once they are "sure" about the majority color of the jar, in the latter, the subject watches the whole sequence and rates the probability of one of the jars after each bead.

The best-replicated finding in this literature is that subjects with schizophrenia tend to stop the sequence after only one or two beads compared to healthy volunteers (Hedge's $g = -0.52$ (95% CI, $-0.69$ to $-0.36$)) or psychiatric controls (Hedge's $g = -0.58$ (95% CI, $-0.8$ to $-0.35$)) (Dudley et al., 2016): a phenomenon known as the JTC bias. Whilst it is more present in schizophrenic subjects with delusions than without (Hedge's $g = -0.29$ (95% CI, $-0.48$ to $-0.09$)), the JTC bias only shows a small and nonsignificant association with delusional severity, so its relationship to delusions is unclear. However, if the JTC bias principally contributes to the formation rather than the maintenance of delusions perhaps this is to be expected. Studies with sufficient power (i.e., sample size), stage of illness as well as variation in symptom status will be needed to test this hypothesis.

The usual measures in the PE version of the beads task are the "draws to certainty" (how many beads are seen before the subjects rates a jar as being $>x\%$ likely) and the response to disconfirmatory evidence, e.g., how much a subject adjusts their belief on the fourth bead in the sequence B-B-B-R. Each of these measures has been found to be more pronounced in individuals with schizophrenia, although less so than the JTC bias, and perhaps not at all if psychiatric controls are used (Fine et al., 2007).

Although many studies have used the beads task, very few have attempted to understand the JTC or other effects in computational terms. This is important because the JTC bias is usually interpreted as a tendency to overweight evidence—e.g., a higher learning rate—but there are many other potential explanations for it, which carefully controlled experiments and modeling could exclude. These include the following:

- a lower decision threshold
- an inability to inhibit a prepotent response
- more stochastic decision-making (measured using decision "temperature")
- a lower perceived cost of making a wrong decision
- a higher perceived cost of sampling

Indeed, the smaller effect sizes of belief updating measures in the PE beads task compared with that of the JTC effect may indicate that one or more of the above alternatives is contributing to the JTC bias in schizophrenia.

Moutoussis et al. (2011) explored whether the last three parameters listed above—i.e., decision temperature $\tau$, cost of wrong decision $C_W$, or cost of sampling $C_S$—could explain the JTC bias in individuals with schizophrenia both with and without active psychosis. This study modeled their belief updating using either an ideal Bayesian observer who looked ahead to see how many draws were left (hence incorporating $\tau$, $C_W$, and $C_S$ into its computation of action values) or a much simpler model that decided (with stochasticity $\tau$) based on the ratio of colors already seen, but didn't look into the future.

The authors found that acutely psychotic subjects had much higher $\tau$ but no differences in $C_W$ and $C_S$ (although $C_W$ and $C_S$ were not experimentally manipulated), and that responses of patients with schizophrenia (either psychotic or in remission) were fit best by the simpler "bead ratio" model than the model that looked ahead to the future. A higher $\tau$ may mean that decisions are truly more stochastic, or it may mean that the source of variability in decision-making has not been captured by the model; in this case, for example, the models did not contain subject-specific learning rate parameters that could govern the "weighting" of evidence.

How does the belief updating of individual with schizophrenia and healthy controls compare to that of an ideal Bayesian observer? Speechley et al. (2010) examined this question using an variation of the PE beads task in which the two jars did not contain identical but opposite proportions of beads (using e.g., 80/20 and 50/50 ratios), and in which subjects rated the probability of each jar, thus abandoning the assumption that the subjective probabilities of the two jars must sum to one. They showed that delusional patients with schizophrenia consistently judged the probability of whichever jar matched the evidence as being higher than nondelusional schizophrenia patients, bipolar patients and healthy controls, yet the probability ratings of the least likely jar were no different in all four groups. No group was closer to the Bayesian observer than any other. This result is even more interesting in the light of the finding that the reliability of the current cognitive strategy is evaluated in different part of PFC to the reliability of alternative strategies (Donoso et al., 2014), indicating that psychosis may affect estimation of the former (similar to confirmation bias, Balzan et al., 2013) but not the latter.

It is difficult to fit individual learning models to DTD beads task data unless the task is repeated many times. To inform their DTD-task interpretation, Averbeck et al. (2011) asked subjects with schizophrenia and controls to perform a sequence learning task with probabilistic feedback (85% correct), which allowed them to estimate how much subjects learned from positive and negative feedback (although the task was not optimized for the latter). They found that the schizophrenic subjects learned *less* from positive feedback than controls, and the less they learned, the more likely they were to show the JTC bias. They took this as evidence that a lower decision threshold—rather than overweighting evidence—is behind the JTC bias, but a more recent study (Collins et al., 2014) may force a reinterpretation of these data (discussed below).

Numerous studies have examined learning about the rewarding value of stimuli in schizophrenia (reviewed in depth elsewhere Deserno et al., 2013; Strauss et al., 2014). A common finding is that subjects with schizophrenia learn more slowly than controls, and show impaired learning from rewards but not from punishments—especially those with high negative symptoms (Gold et al., 2012). Given the finding that subjects with schizophrenia also have diminished reward prediction error-related functional magnetic resonance imaging activity in dopaminergic regions (Radua et al., 2015), many would conclude that striatal reinforcement learning must be impaired in schizophrenia. This may not be the whole story, however.

Collins et al. (2014) explored this question using a task in which subjects had to learn the correct actions to make in response to different stimuli. Their key manipulation was to adjust the stimulus set size in different blocks, from only two up to six. They reasoned that subjects were likely to use a combination of working memory and reinforcement learning to learn the correct responses, with dependence on the latter increasing as set size increased. Therefore, they included a working memory component in their model. Fitting the model to individuals' data, they found that individual with schizophrenia exhibited lower working memory capacity and greater decay rates, but their reinforcement learning parameters (learning and decay rates) were no different to controls'. Crucially, individuals with schizophrenia did not exhibit a more stochastic pattern of responses relative to healthy controls. This important result demonstrates that unless working memory is explicitly included in the computational model, inferences about reinforcement learning parameters in schizophrenia must be treated with caution. In neurobiological terms, this implies that there observed pathology in learning mechanisms in schizophrenia may be mediated by disruptions of neural systems supporting executive computations rather than exclusively striatal circuits (which are canonically implicated in learning deficits). How this relates to delusion formation is unclear, though, as none of the model parameters or their principal components correlated with any symptom measures.

What can this work tell us about the potential mechanisms underlying the formation of delusions? The beads task literature indicates that belief updating is probably increased in schizophrenia, especially in those with delusions—albeit in simple tasks without a working memory load. This increased learning rate may be specific to the currently entertained hypothesis (but it is probably not the sole contributor to the JTC bias). Reward-learning studies that have shown impaired reward learning have often not accounted for working memory problems, although doing so does not reveal parameters that correlate with delusion attributes.

This raises a more fundamental question: even if schizophrenic subjects are conclusively shown to update their beliefs more in the light of confirmatory evidence, is this sufficient to explain the formation of delusional beliefs? One cause for skepticism here is the speed with which some delusions arrive; for example, in delusional perception—a "first rank" schizophrenia symptom—a normal percept is accompanied by a fully formed delusional belief, likewise delusional ideas can just appear in a patient's mind. It is debatable whether any kind of incremental (striatal) learning mechanism would give rise to such phenomena. For explanations of these other symptoms, we should perhaps be looking to abnormal "one-shot" learning processes (Lee et al., 2015), e.g., in HC and PFC, rather than reinforcement learning mechanisms.

## 7.4 MODELING THE MAINTENANCE OF DELUSIONS

The second key quality of delusions, in addition to their departure from intersubjective norms about reasonable beliefs, is their resistance to change, even in the face of compelling counterevidence. For this reason, the "overweighting of evidence," if it exists in schizophrenia, cannot be uniquely explained as a uniform increase in learning rates, because in this case odd beliefs would be revised as quickly as they arose.

"Reversal learning" paradigms are designed to investigate how quickly beliefs are revised in the light of changing evidence (i.e., after learning a given contingency, being slower to learn a reversal in that contingency than you were to learn the original one). Subjects with schizophrenia have a specific impairment in reversal learning, e.g., in the intradimensional/extradimensional (ID/ED) set-shifting task (Elliott et al., 1995; Pantelis et al., 1999), even after controlling for IQ (Leeson et al., 2009). Subjects with schizophrenia are impaired at reversal in both deterministic tasks (in which feedback about the choice is 100% accurate), like the ID/ED task, and probabilistic tasks (Waltz and Gold, 2007) (in which feedback can sometimes be misleading), like the beads task.

Numerous factors could be responsible for reversal learning deficits. In tasks involving explicit rewards and punishments, decreased reversal could be due to learning faster from rewards than from punishments (Maia and Frank, 2017), although the opposite pattern is generally seen in schizophrenia. A decline in a single learning rate over time would also have the same effect on reversal, such a decline could be modeled as the result of an individual bias toward learning from past experience (den Ouden et al., 2013), or in a more normative fashion, as a consequence of declining uncertainty about the contingencies over time (as in a Kalman filter Kalman, 1960 or a hierarchical model in which volatility decreases during the task Mathys et al., 2011). Another plausible hypothesis is that subjects with schizophrenia perseverate (Collins et al., 2014)—i.e., find it difficult to change an established response—although whether this extends beyond behavior to beliefs is another question.

Not all investigations find perseveration in schizophrenia, however, several reversal learning studies (Schlagenhauf et al., 2014; Waltz et al., 2013) have shown that patients have an increased tendency to switch back to the original response that is no longer correct, hence they take longer to reach reversal criterion, but not because they perseverate. Schlagenhauf et al. (2014) performed detailed modeling of subjects' responses in a probabilistic reversal learning task and showed that most controls' response were best fit by a Hidden Markov Model rather than simpler Q-learning models. In effect, this means that controls realized that there was an underlying state in the task—e.g., the left button is most likely rewarded—meaning that in this state the best response is always to choose left, even if sometimes feedback is negative (a Q-learner would update its belief about the best action more in this situation). Only half of the schizophrenic subjects' responses were fitted by this model, however, mirroring previous findings (Moutoussis et al., 2011) that these subjects often do not develop sophisticated models of tasks. Interestingly, the well-fitted patients did not differ from controls in their response stochasticity, but their belief about the volatility of the contingencies was greater. Similarly, a recent study found that subjects with schizophrenia overestimate the probability of contextual change (Kaplan et al., 2016). These are intriguing results, but it is hard to see how they could explain delusional maintenance; if one expects more environmental volatility, one ought to be more willing to update one's beliefs.

If belief reversal really is impaired in schizophrenia, what could be the potential mechanism? The two most likely areas of dysfunction in impaired reversal are the OFC and/or striatum (Robbins, 2007). With respect to neuromodulatory systems, it seems that serotonin (but not dopamine) is crucial for the cortical component of reversal (Clarke et al., 2005; Rogers et al., 1999), whereas dopamine affects the striatal component. Crucially, from the point of view of psychosis research, *elevated* dopamine levels in the striatum reduce reversal ability (Cools et al., 2001, 2007), and this effect appears to be mediated by $D_2$ receptors (Clatworthy et al., 2009; Groman et al., 2011). Indeed, a genetic polymorphism, which is thought to increase tonic striatal dopamine levels (best detected by $D_2$ receptors) impairs reversal, and this effect was modeled (using a model without a volatility component) as a overweighting of past experience (den Ouden et al., 2013). These findings raise the intriguing question of whether—if cognitive representations are more unstable in schizophrenia (Moutoussis et al., 2011; Schlagenhauf et al., 2014)—striatal $D_2$ hyperactivity might be an adaptive change to try to stabilize decision-making, at the expense of reduced flexibility (Adams et al., 2016)? Indeed, patients with worse delusions have lower engagement of the dorsolateral PFC with anterior PFC but greater connectivity between the former and the midbrain (Kaplan et al., 2016).

One should not oversell the potential contribution of a reversal learning impairment to delusion persistence in schizophrenia, however. Aside from cognitive factors such as reversal impairment, confirmation bias in evidence-seeking (Balzan et al., 2013) and a more generalized reduction in IQ, it is likely that effects (both negative and positive) also make huge contributions to delusion persistence; e.g., anxiety can cause someone to ruminate on one particular belief, ignore more complex alternative explanations, and seek refuge in anxiolytic safety behaviors, all of which may reinforce the original anxiogenic belief, creating a vicious circle (Freeman, 2007). In addition, social isolation—common in psychosis—is likely to prevent odd beliefs being revised by others.

# 7.5 CONCLUSIONS AND FUTURE DIRECTIONS

In this chapter we have explored how hierarchical Bayesian models (using predictive coding) can infer the causal structure of the environment. Key to this process is the accurate estimation of precision (inverse variance, or uncertainty) at each level of the model—failure to do so will change inference. Precision determines the influence of one probability distribution over another in Bayesian inference, and hence it is likely encoded by synaptic gain, which determines the influence one neuronal population can have over another. There is ample evidence of neuro-modulatory and oscillatory abnormalities, which affect synaptic gain in schizophrenia and also of an elevated E/I balance, both of which would decrease the precision of probabilistic beliefs in higher-hierarchical areas. This loss of high-level precision would lead to increased belief updating from new evidence, which is found in both perceptual and cognitive tasks in schizophrenia.

How this model relates to the formation and maintenance of delusional beliefs, and to striatal $D_2$ receptor activity, is less clear. There is certainly a relationship between JTC and delusions, but the former is only partially explained by an increased learning rate, and the latter sometimes arrive too quickly to be accounted for by any incremental learning process. Furthermore, explanations of delusion formation that rely on an increased learning rate run into difficulty when trying to account for their persistence in the face of contradictory evidence. One possibility is that people with schizophrenia have an additional impairment in reversal learning (although this may sometimes be due to too much switching rather than not enough!). Another is that the strong effect (or persistent prediction errors, Corlett et al., 2009) associated with many delusional beliefs leads to rumination and gradual reinforcement of the belief. Investigating how affect and hierarchical inference interact (Browning et al., 2015) is a crucial question for computational psychiatry research to explore.

The positive symptoms of schizophrenia are strongly associated with the elevated presynaptic availability of dopamine in the dorsal (associative) striatum (Howes and Kapur, 2009), and are reduced by $D_2$ receptor antagonists (although neither is always the case, Demjaha et al., 2012). This indicates that psychosis is characterized (in most) by increased tonic dopamine levels at $D_2$ receptors (and perhaps disordered phasic dopamine release at $D_1$ receptors); $D_1$ receptors are most sensitive to phasic bursts, whereas tonic activity and phasic pauses are best detected by $D_2$ receptors (Dreyer et al., 2010). Whilst it is known that increased E/I balance in PFC (from ketamine administration) can potentiate amphetamine-induced dopamine release (Kegeles et al., 2000), the computational reason for this remains unknown.

Maia and Frank (2017) have developed the most thorough attempt to explain delusions via a computational account of disordered striatal dopamine transmission (Maia and Frank, 2017), in which "spontaneous" (i.e., inappropriate) reward prediction errors (and a loss of appropriate reward prediction errors) cause aberrant valuation of states in the environment and aberrant gating of thoughts or actions which otherwise might not be entertained. This detailed account provides a compelling explanation for many reward learning abnormalities in schizophrenia, and also for negative symptoms (although there are also cortical contributions to each Strauss et al., 2014), but aspects of delusions remain to be accounted, for example, their existence despite (apparently) normal striatal dopamine function in some cases, and their distinctive themes—of persecution, surveillance, self-reference, etc.—which often seem beyond the realm of aberrant reward learning.

In an alternative account of the computational function of striatal dopamine, Friston et al. (2013) have proposed that it encodes the precision of beliefs about goals and actions (in its tonic signaling) and the (phasic) updates to this precision during variational message passing. Thus, abnormal inference in cortex (perhaps due to imbalanced hierarchical precision encoding) could then be reflected in and perhaps compounded by abnormal precision encoding in the striatum. This account provides an important (bidirectional) link between cortical and striatal (dys)function, and is consistent with hyperdopaminergia being sufficient but not necessary for abnormal inference (FitzGerald et al., 2015) and being associated with beliefs of unduly high precision, but much more work is required for it to explain delusions themselves. It would also be nice to have a computational explanation of why so many other risk factors for schizophrenia—e.g., social isolation or subordination, prenatal or perinatal adversity, and acute stress—cause dopamine hyperactivity (Howes and Kapur, 2009).

# References

Abi-Saab, W.M., D'Souza, D.C., Moghaddam, B., Krystal, J.H., 1998. The NMDA antagonist model for schizophrenia: promise and pitfalls. Pharmacopsychiatry 31 (Suppl. 2), 104—109.

Adams, R.A., Stephan, K.E., Brown, H.R., Frith, C.D., Friston, K.J., 2013. The computational anatomy of psychosis. Front. Psychiatry 4, 47.

Adams, R.A., Huys, Q.J.M., Roiser, J.P., January 2016. Computational psychiatry: towards a mathematically informed understanding of mental illness. J. Neurol. Neurosurg. Psychiatry 87 (1), 53—63.

Anticevic, A., Cole, M.W., Repovs, G., et al., 2014. Characterizing thalamo-cortical disturbances in schizophrenia and bipolar illness. Cereb. Cortex 24 (12), 3116—3130.

Anticevic, A., Hu, X., Xiao, Y., et al., 2015. Early-course unmedicated schizophrenia patients exhibit elevated prefrontal connectivity associated with longitudinal change. J. Neurosci. 35 (1), 267—286.

Averbeck, B.B., Evans, S., Chouhan, V., Bristow, E., Shergill, S.S., April 2011. Probabilistic learning and inference in schizophrenia. Schizophr. Res. 127 (1—3), 115—122.

Balzan, R., Delfabbro, P., Galletly, C., Woodward, T., 2013. Confirmation biases across the psychosis continuum: the contribution of hypersalient evidence-hypothesis matches. Br. J. Clin. Psychol. 52 (1), 53—69.

Botvinick, M.M., Niv, Y., Barto, A.C., 2009. Hierarchically organized behavior and its neural foundations: a reinforcement learning perspective. Cognition 113 (3), 262—280.

Browning, M., Behrens, T.E., Jocham, G., O'Reilly, J.X., Bishop, S.J., 2015. Anxious individuals have difficulty learning the causal statistics of aversive environments. Nat. Neurosci. 18 (4), 590−596.

Clarke, H.F., Walker, S.C., Crofts, H.S., Dalley, J.W., Robbins, T.W., Roberts, A.C., 2005. Prefrontal serotonin depletion affects reversal learning but not attentional set shifting. J. Neurosci. 25 (2), 532−538.

Clatworthy, P.L., Lewis, S.J.G., Brichard, L., et al., 2009. Dopamine release in dissociable striatal subregions predicts the different effects of oral methylphenidate on reversal learning and spatial working memory. J. Neurosci. 29 (15), 4690−4696.

Collins, A.G.E., Brown, J.K., Gold, J.M., Waltz, J.A., Frank, M.J., 2014. Working memory contributions to reinforcement learning impairments in schizophrenia. J. Neurosci. 34 (41), 13747−13756.

Cools, R., Barker, R.A., Sahakian, B.J., Robbins, T.W., 2001. Enhanced or impaired cognitive function in Parkinson's disease as a function of dopaminergic medication and task demands. Cereb. Cortex 11 (12), 1136−1143. N.Y. 1991.

Cools, R., Lewis, S.J.G., Clark, L., Barker, R.A., Robbins, T.W., 2007. L-DOPA disrupts activity in the nucleus accumbens during reversal learning in Parkinson's disease. Neuropsychopharmacology 32 (1), 180−189.

Corlett, P.R., Fletcher, P.C., October 2014. Computational psychiatry: a Rosetta Stone linking the brain to mental illness. Lancet Psychiatry 1 (5), 399−402.

Corlett, P.R., Krystal, J.H., Taylor, J.R., Fletcher, P.C., 2009. Why do delusions persist? Front. Hum. Neurosci. 3, 12.

Demjaha, A., Murray, R.M., McGuire, P.K., Kapur, S., Howes, O.D., 2012. Dopamine synthesis capacity in patients with treatment-resistant schizophrenia. Am. J. Psychiatry 169 (11), 1203−1210.

den Ouden, H.E.M., Daw, N.D., Fernandez, G., et al., 2013. Dissociable effects of dopamine and serotonin on reversal learning. Neuron 80 (4), 1090−1100.

Deserno, L., Boehme, R., Heinz, A., Schlagenhauf, F., 2013. Reinforcement learning and dopamine in schizophrenia: dimensions of symptoms or specific features of a disease group? Front. Psychiatry 4, 172.

Dickinson, D., Ramsey, M.E., Gold, J.M., 2007. Overlooking the obvious: a meta-analytic comparison of digit symbol coding tasks and other cognitive measures in schizophrenia. Arch. Gen. Psychiatry 64 (5), 532−542.

Dima, D., Roiser, J.P., Dietrich, D.E., et al., 2009. Understanding why patients with schizophrenia do not perceive the hollow-mask illusion using dynamic causal modelling. Neuroimage 46 (4), 1180−1186.

Donoso, M., Collins, A.G.E., Koechlin, E., 2014. Human cognition. Foundations of human reasoning in the prefrontal cortex. Science 344 (6191), 1481−1486.

Dreyer, J.K., Herrik, K.F., Berg, R.W., Hounsgaard, J.D., 2010. Influence of phasic and tonic dopamine release on receptor activation. J. Neurosci. 30 (42), 14273−14283.

Dudley, R., Taylor, P., Wickham, S., Hutton, P., 2016. Psychosis, delusions and the "jumping to conclusions" reasoning bias: a systematic review and meta-analysis. Schizophr. Bull. 42 (3), 652−665.

Elliott, R., McKenna, P.J., Robbins, T.W., Sahakian, B.J., 1995. Neuropsychological evidence for frontostriatal dysfunction in schizophrenia. Psychol. Med. 25 (3), 619−630.

Feldman, H., Friston, K.J., 2010. Attention, uncertainty, and free-energy. Front. Hum. Neurosci. 4, 215.

Fine, C., Gardner, M., Craigie, J., Gold, I., 2007. Hopping, skipping or jumping to conclusions? Clarifying the role of the JTC bias in delusions. Cogn. Neuropsychiatry 12 (1), 46−77.

FitzGerald, T.H.B., Dolan, R.J., Friston, K., 2015. Dopamine, reward learning, and active inference. Front. Comput. Neurosci. 9, 136.

Freeman, D., 2007. Suspicious minds: the psychology of persecutory delusions. Clin. Psychol. Rev. 27 (4), 425–457.

Fries, P., 2005. A mechanism for cognitive dynamics: neuronal communication through neuronal coherence. Trends Cogn. Sci. 9 (10), 474–480.

Friston, K.J., Schwartenbeck, P., Fitzgerald, T., Moutoussis, M., Behrens, T., Dolan, R.J., 2013. The anatomy of choice: active inference and agency. Front. Hum. Neurosci. 7, 598.

Friston, K.J., 2005. A theory of cortical responses. Philos. Trans. R Soc. Lond. B Biol. Sci. 360 (1456), 815–836.

Friston, K.J., 2008. Hierarchical models in the brain. PLoS Comput. Biol. 4 (11), e1000211.

Garety, P.A., Hemsley, D.R., Wessely, S., 1991. Reasoning in deluded schizophrenic and paranoid patients. Biases in performance on a probabilistic inference task. J. Nerv. Ment. Dis. 179 (4), 194–201.

Gold, J.M., Waltz, J.A., Matveeva, T.M., et al., 2012. Negative symptoms and the failure to represent the expected reward value of actions: behavioral and computational modeling evidence. Arch. Gen. Psychiatry 69 (2), 129–138.

Goldman-Rakic, P.S., Castner, S.A., Svensson, T.H., Siever, L.J., Williams, G.V., 2004. Targeting the dopamine D1 receptor in schizophrenia: insights for cognitive dysfunction. Psychopharmacology (Berl.) 174 (1), 3–16.

Groman, S.M., Lee, B., London, E.D., et al., 2011. Dorsal striatal D2-like receptor availability covaries with sensitivity to positive reinforcement during discrimination learning. J. Neurosci. 31 (20), 7291–7299.

Hong, L.E., Turano, K.A., O'Neill, H., et al., 2008. Refining the predictive pursuit endophenotype in schizophrenia. Biol. Psychiatry 63 (5), 458–464.

Howes, O.D., Kapur, S., 2009. The dopamine hypothesis of schizophrenia: version III—the final common pathway. Schizophr. Bull. 35 (3), 549–562.

Huys, Q.J.M., Lally, N., Faulkner, P., et al., March 10, 2015. The interplay of approximate planning strategies. Proc. Natl. Acad. Sci. U.S.A. 112 (10), 3098–3103.

Jardri, R., Denève, S., 2013. Circular inferences in schizophrenia. Brain J. Neurol. 136 (Pt 11), 3227–3241.

Kalman, R.A., 1960. New approach to linear filtering and prediction problems. Trans. ASME J. Basic Eng. 82 (Series D), 35–45.

Kaplan, C.M., Saha, D., Molina, J.L., et al., 2016. Estimating changing contexts in schizophrenia. Brain J. Neurol. 139 (Pt 7), 2082–2095.

Kegeles, L.S., Abi-Dargham, A., Zea-Ponce, Y., et al., 2000. Modulation of amphetamine-induced striatal dopamine release by ketamine in humans: implications for schizophrenia. Biol. Psychiatry 48 (7), 627–640.

Knill, D.C., Pouget, A., 2004. The Bayesian brain: the role of uncertainty in neural coding and computation. Trends Neurosci. 27 (12), 712–719.

Lee, S.W., O'Doherty, J.P., Shimojo, S., 2015. Neural computations mediating one-shot learning in the human brain. PLoS Biol. 13 (4), e1002137.

Leeson, V.C., Robbins, T.W., Matheson, E., et al., 2009. Discrimination learning, reversal, and set-shifting in first-episode schizophrenia: stability over six years and specific associations with medication type and disorganization syndrome. Biol. Psychiatry 66 (6), 586–593.

Lewis, D.A., Curley, A.A., Glausier, J.R., Volk, D.W., 2012. Cortical parvalbumin interneurons and cognitive dysfunction in schizophrenia. Trends Neurosci. 35 (1), 57–67.

Maia, T.V., Frank, M.J., January 1, 2017. An integrative perspective on the role of dopamine in schizophrenia. Biol. Psychiatry 81 (1), 52–66.

Mathys, C., Daunizeau, J., Friston, K.J., Stephan, K.E.A., 2011. Bayesian foundation for individual learning under uncertainty. Front. Hum. Neurosci. 5, 39.

Mayer, J.S., Park, S., 2012. Working memory encoding and false memory in schizophrenia and bipolar disorder in a spatial delayed response task. J. Abnorm. Psychol. 121 (3), 784–794.

Mirian, D., Heinrichs, R.W., McDermid Vaz, S., 2011. Exploring logical reasoning abilities in schizophrenia patients. Schizophr. Res. 127 (1–3), 178–180.

Moutoussis, M., Bentall, R.P., El-Deredy, W., Dayan, P., 2011. Bayesian modelling of jumping-to-conclusions bias in delusional patients. Cogn. Neuropsychiatry 16 (5), 422–447.

Murray, J.D., Anticevic, A., Gancsos, M., et al., 2014. Linking microcircuit dysfunction to cognitive impairment: effects of disinhibition associated with schizophrenia in a cortical working memory model. Cereb. Cortex 24 (4), 859–872. N.Y.N 1991.

Network and Pathway Analysis Subgroup of the Psychiatric Genomics Consortium, International Inflammatory Bowel Disease Genetics Consortium (IIBDGC), 2015. Psychiatric genome-wide association study analyses implicate neuronal, immune and histone pathways. Nat. Neurosci. 18 (2), 199–209.

Olney, J.W., Farber, N.B., 1995. Glutamate receptor dysfunction and schizophrenia. Arch. Gen. Psychiatry 52 (12), 998–1007.

Owen, G.S., Cutting, J., David, A.S., 2007. Are people with schizophrenia more logical than healthy volunteers? Br. J. Psychiatry J. Ment. Sci. 191, 453–454.

Pantelis, C., Barber, F.Z., Barnes, T.R., Nelson, H.E., Owen, A.M., Robbins, T.W., 1999. Comparison of set-shifting ability in patients with chronic schizophrenia and frontal lobe damage. Schizophr. Res. 37 (3), 251–270.

Pareés, I., Brown, H., Nuruki, A., et al., 2014. Loss of sensory attenuation in patients with functional (psychogenic) movement disorders. Brain J. Neurol. 137 (Pt 11), 2916–2921.

Payzan-LeNestour, E., Bossaerts, P., 2011. Risk, unexpected uncertainty, and estimation uncertainty: Bayesian learning in unstable settings. PLoS Comput. Biol. 7 (1), e1001048.

Pellicano, E., Burr, D., 2012. When the world becomes "too real": a Bayesian explanation of autistic perception. Trends Cogn. Sci. 16 (10), 504–510.

Radua, J., Schmidt, A., Borgwardt, S., et al., 2015. Ventral striatal activation during reward processing in psychosis: a neurofunctional meta-analysis. JAMA Psychiatry 72 (12), 1243–1251.

Rao, R.P., Ballard, D.H., 1999. Predictive coding in the visual cortex: a functional interpretation of some extra-classical receptive-field effects. Nat. Neurosci. 2 (1), 79–87.

Robbins, T.W., 2007. Shifting and stopping: fronto-striatal substrates, neurochemical modulation and clinical implications. Philos. Trans. R Soc. Lond. B Biol. Sci. 362 (1481), 917–932.

Rogers, R.D., Blackshaw, A.J., Middleton, H.C., et al., 1999. Tryptophan depletion impairs stimulus-reward learning while methylphenidate disrupts attentional control in healthy young adults: implications for the monoaminergic basis of impulsive behaviour. Psychopharmacology (Berl.) 146 (4), 482–491.

Schlagenhauf, F., Huys, Q.J.M., Deserno, L., et al., April 1, 2014. Striatal dysfunction during reversal learning in unmedicated schizophrenia patients. Neuroimage 89, 171–180.

Schmack, K., Gòmez-Carrillo de Castro, A., Rothkirch, M., et al., 2013. Delusions and the role of beliefs in perceptual inference. J. Neurosci. 33 (34), 13701–13712.

Shergill, S.S., Samson, G., Bays, P.M., Frith, C.D., Wolpert, D.M., 2005. Evidence for sensory prediction deficits in schizophrenia. Am. J. Psychiatry 162 (12), 2384–2386.

Silverstein, S.M., Keane, B.P., 2011. Perceptual organization impairment in schizophrenia and associated brain mechanisms: review of research from 2005 to 2010. Schizophr. Bull. 37 (4), 690–699.

Speechley, W.J., Whitman, J.C., Woodward, T.S., 2010. The contribution of hypersalience to the "jumping to conclusions" bias associated with delusions in schizophrenia. J. Psychiatry Neurosci. JPN 35 (1), 7–17.

Stephan, K.E., Baldeweg, T., Friston, K.J., 2006. Synaptic plasticity and dysconnection in schizophrenia. Biol. Psychiatry 59 (10), 929–939.

Strauss, G.P., Waltz, J.A., Gold, J.M., 2014. A review of reward processing and motivational impairment in schizophrenia. Schizophr. Bull. 40 (Suppl. 2), S107–S116.

Teufel, C., Subramaniam, N., Dobler, V., et al., 2015. Shift toward prior knowledge confers a perceptual advantage in early psychosis and psychosis-prone healthy individuals. Proc. Natl. Acad. Sci. U.S.A. 112 (43), 13401–13406.

Turetsky, B.I., Calkins, M.E., Light, G.A., Olincy, A., Radant, A.D., Swerdlow, N.R., 2007. Neurophysiological endophenotypes of schizophrenia: the viability of selected candidate measures. Schizophr. Bull. 33 (1), 69–94.

Waltz, J.A., Gold, J.M., 2007. Probabilistic reversal learning impairments in schizophrenia: further evidence of orbitofrontal dysfunction. Schizophr. Res. 93 (1–3), 296–303.

Waltz, J.A., Kasanova, Z., Ross, T.J., et al., 2013. The roles of reward, default, and executive control networks in set-shifting impairments in schizophrenia. PLoS One 8 (2), e57257.

Yu, A.J., Dayan, P., 2005. Uncertainty, neuromodulation, and attention. Neuron 46 (4), 681–692.

Yuille, A., Kersten, D., 2006. Vision as Bayesian inference: analysis by synthesis? Trends Cogn. Sci. 10 (7), 301–308.

# CHARACTERIZING COMPLEX PSYCHIATRIC SYMPTOMS VIA MATHEMATICAL MODELS

# 8

# A Case Study in Computational Psychiatry: Addiction as Failure Modes of the Decision-Making System

Cody J. Walters, A.D. Redish

University of Minnesota, Minneapolis, MN, United States

## OUTLINE

Because addiction is so hard to define, the DSM-IV defined drug dependence and avoided the word *addiction* (DSM-IV-TR, 2000). However, more recent studies have suggested that addiction-like behaviors can underlie nondrug decision problems as well (Holden, 2001; Schüll, 2012;

*Computational Psychiatry*
http://dx.doi.org/10.1016/B978-0-12-809825-7.00008-0

**199**

Robbins and Clark, 2015). But then we run into the problem of whether all continued behaviors are addictions. Do we really want to say that Brett Favre was "addicted" to football because he continued playing long after the game had damaged his body? Do we really want to say that Osip Mandelstam was "addicted" to poetry because he continued to write even after Stalin had sent him to the Gulag (where he eventually died)? To avoid this difficult definition, we will unask the question and instead concentrate on specific decision-making errors and relate that to problematic behaviors often categorized as addiction (Redish et al., 2008; Heyman, 2009, 2013; Redish, 2013).

Current models of psychiatry suggest that psychiatric disorders should be defined in terms of "harmful dysfunction" (Wakefield, 2007). This definition includes a scientific component (dysfunction) and a sociological component (harm). For example, illiteracy is harmful but is not usually considered a brain dysfunction. (On the other hand, dyslexia is both harmful and a brain dysfunction (Norton et al., 2015; Jaffe-Dax et al., 2015).) Synesthesia is due to a brain dysfunction but is not generally considered harmful (Cytowic, 1998). Treatment needs to be predicated on fixing those things that are harmful, but the appropriate treatment depends on the dysfunction. For example, most clinicians do not feel a need to treat synesthesia, but both dyslexia and illiteracy need treatment. Both of these problems require treatment, but because the causes are different, treatments for illiteracy and dyslexia will likely need to be different. In this chapter, we will make the case that addiction is a symptom, not a disease, and that because the underlying causes (the underlying dysfunctions) for addiction are varied the necessary treatments must be varied (Redish et al., 2008). We will make the case that rather than categorizing subjects in terms of their addiction (cocaine addiction, heroin addiction, gambling addiction), we should be defining them in terms of their decision-making dysfunctions (overvaluation, errors in expectation, reactions to anxiety). Lastly, we will argue that treatments should be guided by identifying the underlying dysfunction in an addict's decision-making circuitry to allow clinicians to individualize treatments.

A key concept that we will build this chapter on is that of a *vulnerability* or *failure mode*—a breakdown in a process due to a malfunctioning component. These terms come from the field of reliability engineering where one tries to identify the underlying breakdown that has caused a system-wide failure. A flat tire, for example, is a *failure mode* of automobiles (and bicycles) because a tire is a thin rubber tube filled with air. If that tube becomes punctured, then the air leaks out and the car is no longer riding on the normal air cushion. On the other hand, tank treads (which do not ride on air) are not susceptible to going flat (although they are vulnerable to being split). Just as cars and tanks have different failure modes that depend on their underlying mechanisms, so too do decision-making systems.

# 8.1 THE MACHINERY OF DECISION-MAKING

To understand how decision-making systems can go wrong, one needs to understand the fundamental mechanisms by which they work. Ultimately, decisions are about interactions with the world or changes in one's behavior that affect the world. For completeness, we will include both actions as visceral changes (heart rate, thermal regulation, hormone levels) as well as external, physical actions (pushing a button, lighting a cigarette, going to some location, signing on a dotted line). By this definition, ultimately, all decisions entail taking an action. At the point where we have defined decision as action selection, we are now in the domain of computational information processing—all decisions are, ultimately, a consequence of processing information about one's present circumstances (perception), information about one's past (memory), and information about one's needs/goals (motivation); however, as we will see below, these components do not have to be explicitly represented to be a part of the process—sometimes they can be hidden within the process itself.

Importantly, how information is stored changes how easy it is to access and how it generalizes to new situations (Redish and Mizumori, 2015). For example, if you are looking for a specific book to cite on a topic, it will be easiest to find if your bookshelf is sorted by a library catalog system, such as the Dewey decimal or Library of Congress system, rather than if you have sorted your books by size. However, if you are looking for a book to level out the table with a short leg, then sorting by size is going to get you to your target faster. This point is one of the main discoveries of computer science in the last century—data structures matter (Cormen et al., 1990).

In the same way that data structures matter in a digital computer, so too do the information processes that are being used to select an action. Current taxonomies have suggested that there are four key action-selection information processing systems that store (and generalize) information differently (Rangel et al., 2008; van der Meer et al., 2012): **Reflexes, Pavlovian action-selection systems, Procedural decision-making systems**, and **Deliberative decision-making systems**. Each of these systems uses different computational information processing mechanisms through different brain structures and thus has different failure modes (Montague et al., 2012; Redish, 2013).

- **Reflexes** select actions based on immediately available perceptions and have stored an appropriate action in the spinal circuitry (Sherrington, 1906; Eaton, 1984), which was learned by the species through genetic variation and selection (a genetic learning algorithm). Both the memory and the motivational components are hardwired into the circuitry through genetically controlled wiring.

- **Pavlovian action-selection systems** entail species-specific actions that one learns when to release (Rangel et al., 2008; van der Meer et al., 2012). At first, these relationships are hardwired (salivate when presented with food), but with experience, one can learn to take these actions in response to predictive cues (salivate when you hear the dinner bell, Pavlov, 1927). Here, the motivational components remain in the circuitry through genetic learning, but the predictive stimuli can be learned (perception/memory) (Rescorla, 1988; Domjan, 1998).
- **Procedural action-selection systems** entail learning an arbitrary action sequence that one can release in an appropriate situation (hitting a fastball) (Squire, 1987; Mishkin and Appenzeller, 1987; Saint-Cyr et al., 1988). Here the motivational components are cached in the circuitry once learned, but they are learned through individual experience. Importantly, Procedural systems require learning in the perceptual system (situation recognition) that goes well beyond stimulus recognition (such as learning to differentiate a fastball from a curveball and being able to identify where and when the ball will cross the plate, McClelland and Rogers, 2003; Redish et al., 2007; Gershman and Niv, 2010).
- Finally, **Deliberative action-selection systems** entail an explicit imagination, evaluation, and planning process by which one creates a simulated (hypothetical) future (imagining the consequences of one's actions) and then evaluates that simulated future to select the best action (Gilbert and Wilson, 2007; Buckner and Carroll, 2007; Johnson et al., 2007). Interestingly, current models of imagination suggest that the same perceptual systems are reused for imagination (Kosslyn, 1994; O'Craven and Kanwisher, 2000), which implies that deliberation will depend on the same situation recognition processes as procedural learning (Pearson et al., 2015). Similarly, current models suggest that the deliberative process uses structures originally evolved to measure current rewards and punishments (this cake tastes good, that wool shirt is itchy) to evaluate the imagined worlds (Andrade and Ariely, 2009; Phelps et al., 2014). This explains why current emotional states can affect one's decisions about the future (like why you buy more food at the grocery store when you are hungry).

## 8.2 ADDICTION AS FAILURE MODES OF DECISION-MAKING SYSTEMS

A corollary of having multiple decision-making systems is that there are multiple ways for those systems to fail. Failure can occur at multiple targets in any given decision-making system, and each failure point can generate a subtly different behavioral phenotype (Table 8.1).

TABLE 8.1    Some Failure Modes of the Decision-Making System That Can Lead to Addiction. Obviously Incomplete

| Failure-Point | Clinical Consequence |
|---|---|
| Changing allostatic set points (Koob and Le Moal, 2006; Koob and Volkow, 2010) | Physiological needs, craving |
| Cue-outcome associations elicit prewired visceral actions (Damasio, 1994; Bechara and Damasio, 2002; Bechara, 2005) | Incorrect action selection, craving |
| Escape from negative emotions (Koob, 2009) | Incorrect action selection |
| Mimicking reward (Volkow et al., 2002; Wise, 2005; Dezfouli et al., 2009) | Incorrect action selection, craving |
| Errors in expected outcomes (Goldman et al., 1999; Jones et al., 2001; Redish and Johnson, 2007) | Incorrect action selection |
| Increased likelihood of retrieving a specific expected action-outcome path (Redish and Johnson, 2007) | Obsession |
| Overvaluation of expected outcomes (Robinson and Berridge, 2001, 2003) | Incorrect action selection |
| Overvaluation of learned actions (Di Chiara, 1999; Redish, 2004) | Automated, robotic drug use |
| Timing errors (Ross, 2008) | Preferences for unpredictable events |
| Overfast discounting processes (Bickel and Marsch, 2001; Bickel and Mueller, 2009) | Impulsivity |
| Changes in learning rates (Franken et al., 2005; Gutkin et al., 2006; Redish et al., 2008; Piray et al., 2010) | Excess drug-related cue associations |
| Selective inhibition of the deliberative system (Bernheim and Rangel, 2004; Bechara, 2005; Bickel et al., 2008, 2012; Baumeister and Tierney, 2011) | Fast development of habit learning |
| Selective excitation of the habit system (Everitt and Robbins, 2005; Bickel et al., 2008; Keramati and Gutkin, 2013) | Fast development of habit learning |
| Misclassification of situations: overcategorization (Redish et al., 2007) | Illusion of control, hindsight bias |
| Misclassification of situations: overgeneralization (Redish et al., 2007) | Perseveration in the face of losses |

*Modified from Redish, A.D., 2013. The Mind within the Brain: How We Make Decisions and How Those Decisions Go Wrong, Oxford Univ. Press, Oxford, UK.*

Biological organisms, for instance, actively regulate crucial biological parameters homeostatically (Mayr, 1998). Although at any given moment there is a set value that the organism will attempt to maintain, this set point can and does vary as a function of context (allostasis, Koob and Le Moal, 2006). Drugs of abuse are capable of altering an individual's natural set point and thus changing the biological needs of an organism (Meyer and Mirin, 1979; Benowitz, 1996; Koob and Le Moal, 2006). Cessation of drug use can thus disrupt the new drug-induced set point and result in withdrawal. These reflex-driven withdrawal symptoms would lead to highly negative sensations that require relief, which can drive drug seeking from multiple decision systems, including both Pavlovian and Deliberative. Importantly, however, there are other failure modes that can also drive drug seeking long after withdrawal symptoms have been eliminated; withdrawal and craving are dissociable (Childress et al., 1988).

A second well-studied failure mode can arise from cues that have come to predict upcoming drug administration, which can activate compensatory mechanisms in an addict. Heroin addiction offers a striking illustration of this system at work (Meyer and Mirin, 1979): When an addict prepares to administer the drug in the same setting that the drug is typically taken, physiological mechanisms (enzyme changes, modulation of receptor kinetics) will prepare the user's body for an upcoming dose and thus temporarily provide the individual with heightened tolerance. However, if the drug is taken in a novel setting the user is liable to overdose due to a failure of these compensatory Pavlovian mechanisms to provide that conditioned tolerance. Similarly, alcoholics have reduced alcohol-related coordination deficits in bars and other places where they expect to drink than in nonalcohol-associated environments (such as offices) (Hunt, 1998).

Because deliberative systems depend on evaluation circuits that evolved to evaluate ongoing needs (Phelps et al., 2014; Redish, 2016), expectations of future outcomes can depend on immediate needs. Thus, the compensation processes that occur on cue delivery can drive positive evaluations of drug-related outcomes (providing relief from the allostatic shifts, Koob, 2009). Because the recall of memory from a search process depends on recall and framing components (Redish and Johnson, 2007; Winstanley et al., 2012), it can also be guided toward these reminded outcomes. For example, video poker machines at the entrance to a grocery store can cue a whole imagined scenario of potential game playing (Schüll, 2012). This is a process termed "Pavlovian-to-Instrumental" transfer (Kruse et al., 1983; Talmi et al., 2008).

Moving beyond cue associations, the ability to encode refined cached action chains (i.e., habits) that are released in the appropriate situation is crucial to many forms of expertise (Graybiel, 1995; Klein, 1999). Current theories suggest that this system entails the recognition of situations and the release of cached actions (Daw et al., 2005; Dezfouli and Balleine, 2012; van der Meer et al., 2012). However, these learned situation-action sequences are inflexible (because they evolved to respond quickly) and can turn maladaptive but well-practiced behaviors into tenacious habits. Whereas Pavlovian-to-Instrumental transfer can allow cues to increase the likelihood of deliberative systems to drive behavior toward drug seeking, a failure in the cached action system will make it such that, on recognizing and categorizing a situation, an inflexible and automated (potentially drug-related) action sequence will be released. For example, a smoker who has made it a matter of mindless routine to light up a cigarette first thing every morning (Tiffany, 1990) or the video poker player who gets lost in the flow of the game (Schüll, 2012) are two examples of these learned situation-action procedural mechanisms gone awry.

Higher level cognitive dysfunctions are also major contributors to drug abuse and relapse. To plan for the future, for example, an agent must evaluate available actions and their expected outcomes (Redish, 2016). Drugs of abuse often disrupt this planning and evaluation process, which leads to distorted outcome expectations (Goldman et al., 1987, 1999; Jones et al., 2001; Oscar-Berman and Marinkovic, 2003). The orbitofrontal cortex and nucleus accumbens are both key structures implicated in outcome evaluation (O'Doherty, 2004; McDannald et al., 2011; van der Meer et al., 2012) and a failure to receive, process, or generate the appropriate signal in these structures would negatively impact behavior because of a mis-valuation of expected outcomes. Both orbitofrontal cortex and nucleus accumbens are often disrupted in drug users (Carelli and Wondolowski, 2003; Schoenbaum et al., 2006; Kourrich and Thomas, 2009; Koob and Volkow, 2010).

Many theories suggest that the evaluation of rewards in some systems, particularly the Pavlovian and Procedural systems, is due to changes in dopamine release (Montague et al., 1996; Schultz et al., 1997). Dopamine signaling increases to unexpected rewards, and dopamine neurons shift their firing to earlier cues that reliably predict those rewards (Schultz and Dickinson, 2000). Correctly predicted rewards produce flat rates of dopamine spiking, whereas expected but undelivered rewards produce a decrease in dopamine signaling. These three components mean that dopamine could signal "reward prediction errors," which can be used to train reactive decision systems such as Pavlovian or Procedural systems by driving estimates of value in the direction of the predicted error (Rescorla and Wagner, 1972; Montague et al., 1996; Sutton and Barto, 1998; Schultz and Dickinson, 2000). Redish (2004) noted that if a pharmacological agent (such as a drug) provided dopamine in a way that bypassed the normal neural calculations, it would lead to addictive behaviors because the value would be driven to infinity. (Of course, neuronal representations would have to renormalize value, which could explain why nondrug rewards can lose their value in the face of extensive drug experiences, Goldstein, 2000; Heyman, 2009.) Importantly, this would only be one of many potential failure modes that could lead to addictive behaviors (Redish et al., 2008), and dopamine neurons signal other information as well as prediction errors (Bromberg-Martin et al., 2010). Nevertheless, the excess reward prediction error theory predicted that drugs of abuse should not show Kamin blocking, a phenomenon whereby predictive cues are not associated with already predicted stimuli (Redish, 2004). Interestingly, several subsequent experiments found that animals do show Kamin blocking using cocaine delivery as a reward (Panlilio et al., 2007; Marks et al., 2010). However, a further study found that while most animals showed Kamin blocking in a nicotine access experiment, the subset of animals that showed uncontrollable nicotine seeking did not (Jaffe et al., 2014). This elucidates one of the main points of this chapter: drug seeking can occur due to many potential failure modes; different individuals may have different reasons for their drug seeking. Treatment will need to identify the active failure mode to successfully treat addicts.

The ability to mentally construct imagined futures and play out competing scenarios to predict the anticipated value of a potential action confers a great advantage when making important "one-time-only" decisions (e.g., mentally simulating and comparing which of two job offers to accept, Gilbert and Wilson, 2007; Redish, 2016). However, these computational processes also have failure modes endemic to them. For example, either a miscalculation of the anticipated outcome or a misevaluation of a correctly anticipated outcome would lead to dysfunctional decision-making. The former can drive obsession and craving (Redish and Johnson, 2007), whereas the latter can drive decisions that will lead to negative outcomes. These can be seen in the heroin user looking for the orgasmic high of the first hit (Meyer and Mirin, 1979), the gambler trying to recreate the one time they won big at the machine (Lesieur, 1977; Custer, 1984) or the smoker underestimating the likelihood of getting cancer (Weinstein et al., 2005).

The etiology of addiction is not always driven by deficits in reward networks. Indeed, substance abuse has a high comorbidity rate with neuropsychiatric states. A well-studied example is that of alcoholism and its relation to anxiety disorders and major depression. In alcohol rehabilitation clinics alone, 50% of patients are diagnosed with either an anxiety disorder or major depression and these patients are twice as likely as their noncomorbid counterparts to relapse after leaving the clinic (Hobbs et al., 2011; Schadé et al., 2005). Although the comorbidity of psychiatric disease and substance abuse is established, the causal relation between the two is a matter of debate. Intriguingly, a metaanalysis of epidemiological surveys and field studies found that comorbidity of alcoholism with anxiety was dependent on the type of anxiety disorder that the patient had—while agoraphobia and social anxiety were found to be risk factors for developing alcoholism, panic disorder and generalized anxiety were found to result from alcoholism (Hall, 1990). It would appear then that the causal relationship between anxiety and pathological alcohol consumption is bidirectional: pathological anxiety is a risk factor for alcohol abuse, but long-term alcohol use has the potential to induce pathological anxiety (Kushner et al., 2000).

A common argument for anxiety driving alcohol abuse is that individuals suffering from pathological anxiety might resort to alcohol as a means of self-medication (Quitkin et al., 1972). It has been suggested that the pharmacological profile of alcohol is such that it alleviates anxiety in a similar fashion as commonly prescribed anxiolytic compounds such as benzodiazepines and barbiturates (Liljequist and Engel, 1984). In support of this view, it has been shown that cross-tolerance occurs with alcohol and anxiolytics, thus highlighting a potential shared mechanism.

In contrast to the anxiolytic effects seen with acute alcohol administration, chronic alcohol use produces long-term changes in $GABA_A$ inhibitory receptors and in NMDA-sensitive glutamatergic receptors (Valenzuela and Harris, 1997; Littleton, 1998; Hunt, 1998) and is anxiogenic (Coffman and Petty, 1985; Tran et al., 1981). One explanation for why alcohol abuse could result in the development of an anxiety disorder hinges on the effect alcohol has on Deliberative and Pavlovian systems. Alcohol, for example, specifically impairs hippocampal and prefrontal function (Hunt, 1998; White, 2003; Oscar-Berman and Marinkovic, 2003), which could shift the balance from Deliberative to more Pavlovian and Procedural systems. Importantly, early alcohol use could depend on cognitive and social expectations (Goldman et al., 1987, 1999; Bobo and Husten, 2000), whereas later use may depend on dysfunctions in Pavlovian and Procedural systems (Dickinson et al., 2002; Oei and Baldwin, 2002).

It is important to note that the various decision-making mechanisms identified above interact to generate behavior. Failure in one system can affect other systems (such as evaluation errors affecting deliberative systems), but also a working system can be used to drive behavior when another system is dysfunctional. For example, social factors certainly play a part in the potential for alcoholism to induce anxiety, with social ramifications of alcohol abuse such as divorce and unemployment undoubtedly acting as potent anxiogenic stressors. We will see additional examples below.

## 8.3 BEYOND SIMPLE FAILURE MODES

Neuropsychiatric symptoms often result in unhealthy and unsafe behaviors that themselves drive the expression of new symptoms (Borsboom and Cramer, 2013). Symptoms that tend to cooccur can then causally influence one another (sleep loss → fatigue → loss of interest, etc.). As such, an initial external event (e.g., a debilitating physical injury) is capable of triggering a symptom network (e.g., injury → stress → depressed mood → insomnia → impaired attention → etc.). Once activated, a symptom network might itself be diagnosed as a mental disorder. The degree to which neuropsychiatric states are the result of internally driven defects in the neural circuitry or the externally imposed ramifications of initial symptoms is a topic of debate. As has often been found with such scientific debates (e.g., nature vs. nurture), it is likely a combination of the two.

## 8.4 RELIABILITY ENGINEERING

Broadly, reliability engineering refers to a collection of methods designed to minimize the likelihood of a system failing. To address this issue, one identifies the potential failures of the components and asks how those potential failure modes would affect the function of the system as a whole. This deductive, top-down approach is known as fault tree analysis (MacDonald et al., 2016). In fault tree analysis, the relationship between elements in the system and the ramifications of a failure in any one element on the system as a whole are evaluated using probabilistic causal networks (Pearl, 1988, 2009).

Applying this systems engineering outlook to the nervous system has recently emerged as a valuable tool capable of providing insight into the etiology of mental illnesses (MacDonald et al., 2016; Redish and Gordon, 2016). Specifically, by identifying the relationship between neural circuits and the points at which they are susceptible to failure, reliability engineering offers a toolbox of techniques for predicting the underlying cause of a neuropsychiatric disease (Flagel et al., 2016). As a result, more effective and individualized treatments can then be designed to address an individual's specific constellation of network failures that underpin their specific neuropsychiatric phenotype.

Just as with any other clinical condition, neuropsychiatric disorders are often identified on the basis of outwardly observable symptoms that are thought to reflect an underlying physiological deficit. The unobservable biological dysfunctions that generate the observable symptoms are known as latent variables because, despite being the direct cause of the pathological symptoms, they are hidden from view.

The aim of generating a fault tree for a given neuropsychiatric disease is then to identify all the distinct latent variables (and relations between them) contributing to the disease state. With this tool, a clinician can understand the pattern of potential dysfunctional components in the system that could result in a given patient's symptoms. Seeing as there are multiple combinations of latent variable defects that can result in the same symptomatology, an inductive principle that could be used to make increasingly accurate predictions about the most likely cause of a given disorder would be useful. Such a principle for making claims about uncertain variables does in fact exist and can be found in the mathematical rules governing Bayesian inference (Pearl, 1988, 2009).

## 8.5 IMPLICATIONS FOR TREATMENT

The goal of treatment is to reduce the harm underlying the "harmful dysfunction" discussed in the opening of this chapter. For example, a number of treatments have been aimed at attempting to reduce identifiable dysfunctions occurring in addicts, such as treatments to mitigate the effect of heroin on the mu-opioid receptor in some heroin users (Meyer and Mirin, 1979), treatments to slowly ramp down the changed set point in some nicotine users (Hanson et al., 2003), as well as treatments that make imbibing alcohol unpleasant (Wright and Moore, 1990), and treatments to extinguish the cigarette-nicotine association with denicotinized cigarettes (Johnson et al., 2004; Buchhalter et al., 2005). By identifying the underlying failure modes, we can move toward personalized treatments that attempt to address the individual dysfunctional mechanisms that are active in any given user.

However, it is also possible to provide compensatory mechanisms that can alleviate the harm, even without treating the dysfunction itself. For example, eyeglasses reliably treat the harmful dysfunction of nearsightedness without actually repairing the dysfunctional lenses. Presumably, the reason that organisms evolved multiple decision-making systems is that they make more optimal decisions under different conditions. By switching to the most effective system in a given situation, an organism could outcompete other organisms trying to use a single information processing algorithm for all situations. Just as we saw in the shelving example above, where it was easier to find the next book in a series if it was organized by author, if we had multiple indices that provided pointers to where the book was, perhaps one index of subject classification and another of size, then we could use the subject classification index when we wanted to locate a book on a topic and the size index when we needed a book to stabilize the table. Which decision-making algorithm will control behavior depends on a number of incompletely understood factors, but one factor is the situation that one is in. This means that it could be possible to change the situation and, as a result of this change in situation, change one's addictive behavior.

For example, one strategy for coping with addiction is to eliminate exposure to cues that are known to trigger maladaptive behavior by precommitment (Ainslie, 1992; Kurth-Nelson and Redish, 2012a). Crucial to this method is the notion of shifting between decision-making systems (Kurth-Nelson and Redish, 2010; Kurzban, 2010; Redish, 2013). By preventing oneself from having the option to engage in addictive behavior ahead of time, an addict can precommit to a choice (say via the Deliberative system) that allows them to avoid placing themselves in temptation's path (that might trigger a Pavlovian and Procedural action). For example, an alcoholic might decide ahead of time to avoid walking down the street that has a liquor store on the way home from work or a gambler might avoid driving by a casino. (This can be difficult if there are video poker machines in every store, even the grocery store, Schüll, 2012.) This strategy of precommitment is a commonly employed and often effective method for minimizing exposure to cues that trigger impulsive and addictive behaviors. Fundamentally, it depends on the existence of multiple value functions, such as would occur with multiple, competing decision-making systems (Kurth-Nelson and Redish, 2010).

A related strategy for overcoming compulsive behavior is known as bundling (Ainslie, 1992; Kurth-Nelson and Redish, 2012b). Bundling entails changing the space of potential outcomes, usually by looking beyond a single choice. Effectively, bundling is a way of saying "doing this will lead to that." For example, an alcoholic acknowledging that there is no such thing as "just one drink" realizes that if they choose to drink, they will end up drinking to excess. This knowledge changes the value of the two options (drink or do not) relative to having a third (now unavailable) option of drinking "just one drink." This simple reestimation of the space of potential outcomes can help individuals step out of the vicious cycle of distorted expectations and destructive behavior.

Another commonly employed method for breaking addictive behavior is called contingency management (Higgins et al., 2002; Petry, 2011). Contingency management introduces a reward system that serves as an alternative to the reward that an addict would obtain from engaging in their maladaptive behavior. Addicts, by remaining abstinent, earn prizes or credits that can be used to purchase goods, and it has been shown that this promise of future reward will often incentivize them to remain abstinent.

Although current hypotheses describe the reasons for contingency management's success in terms of alternate reinforcers and lost opportunity costs (Higgins et al., 2004; Bickel et al., 2007; Packer et al., 2012),

Regier and Redish (2015) did a comparison of expected decreases in drug use relative to the actual alternative compensations provided in contingency management. We found that contingency management worked much better than expected.

One hypothesis to explain this surprising effectiveness of contingency management is that the promise of earning a delayed reinforcer if the addict remains abstinent nudges the individual into using deliberative processes rather than more reactive processes. This hypothesis predicts that contingency management success rates will be positively correlated with the integrity of the abilities (such as executive function processes) that underlie deliberation. This implies that cognitive tests that measure the viability of an individual's Deliberative system could be used to determine whether a patient is a good candidate for contingency management treatment. Furthermore, if the patient's Deliberative system was also compromised, perhaps executive function training could be used beforehand to prepare a patient for contingency management.

## 8.6 CONCLUSIONS

Multiple decision-making systems coexist and interact with one another to generate complex behavior. These decision-making systems and their interactions are vulnerable to distinct failure modes that can provide multiple paths to addiction. A computational understanding of decision-making circuitry offers the promise of a powerful tool that can be of tremendous value to both researchers and clinicians. Clinically, a more thorough appreciation of addiction mechanisms, from the underlying computations that neural circuits perform to how deficits in those neural circuits relate to clinical phenotypes, can inform the design of more effective treatments. A deeper understanding of which decision-making system vulnerabilities give rise to which clinical phenotypes will lead to more accurate methods for identifying, categorizing, and treating addictive dysfunction in an increasingly meaningful and patient-specific fashion.

## References

Ainslie, G., 1992. Picoeconomics. Cambridge Univ. Press.
Andrade, E.B., Ariely, D., 2009. The enduring impact of transient emotions on decision making. Organ. Behav. Hum. Decis. Process. 109, 1—8.
Baumeister, R.F., Tierney, J., 2011. Willpower: Rediscovering the Greatest Human Strength. Penguin Press.
Bechara, A., 2005. Decision making, impulse control and loss of willpower to resist drugs: a neurocognitive perspective. Nat. Neurosci. 8, 1458—1463.
Bechara, A., Damasio, H., 2002. Decision-making and addiction (part I): impaired activation of somatic states in substance dependent individuals when pondering decisions with negative future consequences. Neuropsychologia 40, 1675—1689.

Benowitz, N.L., 1996. Pharmacology of nicotine: addiction and therapeutics. Annu. Rev. Pharmacol. Toxicol. 36, 597–613.

Bernheim, B.D., Rangel, A., 2004. Addiction and cue-triggered decision processes. Am. Econ. Rev. 94, 1558–1590.

Bickel, W.K., Jarmolowicz, D.P., Mueller, E.T., Gatchalian, K.M., McClure, S.M., 2012. Are executive function and impulsivity antipodes? A conceptual reconstruction with special reference to addiction. Psychopharmacology 221, 361–387.

Bickel, W.K., Marsch, L.A., 2001. Toward a behavioral economic understanding of drug dependence: delay discounting processes. Addiction 96, 73–86.

Bickel, W.K., Miller, M.L., Yi, R., Kowal, B.P., Lindquist, D.M., Pitcock, J.A., 2007. Behavioral and neuroeconomics of drug addiction: competing neural systems and temporal discounting processes. Drug Alcohol Depend. 90, S85–S91.

Bickel, W.K., Mueller, E.T., 2009. Toward the study of trans-disease processes: a novel approach with special reference to the study of co-morbidity. J. Dual Diagn. 5, 131–138.

Bickel, W.K., Yi, R., Kowal, B.P., Gatchalian, K.M., 2008. Cigarette smokers discount past and future rewards symmetrically and more than controls: is discounting a measure of impulsivity? Drug Alcohol Depend. 96, 256–262.

Bobo, J.K., Husten, C., 2000. Sociocultural influences on smoking and drinking. Alcohol Res. Health 24, 225–232.

Borsboom, D., Cramer, A.O., 2013. Network analysis: an integrative approach to the structure of psychopathology. Annu. Rev. Clin. Psychol. 9, 91–121.

Bromberg-Martin, E.S., Matsumoto, M., Nakahar, H., Hikosaka, O., 2010. Multiple timescales of memory in lateral habenula and dopamine neurons. Neuron 67, 499–510.

Buchhalter, A.R., Acosta, M.C., Evans, S.E., Breland, A.B., Eissenberg, T., 2005. Tobacco abstinence symptom suppression: the role played by the smoking-related stimuli that are delivered by denicotinized cigarettes. Addiction 100, 550–559.

Buckner, R.L., Carroll, D.C., 2007. Self-projection and the brain. Trends Cogn. Sci. 11, 49–57.

Carelli, R.M., Wondolowski, J., 2003. Selective encoding of cocaine versus natural rewards by nucleus accumbens neurons is not related to chronic drug exposure. J. Neurosci. 23, 11214–11223.

Childress, A.R., McLellan, A.T., Ehrman, R., O'Brien, C.P., 1988. Classically conditioned responses in opioid and cocaine dependence: a role in relapse? NIDA Res. Monogr. 84, 25–43.

Coffman, J.A., Petty, F., 1985. Plasma GABA levels in chronic alcoholics. Am. J. Psychiatry 142, 1204–1205.

Cormen, T.H., Leiserson, C.E., Rivest, R.L., 1990. Introduction to Algorithms. MIT Press and McGraw-Hill, Cambridge, MA and New York, NY.

Custer, R.L., 1984. Profile of the pathological gambler. J. Clin. Psychiatry 45, 35–38.

Cytowic, R.E., 1998. The Man Who Tasted Shapes. MIT Press.

Damasio, A., 1994. Descartes' Error: Emotion, Reason, and the Human Brain. Quill Press.

Daw, N.D., Niv, Y., Dayan, P., 2005. Uncertainty-based competition between prefrontal and dorsolateral striatal systems for behavioral control. Nat. Neurosci. 8, 1704–1711.

Dezfouli, A., Balleine, B., 2012. Habits, action sequences and reinforcement learning. Eur. J. Neurosci. 35, 1036–1051.

Dezfouli, A., Piray, P., Keramati, M.M., Ekhtiari, H., Lucas, C., Mokri, A., 2009. A neurocomputational model for cocaine addiction. Neural Comput. 21, 2869–2893.

Di Chiara, G., 1999. Drug addiction as dopamine-dependent associative learning disorder. Eur. J. Pharmacol. 375, 13–30.

Dickinson, A., Wood, N., Smith, J.W., 2002. Alcohol seeking by rats: action or habit? Q. J. Exp. Psychol. Sect. B 55, 331–348.

Domjan, M., 1998. The Principles of Learning and Behavior, fourth ed. Brooks/Cole.

DSM-IV-TR, 2000. Diagnostic and Statistical Manual of Mental Disorders. American Psychiatric Association.

Eaton, R.C. (Ed.), 1984. Neural Mechanisms of Startle Behavior. Springer.

Everitt, B.J., Robbins, T.W., 2005. Neural systems of reinforcement for drug addiction: from actions to habits to compulsion. Nat. Neurosci. 8, 1481−1489.

Flagel, S.B., Pine, D.S., Ahmari, S.E., First, M.B., Friston, K.J., Mathys, C., Redish, A.D., Schmack, K., Smoller, J., Thapar, A., 2016. A novel framework for improving psychiatric diagnostic nosology. In: Redish, A.D., Gordon, J.A. (Eds.), Computational Psychiatry: New Perspectives on Mental Illness. MIT Press, Cambridge, MA. Strüngmann Forum Reports.

Franken, I.H., Booij, J., van den Brink, W., 2005. The role of dopamine in human addiction: from reward to motivated attention. Eur. J. Pharmacol. 526, 199−206.

Gershman, S.J., Niv, Y., 2010. Learning latent structure: carving nature at its joints. Curr. Opin. Neurobiol. 20, 251−256.

Gilbert, D.T., Wilson, T.D., 2007. Prospection: experiencing the future. Science 317, 1351−1354.

Goldman, M.S., Boca, F.K.D., Darkes, J., 1999. Alcohol expectancy theory: the application of cognitive neuroscience. In: Leonard, K.E., Blane, H.T. (Eds.), Psychological Theories of Drinking and Alcoholism. Guilford, pp. 203−246.

Goldman, M.S., Brown, S.A., Christiansen, B.A., 1987. Expectancy theory: thinking about drinking. In: Blaine, H.T., Leonard, K.E. (Eds.), Psychological Theories of Drinking and Alcoholism. Guilford, New York, pp. 181−226.

Goldstein, A., 2000. Addiction: From biology to Drug Policy. Oxford, New York.

Graybiel, A.M., 1995. Building action repertoires: memory and learning functions of the basal ganglia. Curr. Opin. Neurobiol. 5, 733−741.

Gutkin, B.S., Dehaene, S., Changeux, J.P., 2006. A neurocomputational hypothesis for nicotine addiction. Proc. Natl. Acad. Sci. U.S.A. 103, 1106−1111.

Hall, M., 1990. The relation between alcohol problems and the anxiety disorders. Am. J. Psychiatry 1, 685.

Hanson, K., Allen, S., Jensen, S., Hatsukami, D., 2003. Treatment of adolescent smokers with the nicotine patch. Nicotine Tob. Res. 5, 515−526.

Heyman, G., 2009. Addiction: A Disorder of Choice. Harvard.

Heyman, G.M., 2013. Addiction and choice: theory and new data. Front. Psychiatry 4.

Higgins, S.T., Alessi, S.M., Dantona, R.L., 2002. Voucher-based incentives: a substance abuse treatment innovation. Addict. Behav. 27, 887−910.

Higgins, S.T., Heil, S.H., Lussier, J.P., 2004. Clinical implications of reinforcement as a determinant of substance use disorders. Annu. Rev. Psychol. 55, 431−461.

Hobbs, J.D., Kushner, M.G., Lee, S.S., Reardon, S.M., Maurer, E.W., 2011. Meta-analysis of supplemental treatment for depressive and anxiety disorders in patients being treated for alcohol dependence. Am. J. Addict. 20, 319−329.

Holden, C., 2001. 'behavioral' addictions: do they exist? Science 294, 980−982.

Hunt, W.A., 1998. Pharmacology of alcohol. In: Tarter, R.E., Ammerman, R.T., Ott, P.J. (Eds.), Handbook of Substance Abuse: Neurobehavioral Pharmacology. Plenum, New York, pp. 7−22.

Jaffe, A., Pham, J.A.Z., Tarash, I., Getty, S.S., Fanselow, M.S., Jentsch, J.D., 2014. The absence of blocking in nicotine high-responders as a possible factor in the development of nicotine dependence? Open Addict. J. 7, 8−16.

Jaffe-Dax, S., Raviv, O., Jacoby, N., Loewenstein, Y., Ahissar, M., 2015. Towards a computational model of dyslexia. BMC Neurosci. 16, 1.

Johnson, A., van der Meer, M.A.A., Redish, A.D., 2007. Integrating hippocampus and striatum in decision-making. Curr. Opin. Neurobiol. 17, 692−697.

Johnson, M.W., Bickel, W.K., Kirshenbaum, A.P., 2004. Substitutes for tobacco smoking: a behavioral economic analysis of nicotine gum, denicotinized cigarettes, and nicotine-containing cigarettes. Drug Alcohol Depend. 74, 253–264.

Jones, B.T., Corbin, W., Fromme, K., 2001. A review of expectancy theory and alcohol consumption. Addiction 96, 57–72.

Keramati, M., Gutkin, B., 2013. Imbalanced decision hierarchy in addicts emerging from drug-hijacked dopamine spiraling circuit. PLoS One 8, e61489.

Klein, G., 1999. Sources of Power: How People Make Decisions. MIT Press.

Koob, G.F., 2009. Neurobiological substrates for the dark side of compulsivity in addiction. Neuropharmacology 56, 18–31.

Koob, G.F., Le Moal, M., 2006. Neurobiology of Addiction. Elsevier Academic Press.

Koob, G.F., Volkow, N.D., 2010. Neurocircuitry of addiction. Neuropsychopharmacology 35, 217–238.

Kosslyn, S.M., 1994. Image and Brain. MIT Press.

Kourrich, S., Thomas, M.J., 2009. Similar neurons, opposite adaptations: psychostimulant experience differentially alters firing properties in accumbens core versus shell. J. Neurosci. 29, 12275–12283.

Kruse, J.M., Overmier, J.B., Konz, W.A., Rokke, E., 1983. Pavlovian conditioned stimulus effects upon instrumental choice behavior are reinforcer specific. Learn. Motiv. 14, 165–181.

Kurth-Nelson, Z., Redish, A.D., 2010. A reinforcement learning model of pre-commitment in decision making. Front. Behav. Neurosci. 4, 184.

Kurth-Nelson, Z., Redish, A.D., 2012a. Don't let me do that! Models of precommitment. Front. Neurosci. 6, 138.

Kurth-Nelson, Z., Redish, A.D., 2012b. Modeling decision-making systems in addiction. In: Gutkin, B., Ahmed, S.H. (Eds.), Computational Neuroscience of Drug Addiction. Springer, pp. 163–188 (Chapter 6).

Kurzban, R., 2010. Why Everyone (Else) is a Hypocrite. Princeton.

Kushner, M.G., Abrams, K., Borchardt, C., 2000. The relationship between anxiety disorders and alcohol use disorders: a review of major perspectives and findings. Clin. Psychol. Rev. 20, 149–171.

Lesieur, H., 1977. The Chase : Career of the Compulsive Gambler. Anchor Press.

Liljequist, S., Engel, J.A., 1984. The effects of GABA and benzodiazepine receptor antagonists on the anti-conflict actions of diazepam or ethanol. Pharmacol. Biochem. Behav. 21, 521–525.

Littleton, J., 1998. Neurochemical mechanisms underlying alcohol withdrawal. Alcohol Res. Health 22, 13–24.

MacDonald, A.W., Zick, J.L., Netoff, T.I., Chafee, M.V., 2016. The computation of collapse: can reliability engineering shed light on mental illness? In: Redish, A.D., Gordon, J.A. (Eds.), Computational Psychiatry: New Perspectives on Mental Illness. MIT Press, Cambridge, MA. Strüngmann Forum Reports.

Marks, K.R., Kearns, D.N., Christensen, C.J., Silberberg, A., Weissa, S.J., 2010. Learning that a cocaine reward is smaller than expected: a test of Redish's computational model of addiction. Behav. Brain Res. 212, 204–207.

Mayr, E., 1998. This is Biology: The Science of the Living World. Belknap Press.

McClelland, J.L., Rogers, T.T., 2003. The parallel distributed processing approach to semantic cognition. Nat. Rev. Neurosci. 4, 310–322.

McDannald, M.A., Lucantonio, F., Burke, K.A., Niv, Y., Schoenbaum, G., 2011. Ventral striatum and orbitofrontal cortex are both required for model-based, but not model-free, reinforcement learning. J. Neurosci. 31, 2700–2705.

Meyer, R., Mirin, S., 1979. The Heroin Stimulus. Plenum, New York.

Mishkin, M., Appenzeller, T., 1987. The anatomy of memory. Sci. Am. 256, 80–89.

Montague, P.R., Dayan, P., Sejnowski, T.J., 1996. A framework for mesencephalic dopamine systems based on predictive Hebbian learning. J. Neurosci. 16, 1936–1947.

Montague, P.R., Dolan, R.J., Friston, K.J., Dayan, P., 2012. Computational psychiatry. Trends Cogn. Sci. 16, 72–80.

Norton, E.S., Beach, S.D., Gabrieli, J.D., 2015. Neurobiology of dyslexia. Curr. Opin. Neurobiol. 30, 73–78.

O'Craven, K.M., Kanwisher, N., 2000. Mental imagery of faces and places activates corresponding stimulus-specific brain regions. J. Cogn. Neurosci. 12, 1013–1023.

O'Doherty, J.P., 2004. Reward representations and reward-related learning in the human brain: insights from neuroimaging. Curr. Opin. Neurobiol. 14, 769–776.

Oei, T.P.S., Baldwin, A.R., 2002. Expectancy theory: a two-process model of alcohol use and abuse. J. Stud. Alcohol 55, 525–534.

Oscar-Berman, M., Marinkovic, K., 2003. Alcoholism and the brain: an overview. Alcohol Res. Health 27, 125–134.

Packer, R.R., Howell, D.N., McPherson, S., Roll, J.M., 2012. Investigating reinforcer magnitude and reinforcer delay: a contingency management analog study. Exp. Clin. Psychopharmacol. 20, 287.

Panlilio, L.V., Thorndike, E.B., Schindler, C.W., 2007. Blocking of conditioning to a cocaine-paired stimulus: testing the hypothesis that cocaine perpetually produces a signal of larger-than-expected reward. Pharmacol. Biochem. Behav. 86.

Pavlov, I., 1927. Conditioned Reflexes. Oxford Univ. Press.

Pearl, J., 1988. Probabilistic Reasoning in Intelligent Systems: Networks of Plausible Inference. Morgan Kaufmann.

Pearl, J., 2009. Causality: Models, Reasoning and Inference. Cambridge University Press.

Pearson, J., Naselaris, T., Holmes, E.A., Kosslyn, S.M., 2015. Mental imagery: functional mechanisms and clinical applications. Trends Cogn. Sci. 19, 590–602.

Petry, N.M., 2011. Contingency Management for Substance Abuse Treatment: A Guide to Implementing This Evidence-Based Practice. Routledge.

Phelps, E., Lempert, K.M., Sokol-Hessner, P., 2014. Emotion and decision making: multiple modulatory circuits. Annu. Rev. Neurosci. 37, 263–287.

Piray, P., Keramati, M.M., Dezfouli, A., Lucas, C., 2010. Individual differences in nucleus accumbens dopamine receptors predict development of addiction-like behavior: a computational approach. Neural Comput. 22, 2334–2368.

Quitkin, F.M., Rifkin, A., Kaplan, J., Klein, D.F., 1972. Phobic anxiety syndrome complicated by drug dependence and addiction: a treatable form of drug abuse. Arch. Gen. Psychiatry 27, 159–162.

Rangel, A., Camerer, C., Montague, P.R., 2008. A framework for studying the neurobiology of value-based decision making. Nat. Rev. Neurosci. 9, 545–556.

Redish, A.D., 2004. Addiction as a computational process gone awry. Science 306, 1944–1947.

Redish, A.D., 2013. The Mind within the Brain: How We Make Decisions and How Those Decisions Go Wrong. Oxford Univ. Press, Oxford, UK.

Redish, A.D., 2016. Vicarious trial and error. Nat. Rev. Neurosci. 17, 147–159.

Redish, A.D., Gordon, J.A. (Eds.), 2016. Computational Psychiatry: New Perspectives on Mental Illness. MIT Press, Cambridge, MA. Strüngmann Forum Reports.

Redish, A.D., Jensen, S., Johnson, A., 2008. A unified framework for addiction: vulnerabilities in the decision process. Behav. Brain Sci. 31, 415–487.

Redish, A.D., Jensen, S., Johnson, A., Kurth-Nelson, Z., 2007. Reconciling reinforcement learning models with behavioral extinction and renewal: implications for addiction, relapse, and problem gambling. Psychol. Rev. 114, 784–805.

Redish, A.D., Johnson, A., 2007. A computational model of craving and obsession. Ann. N.Y. Acad. Sci. 1104, 324–339.

Redish, A.D., Mizumori, S.J.Y., 2015. Memory and decision making. Neurobiol. Learn. Mem. 117, 1–3.

Regier, P.S., Redish, A.D., 2015. Contingency management and deliberative decision-making processes. Front. Psychiatry 6, 0076.

Rescorla, R.A., 1988. Pavlovian conditioning: it's not what you think it is. Am. Psychol. 43, 151–160.

Rescorla, R.A., Wagner, A.R., 1972. A theory of Pavlovian conditioning: variations in the effectiveness of reinforcement and nonreinforcement. In: Black, A.H., Prokesy, W.F. (Eds.), Classical Conditioning II: Current Research and Theory. Appleton Century Crofts, New York, pp. 64–99.

Robbins, T.W., Clark, L., 2015. Behavioral addictions. Curr. Opin. Neurobiol. 30, 66–72.

Robinson, T.E., Berridge, K.C., 2001. Mechanisms of action of addictive stimuli: incentive-sensitization and addiction. Addiction 96, 103–114.

Robinson, T.E., Berridge, K.C., 2003. Addiction. Annu. Rev. Psychol. 54, 25–53.

Ross, D., 2008. Timing models of reward learning and core addictive processes in the brain. Behav. Brain Sci. 31, 457–458.

Saint-Cyr, J.A., Taylor, A.E., Lang, A.E., 1988. Procedural learning and neostriatal dysfunction in man. Brain 111, 941–959.

Schadé, A., Marquenie, L.A., Balkom, A.J., Koeter, M.W., Beurs, E., Brink, W., Dyck, R., 2005. The effectiveness of anxiety treatment on alcohol-dependent patients with a comorbid phobic disorder: a randomized controlled trial. Alcohol. Clin. Exp. Res. 29, 794–800.

Schoenbaum, G., Roesch, M., Stalnaker, T.A., 2006. Orbitofrontal cortex, decision making, and drug addiction. Trends Neurosci. 29, 116–124.

Schüll, N.D., 2012. Addiction by Design: Machine Gambling in Las Vegas. Princeton University Press.

Schultz, W., Dayan, P., Montague, P.R., 1997. A neural substrate of prediction and reward. Science 275, 1593–1599.

Schultz, W., Dickinson, A., 2000. Neuronal coding of prediction errors. Annu. Rev. Neurosci. 23, 473–500.

Sherrington, C.S., 1906. The Integrative Action of the Nervous System. Yale.

Squire, L.R., 1987. Memory and Brain. Oxford University Press, New York.

Sutton, R.S., Barto, A.G., 1998. Reinforcement Learning: An Introduction. MIT Press, Cambridge, MA.

Talmi, D., Seymour, B., Dayan, P., Dolan, R.J., 2008. Human Pavlovian-instrumental transfer. J. Neurosci. 28, 360–368.

Tiffany, S.T., 1990. A cognitive model of drug urges and drug-use behavior: role of automatic and nonautomatic processes. Psychol. Rev. 97, 147–168.

Tran, V.T., Snyder, S.H., Major, L.F., Hawley, R.J., 1981. GABA receptors are increased in brains of alcoholics. Ann. Neurol. 9, 289–292.

Valenzuela, C.F., Harris, R.A., 1997. Alcohol: neurobiology. In: Lowinson, J.H., Ruiz, P., Millman, R.B., Langrod, J.G. (Eds.), Substance Abuse: A Comprehensive Textbook. Williams and Wilkins, Baltimore, pp. 119–142.

van der Meer, M.A.A., Kurth-Nelson, Z., Redish, A.D., 2012. Information processing in decision-making systems. Neuroscientist 18, 342–359.

Volkow, N.D., Fowler, J.S., Wang, G.J., 2002. Role of dopamine in drug reinforcement and addiction in humans: results from imaging studies. Behav. Pharmacol. 13, 355–366.

Wakefield, J.C., 2007. The concept of mental disorder: diagnostic implications of the harmful dysfunction analysis. World Psychiatry 6, 149.

Weinstein, N.D., Marcus, S.E., Moser, R.P., 2005. Smokers unrealistic optimism about their risk. Tob. Control 14, 55–59.

White, A.M., 2003. What happened? Alcohol, memory blackouts, and the brain. Alcohol Res. Health 27, 186–196.

Winstanley, C.A., Balleine, B.W., Brown, J.W., Bü chel, C., Cools, R., Durstewitz, D., O'Doherty, J.P., Pennartz, C.M., Redish, A.D., Seamans, J.K., Robbins, T.W., 2012. Search, goals, and the brain. In: Hills, T., McNamara, J., Raaijmakers, J., Robbins, T., Todd, P.M. (Eds.), Cognitive Search. MIT Press, pp. 125–156. Ernst Strüngmann Forum Discussion Series.

Wise, R.A., 2005. Forebrain substrates of reward and motivation. J. Comp. Neurol. 493, 115–121.

Wright, C., Moore, R.D., 1990. Disulfiram treatment of alcoholism. Am. J. Med. 88, 647–655.

# Modeling Negative Symptoms in Schizophrenia

*Matthew A. Albrecht[1,2], James A. Waltz[1], Michael J. Frank[3], James M. Gold[1]*

[1] University of Maryland School of Medicine, Baltimore, MD, United States; [2] Curtin University, Perth, WA, Australia; [3] Brown University, Providence, RI, United States

## OUTLINE

*Computational Psychiatry*
http://dx.doi.org/10.1016/B978-0-12-809825-7.00009-2

*How modeling can inform reward-related decision tasks to uncover the computational bases of decision-making impairments in schizophrenia in relation to negative symptoms, which has implications for understanding motivational disturbances in other affective disorders such as major depression.*

# 9.1  INTRODUCTION: NEGATIVE SYMPTOMS IN SCHIZOPHRENIA

Negative symptoms are a cluster of self-reports and behaviors characterized by the reduction of normal activity or emotion, the origin of which remains unknown. The expressed symptoms superficially overlap with much of the classic depressive disorder phenotype and include motivational deficits, anhedonia, avolition, and reduced affective expression. These symptoms in the context of schizophrenia have seen a steadily increasing research focus, including more direct attempts at investigating treatment options in a hope to alleviate the poor functional outcome (Norman et al., 2000; Evans et al., 2004), poor quality of life (Katschnig, 2000), and low rate of recovery (Strauss et al., 2010). Unfortunately, there is currently no Food and Drug Administration approved drug for the treatment of negative symptoms. This is partly due to a function of our poor understanding of the etiology of negative symptoms, which is centered on clinical observations and phenomenological interrogation.

Some avenues of phenomenological interrogation may have led us astray, having a significant impact on our ability to understand negative symptoms and impaired our ability to develop valid preclinical models, highlighting the need for more formal models of symptom etiology. As an example of where we may have been led astray, it would appear reasonable at first to assume that the negative symptom anhedonia represents an emotional construct of diminished pleasure and therefore diminished in the moment pleasure would be reduced in people exhibiting this negative symptom. However, while patients may report less pleasure relative to controls when reporting on noncurrent feelings, they report similar levels of in the moment/current positive emotion similar to healthy subjects (Kring and Moran, 2008; Cohen and Minor, 2010). It is the former that appears to drive ratings on clinical symptom expression scales (Strauss and Gold, 2012), while we have developed much of our beliefs and animal models on the latter (e.g., sucrose preference test (Papp et al., 1991)). As a consequence, more recent thoughts on anhedonia suggest that it is likely due to a combination of downgrading the hedonic experience of past events and predictions about future events to be more pessimistic than may be the case (Strauss and Gold, 2012). While patient reports of the hedonic quality of past experiences and subsequent behaviors may contribute to nonadaptive behaviors in the future, they would still appear to be entirely rational taking into account a likely history of patient experience. That is, people with schizophrenia are likely to have significantly more negative experiences over their lifetime compared with others, especially if their experiences are interrupted by, or are predicted to be interrupted by, paranoid thoughts or malicious voices.

The refined understanding of negative symptoms was formed through a more extensive and detailed examination of clinical and patient reports, both important sources of information. However, it can sometimes be difficult to advance mechanisms using only subjective self-report sources, highlighting the need to combine more modern perspectives on the clinical/experiential aspects of negative symptoms with a more detailed/objective assessment of functionality based in formally quantified behavioral paradigms. Our research has used behavioral assessments of reinforcement learning (RL) to examine competing models of negative symptom formation. Broadly, patients with schizophrenia have repeatedly shown performance impairments in RL tasks. Intuitively, it can be seen how this could lead to deficits in goal directed action that underpin negative symptoms via straightforward impairments in learning about rewards and then acting on them. More formally, one explanation for these deficits and the propensity to develop negative symptoms could be because there is a failure in signaling reward prediction errors (PEs). A failure to adequately signal reward PEs could happen via two distinct mechanisms, either through a dampened PE response to reward or through noise in the reward PE system such that true reward PEs are drowned out by aberrant signaling (Maia and Frank, 2017). This idea is particularly appealing given the significant role for dopamine systems in motivation and action. Furthermore, it is broadly accepted that dopamine contributes in some way to the clinical profile of positive symptoms of psychosis, and a common underlying neural substrate for schizophrenia possesses a seductive symmetry and parsimony. However, maladaptive value adjustment during RL can happen for any number of reasons besides faulty ventral-striatum reward prediction error signaling.

We will argue below that a critical cognitive pathology in schizophrenia is an inability to effectively and/or flexibly update representations of reward value over time. This specific deficit may more specifically lead to the types of failures in RL performance in schizophrenia, particularly in those patients that present with particularly severe negative symptoms. The key studies supporting this hypothesis were formulated using a computational psychiatry perspective in an attempt to isolate the core computational constructs/modules used by the brain to calculate, store, and act on value. These computational constructs are an abstraction of brain processes where it is assumed that an underlying neural network is associated with a particular calculation. Therefore, these computational models are designed to overlap with neural models thought to generate the expression of behavior. As elaborated below, this framework has led to alternative computational accounts that posit reward-learning deficits in schizophrenia that can arise from alterations within the core dopaminergic RL system itself (Maia and Frank, 2017), or from alterations in more cognitive prefrontal mechanisms that, in the healthy brain, are used to elaborate and enhance learning (Strauss et al., 2011; Gold et al., 2012; Collins et al., 2014; Collins et al., submitted). With a more refined characterization and quantification of the contributions of multiple systems to learning and decision-making, one can begin to design tasks to be more sensitive and specific to disentangle their contributions.

We start with a brief overview of dopamine's role in representing reward value and its anatomical and functional interactions with frontal regions that form a basal-ganglia-cortico-striatal network, as an initial link from brain function to computational abstraction. Following from this, we will describe the application of computational models to RL tasks and their insights into schizophrenia.

## 9.2 DOPAMINE SYSTEMS AND PREDICTION ERRORS

Dopamine neurons in the mesolimbic system seem to play an important role in signaling the value of a stimulus. Increases in phasic firing of dopamine neurons occur in response to the unexpected delivery of a rewarding stimulus. The onset of firing can also be shifted from the rewarding stimulus to a stimulus predictive of that reward, as hypothesized by an extension of classical associative learning schemes to RL algorithms that span temporally extended tasks (Montague et al., 1996). Moreover, if the reward is omitted at the expected time of delivery after a presentation of a reward-predicting stimulus, dopamine neuron firing rate will be suppressed. This simple bidirectional response to value provided a foundation for the idea that dopamine neurons signal a valenced reward PE, their firing rate increasing in response to unexpected reward (or reward prediction), and reducing in response to reward omission, thereby signaling positive and negative PEs, respectively. The magnitude of phasic bursts and dips of DA are correlated with the magnitude of positive and negative PEs, respectively (Bayer et al., 2007; Hart et al., 2014). Dopamine signaled PEs go on to modulate decision-making/action selection by modulating synaptic plasticity of the circuits responsible for action/ behavioral choices. Dopamine can also affect decision-making by modulating choice incentive directly, e.g., positive PEs that increase dopamine can result in a $D_1$-receptor system bias associated with Go-actions adjusting behavior toward riskier options but with higher potential value (Collins and Frank, 2014; Maia and Frank, 2017). Optimal action selection is supported by learning systems situated in frontal cortex, in particular the orbitofrontal cortex (OFC) and medial prefrontal cortex (mPFC). These frontal cortical regions appear to be responsible for calculating, maintaining, and weighing stimulus-action-value associations (Schoenbaum and Roesch, 2005; Frank and Claus, 2006; Fellows and Farah, 2007). Moreover, these frontal systems support rapid and flexible updating of stimulus-action-value calculations compared to the slower updating in the basal ganglia. This ability of OFC to weigh up multiple actions allows simultaneous evaluation of multiple action-outcome scenarios, aiding optimal decision-making. The key limitation here is that action-outcome scenarios are constrained within the capacity of working memory (WM). These two systems play interacting roles in RL and decision-making, complementing each other's strengths and weaknesses. For example, the OFC is able to rapidly update stored values and can directly contrast several competing options, providing an additional top-down influence on decision-making. However, the system is resource-limited particularly by constraints on WM. By contrast, the striatal system contributes a more stable stimulus valuation, and actions are integrated in a more habitual manner. Consequently, the striatal system is not as resource-limited by WM capacity. However, this system can be less flexible to alterations in value and is limited by the actions it can select.

Of course, one of the most influential models of schizophrenia relates to the function of the dopamine system. Computational research has therefore followed this focus, attempting to formalize a relationship between aberrant dopamine signaled positive PEs and the expression of positive symptoms. Computational models of RL can potentially be informative about negative symptoms. The systems that we have discussed above are integral components involved in motivation, pleasure seeking, and action selection; functions that clinically appear to underlie negative symptom expression. Computational modeling is therefore a potentially useful tool to tease apart the separable contributions of these multiple interacting systems (and components within these systems), which may offer key insights.

## 9.3 MODELING IN REWARD-RELATED DECISION TASKS

Computational modeling of RL tasks has at its core the concept of the reward PE; the difference between an outcome that was expected and an outcome that was experienced. Mathematically, this can be expressed as:

$$\delta_t = r_t - V_t$$

where $r_t$ is the experienced reward for trial $t$, $V_t$ is the expected reward value, and $\delta_t$ is the reward PE. This core module will be a critical component of the models described below where we attempt to understand behavior on a trial-by-trial basis. Using the foundational concept of a reward PE, we can build models of increasing complexity to describe the behavioral responses to stimuli. These models can include any number of parameters depending on which computational principles are relevant (and can be identified) when the subject is confronted with a particular task. Of course, the parameters incorporated into the model will depend strongly on the structure of the task. Below we begin with a relatively simple stimulus-selection task that incorporates probabilistic rewards and punishments to shape action selection.

## 9.4 PROBABILISTIC STIMULUS SELECTION—COMBINED ACTOR-CRITIC/ Q-LEARNING

### 9.4.1 Rationale

In a previous report, we had shown that people with schizophrenia demonstrated impaired acquisition in a simple probabilistic RL task (Waltz et al., 2007). The task required participants to select between two stimuli that were reinforced at either 80% (the optimal stimulus) or 20%. They also had to learn to select among two other pairings of stimuli with more probabilistic values (70/30 and 60/40). An example of the structure of this task is described in Fig. 9.1. This kind of task is learned relatively

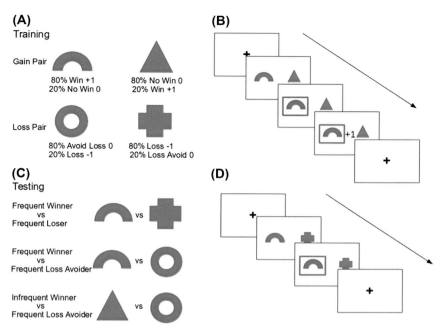

**FIGURE 9.1** Example experimental procedure for the probabilistic stimulus-selection task incorporating rewards and punishments. (A and B) Participants learn to dissociate between pairs of stimuli reinforced or punished at a set probability (in this case 80%). (C and D) Following the training phase, participants are given novel pairings of the previously trained stimuli (or completely novel stimuli).

quickly in control participants who learn to select the most rewarding stimulus of the various pairs. By contrast, patients with schizophrenia showed learning delays at the early stages of training. In a subsequent transfer phase, where the previously trained stimuli were presented to participants without feedback, both controls and patients were able to select the most rewarding stimulus compared with the least rewarding one. Patients did not, however, pick the high-value stimulus (that which would have elicited the largest number of positive PEs) over a neutral novel stimulus as reliably as controls, but they were able to match controls' performance in avoiding the stimulus with lowest reward (that which would have elicited the largest number of negative PEs, and also likely the fewest number of experienced feedback events assuming they learned the task) stimulus with the same neutral novel stimulus.

It was suggested that this pattern of results indicated that patients did not have a problem with learning from negative PEs, they were only selectively impaired at learning from positive PEs. However, while positive PE learning is one mechanism required for subsequent choice of the more rewarding stimulus in this task, it is not the only one. Indeed, while the original task was designed to test a computational model of striatal dopamine function on positive and negative PE learning, subsequent computational analysis based on the same framework suggested that even when positive learning is equivalent across groups, impaired choice of the most rewarding stimulus can arise from a reduced representation of the expected positive value of that action during action selection (Collins and Frank, 2014). In turn, this deficiency can arise from either degraded OFC representations of expected value (EV) (Frank and Claus, 2006), or from reduced striatal DA signaling during the choice period, leading to reduced choice incentive and a greater emphasis on negative outcomes (Collins and Frank, 2014). Both mechanisms have been posited to account for this selective reduction in choice of positively valued options during learning in people with schizophrenia (Gold et al., 2012; Huys et al., 2016). Moreover, patients received less exposure to the high-value stimulus as they were slower to learn the association, increasing the uncertainty of its value. Therefore, a reduced preference for the high-value stimulus compared to controls may also be related to less exposure making the positive EV more uncertain.

To assess some of these hypotheses, we implemented a similar probabilistic RL task that included reward and punishment contingencies to obtain positive and negative reward PEs in both instrumental contexts. That is, in addition to the above (probabilistic) reward versus no-reward condition, a second condition was included where participants learned based on (probabilistic) loss versus no-loss (loss-avoidance). Critically, the same zero outcome would yield both positive and negative PEs depending on gain or loss context. This modification addresses one of the critical confounds in the above experiment; the conflation of reward with positive PE. The previous experiment also did not model behavior to corroborate whether poor PE signaling, or other factors contributing to RL performance, more explicitly contributed to patients performance differences. Therefore, to further examine whether positive PEs or deficits in maintaining EV were potential reasons for patient-control differences, we used a combination of two interacting RL models: a "Q-learning" model and an "Actor-Critic" model. We begin by describing the Q-learning model, a simple two-parameter model that keeps track of state-action value pairs, before introducing the Actor-Critic model. Then the process of combining these two models into a single model with two distinct learning modules will be described. Each of these models is theoretically linked in some way with the basal ganglia system and reward PEs, but each make different predictions about the nature of representations used in reward-based decision-making.

## 9.4.2 Q-Learning

The Q in Q-learning represents the expected utility (Quality) of a state-action pairing. Q-learning models learn about state-action pairs, such that for any given state, multiple possible actions may be entertained. For example, for a given state (or stimulus) $X_1$, an action $Y_1$ possesses an EV 0.5, while action $Y_2$ may possess a value of 0.7. A series of states and possible actions yields a state-action matrix with the dimension of the rows and columns depending on the number of states and actions, respectively e.g.,

|  |  | **Action** | | |
|---|---|---|---|---|
|  |  | $Y_1$ | $Y_2$ | $Y_j$ |
|  | $X_1$ | 0.5 | 0.7 | ... |
| **State** | $X_2$ | 0.2 | 0.4 | ... |
|  | $X_i$ | ... | ... | ... |

The values within the state-action matrix naturally change over time depending on the past history of reinforcements and punishments, and are updated based on a weighted PE, such that:

$$\delta_t = r_t - Q_t(s, a)$$
$$Q_{t+1}(s, a) = Q_t(s, a) + \alpha \delta_t$$

where for trial $t$, $r_t$ gives the obtained reinforcement value for that particular trial (usually given as $+1$, 0, or $-1$ for reward, neutral, and punishment, respectively), $\delta_t$ is the PE for the state-action pair at trial $t$, and $\alpha$ describes the learning rate or the speed in which the state-action pairings update.

For a given new stimulus, action selection for that particular state is made probabilistically based on contrasting the learned Q-values of each candidate action for that state. A transformation of the Q-values into an action probability is often obtained through a softmax function, transforming the Q-values into a probability between 0 and 1. These values reflect the probability of selecting that particular action for that state. The softmax function is given as:

$$p(a|s) = \frac{\exp(\beta Q_t(s, a))}{\sum_{i=1}^{N_a} \exp(\beta Q_t(s, a_i))}$$

where $\beta$ represents the temperature, determining the degree with which differences in Q-values are translated into a more deterministic choice.

### 9.4.3 Actor-Critic Model

The actor-critic model is a two-part model with many of the same features as the Q-learning model. One component, the "Critic" is an evaluator of the state's value. The Critic learns the value of the stimulus without taking into account the possible actions. The second component, the "Actor," is used for action selection and learns stimulus-response weightings for each state-action pair as a function of the critic's evaluation. PEs are generated at the Critic level to update both the state value of the critic and the stimulus-response weights of the Actor. The model is formalized as follows, first the critic PE for each trial is obtained similar to the Q-learning model above for each stimulus (note not a stimulus-action pair):

$$\delta_t^{Critic} = r_t - V(s_t)$$

The critic PE $\delta_t^{Critic}$ then updates the stored value of the state representation modified by the critic's learning rate $\alpha^{Critic}$:

$$V_{t+1}(s) = V_t(s) + \alpha^{Critic}\delta_t^{Critic}$$

The critic's PEs is also used to update the actor's response weighting matrix for that stimulus, by:

$$W_{t+1}(s,a) = W_t(s,a) + \alpha^{Actor}\delta_t^{Critic}$$

where $\alpha^{Actor}$ represents the *Actor's* learning rate. Note also the difference between the stored representations of the critic's value $V_t(s)$ and the *Actor's* weights $W_t(s,a)$, where the former is only representing states and the latter is representing state-action pairings. As can be seen from the formulas, there are two separate learning rates, one for the critic and one for the actor, that represent the speed of updating of each system's values and weights, respectively. The cognitive neuroscience literature has linked the ventral striatum to the critic and the dorsal striatum to the actor, both of which learn from predictions errors signaled by dopamine (O'Doherty et al., 2004). Finally, the action weights are normalized between 0 and 1 by:

$$W_{t+1}(a,s) = \frac{W_{t+1}(s,a)}{\sum_{i=1}^{N_a} W_{t+1}(s,a_i)}$$

## 9.4.4 Combined Actor-Critic/Q-Learning

The above two models represent two unique strategies for learning stimulus-action reinforcement and putatively represent two separate neurological systems. The Q-learning model selects actions by comparing Q-values of state-action pairings and then updates these using a delta rule on the state-action values directly. By contrast, in the actor-critic model, the critic evaluates the overall state without considering competing actions and then feeds this information to the actor that selects actions with strong stimulus-action weights (without directly evaluating the stimulus-action value). The Q-learning model is hypothetically linked to OFC that maintains value representations of the state-action pairs, whereas the actor-critic model is linked more strongly to the basal ganglia system. In the brain, these two systems interact to solve action selection each contributing their unique perspective to the problem at hand. We can model this interaction using a hybrid model, whereby each system is updated independently, with critic PEs and Q-learning PEs calculated for the separate systems and the final action selection is calculated by adding a mixing parameter that apportions the degree to which each system contributes to the final stimulus-action weight, by:

$$q = (1 - mix) \times W_t(s, a) + mix \times Q_t(s, a)$$

where $q$ is the final stimulus-action weight for all possible actions given a stimulus, $mix$ describes the proportion of each model used, and $W_t(s,a)$ and $Q_t(s,a)$ are the actor-critic and Q-learning contributions, respectively, calculated as above. Other more elaborated actor-critic models separate the actor into two opponent processes, capturing asymmetric effects of dopamine on learning and choice based on positive versus negative incentives (Collins and Frank, 2014) and which can be applied to some of the schizophrenia literature (Maia and Frank, 2017), but for brevity we do not detail this model here.

## 9.4.5 Findings

Using this task we found that patients with high number of reported negative symptoms learned best when attempting to avoid punishments, while performing poorest when learning from rewards in a 90% contingency condition compared with both controls and low negative symptom (LNS) patients (Gold et al., 2012). Poor reward-learning performance was further shown in the transfer phase, where high negative symptom (HNS) patients were less likely to select the higher value stimulus, particularly when comparing a frequently winning stimulus with a frequently loss-avoiding stimulus. HNS patients essentially treated the accumulated positive PEs during the reward condition and the loss-avoidance condition with a more similar weight, whereas controls and LNS patients more definitively favored positive PEs from rewarding stimuli over loss-avoidance stimuli. Compared with controls and LNS patients, HNS patients also failed to select the stimulus with the highest EV in the easiest transfer condition comparing the frequently rewarded stimulus with the frequently punished stimulus. A failure to select the more obvious optimal stimulus implicates a general failure in maintaining stimulus value, alternatively it could be due to the reduced exposure to the most frequently rewarded stimulus reducing the certainty surrounding its EV.

The modeling results yielded further information on the cause of the RL performance differences in HNS. In particular, differences between HNS patients and controls were seen in the mixing parameter that describes the ratio of Q-learning to Actor-Critic learning. The mixing proportion in controls was fit at a ratio of $\sim 0.7$ in favor of the Q-learning model, i.e., a greater proportion of their action selection used the model most related to the EV of the choice (thought to be represented in OFC) rather than just the history of PEs. By contrast, in patients with a high burden of negative symptoms, the mix between the Q-learning module and the Actor-Critic module was substantially lower at $\sim 0.4$. LNS patients sat between HNS and controls at $\sim 0.6$. A reduction in the contribution of the Q-learning component to RL in HNS patients implicates a possible mechanism for poor RL performance and its relationship to negative symptoms: a failure of top-down control by the OFC impairs maintenance of EV, particularly for stimuli associated with rewarding outcomes (not necessarily impaired for positive PEs).

## 9.5 TIME CONFLICT—TEMPORAL UTILITY INTEGRATION TASK

### 9.5.1 Rationale

The probabilistic stimulus-selection task described above showed that patients seem to possess difficulties integrating a history of rewarding outcomes and selecting those stimuli over others. Difficulties specific to reward-learning may suggest deficits in striatal DA $D_1$-receptor signaling. The $D_1$ subsystem of the basal ganglia is linked with making actions in response to rewarding outcomes following phasic dopamine bursts. Therefore deficits in this system may be responsible for poor reward Go-learning in the HNS patients. Conversely, there appears to be intact responding to negative PE signaling in HNS patients, which is ostensibly related to the $D_2$ receptor subsystem that is sensitive to phasic dopamine dips and actions within the NoGo pathway. The basal ganglia $D_1$ and $D_2$ subsystems also modulate response times. Following positive PEs and phasic dopamine bursts targeting the $D_1$ subsystem, response times are reduced and Go-learning is enhanced. Following negative PEs and phasic dopamine dips targeting the $D_2$ subsystem, response times slow and NoGo-learning is enhanced.

We can use this phenomenon to investigate the potential role of $D_1$ and $D_2$ systems during RL in people with schizophrenia using a slightly

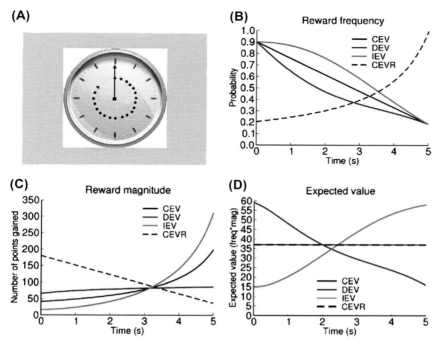

FIGURE 9.2    The time conflict task. (A) Participants are presented with a clock face that turns over a period ~5 s. Reward frequency (B) and magnitude (C) change over the duration of the turning clock hand such that the expected value of a response at a particular time is represented in (D). CEV, constant expected value; CEVR, constant expected value reversed; DEV, decreased expected value; IEV, increased expected value. *Reprinted from Strauss, G.P., Frank, M.J., Waltz, J.A., Kasanova, Z., Herbener, E.S., Gold, J.M., 2011. Deficits in positive rein-forcement learning and uncertainty-driven exploration are associated with distinct aspects of nega-tive symptoms in schizophrenia. Biol. Psychiatry 69, 424–431.*

different approach to the probabilistic selection task. We administered the temporal utility integration task (TUIT) that adjusts reward probabilities and magnitudes as a function of time elapsed (Moustafa et al., 2008). Fig. 9.2 illustrates the experimental procedure for the TUIT. Participants are presented with a clock face with a single arm, where the arm turns over the face of the clock in around 5 s. Participants are required to press a button at any time between the onset of movement of the clock arm and a complete rotation. Points are given to the participant depending on four conditions, where the EV (probability × magnitude) for each condition across the 5 s interval is defined as:

- Decreased expected value (DEV) = early responses yield a higher number of points than expected and should therefore lead to Go-learning/speeding. Faster responses gave more points on average.

- Increased expected value (IEV) = early responses yield a small number of points, which should lead to NoGo-learning and slowed responses. Slower responses gave more points on average.
- Constant expected value (CEV) = reward probability increases, while magnitude decreases as time elapses on the clock
- Constant expected value Reversed (CEVR) = reward probability decreases, while magnitude increases as time elapses on the clock

Using these last two conditions which have constant EV over the clock face, we can determine relative biases in a person's preference for reward probability versus magnitude. Risk averse participants should respond late in CEV and early in CEVR. The first two conditions can yield important information on the relative capabilities of patients to speed or slow their reaction times (RTs).

## 9.5.2 Time Conflict Model

The model for the TUIT begins with the updating of state values as seen in the actor-critic model. Through experience on the task (after pressing the response button at various times) participants develop an EV ($V_t$) for a given moment in time represented on the clock face. As before, the EV is updated by a reward PE:

$$V_{t+1} = V_t + \alpha \delta_t$$

$$\delta_t = r_t - V_t$$

Just like the actor-critic model, the model used for the time conflict experiment assumes that value integration is computed by a critic, and these same PE signals train the actor in the striatum. Because of differential learning mechanisms for approach (Go) and avoid (NoGo) systems, different learning rates ($\alpha$) for positive and negative PEs were used:

$$Go_{t+1}(s,a) = Go_t(s,a) + \alpha^{Go}\delta_t^{+ve}$$

$$NoGo_{t+1}(s,a) = NoGo_t(s,a) + \alpha^{NoGo}\delta_t^{-ve}$$

where $\alpha^{Go}$ and $\alpha^{NoGo}$ describe the amount of speeding and slowing of RTs in response to positive and negative PEs, respectively. As mentioned above, positive PEs are expected to engage an approach-related dopamine $D_1$-dependent speeding effect on RT, while negative PEs are expected to engage an avoidance-related $D_2$-dependent slowing effect on RT.

In addition to the base-RL model used, several further parameters were included to better describe the observed RTs. A parameter $K$, describing the baseline motor response tendency: participants initial RTs generally occurred within the first half of the clock face, participants then slow to explore the remaining clock space. A parameter $\rho$, describing RT adjustment toward RTs obtain more positive PEs. This was estimated using Bayesian updating: participants develop a representation (or prior) of the distribution of positive PEs for fast (RT < median) and slow (RT > median) responses separately over the history of the set, these representations were then updated with new incoming evidence via Bayes rule:

$$P(\theta|\delta_1...\delta_n) \propto P(\delta_1...\delta_n|\theta)P(\theta)$$

where $\theta$, represents the parameters governing the belief distribution of reward PEs for each future response. An explore parameter $\varepsilon$, describing large RT fluctuations during periods of high uncertainty. The magnitude of the uncertainty was derived from the trial-by-trial updated standard deviations derived from beliefs about obtainable rewards yielded from the Bayesian model. The explore parameter was given by:

$$\text{Explore}_t(s) = \varepsilon\left[\sigma_{(\delta|s,a=\text{slow})} - \sigma_{(\delta|s,a=\text{fast})}\right]$$

where $\sigma_{\text{slow}}$ and $\sigma_{\text{fast}}$ are the standard deviations that represent outcome uncertainty after slow and fast relative responses. A parameter $\lambda$, describing the impact of the previous response's RT on current RT. Finally, a parameter $\nu$, describing behavior that is drawn toward the RT that previously gave the largest reward, and which is at least 1 standard deviation above previous rewards. The final TUIT model was given as:

$$RT_t(s) = K + \lambda RT_{t-1}(s) - Go_t(s) + NoGo_t(s) + \rho\left[\mu_t^{\text{slow}}(s) - \mu_t^{\text{fast}}(s)\right]$$
$$+ \nu\left[RT^{\text{best}} - RT^{avg}\right] + \text{Explore}_t(s)$$

### 9.5.3 Findings

As expected, controls demonstrated RT speeding during the DEV condition (Fig. 9.3A; Strauss et al., 2011). RT speeding during the DEV condition was not seen in patients, consistent with impaired Go-learning in patients (Fig. 9.3A). For the remaining conditions, there was little slowing in either group during the IEV condition, and there were no differences between patients and controls on the control conditions (CEV or CEVR). A higher burden of negative symptoms (categorized by a median split) was also associated with greater differences in the DEV condition compared to controls, with LNS patients in between (although, the correlation between scale for the assessment of negative symptoms (SANS) and DEV was not significant). This pattern of results is suggestive of a Go-learning deficit associated with the $D_1$ subsystem in patients with negative symptoms. However, the modeling of the behavior indicated that the largest difference between patients and controls was on the

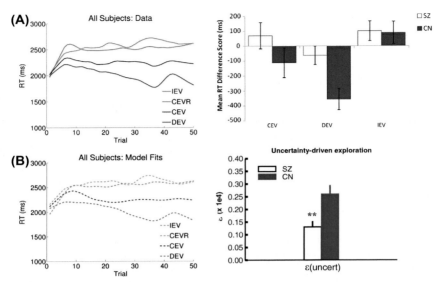

FIGURE 9.3    Results for the temporal utility integration task experiment. (A) Subjects learn to adjust the timing of their responses over trials toward the maximal expected value. This is especially notable for the decreased expected value (DEV) condition where controls show the required speeding of reaction times (RTs) to maximize reward, while patients do not. (B) The modeling replicates faithfully the pattern of response times seen in the data across all subjects. Also shown is the difference between patients and controls in the modeled parameter $\varepsilon$, representing the degree of exploration during uncertainty. This parameter was correlated with scale for the assessment of negative symptoms avolition—anhedonia. CEV, constant expected value; CEVR, constant expected value reversed; CN, control; IEV, increased expected value. *Reprinted from Strauss, G.P., Frank, M.J., Waltz, J.A., Kasanova, Z., Herbener, E.S., Gold, J.M., 2011. Deficits in positive reinforcement learning and uncertainty-driven exploration are associated with distinct aspects of negative symptoms in schizophrenia. Biol. Psychiatry 69, 424—431.*

explore parameter $\varepsilon$ (see Fig. 9.3B) and this parameter correlated with SANS avolition—anhedonia scores (although there were no differences using the median split). As mentioned above, this parameter reflects the degree to which exploration occurs in proportion to uncertainty about reward. With respect to Go- and NoGo-learning rates specifically, there was only weak evidence for a reduction in $\alpha^{Go}$ in patients (no differences in $\alpha^{NoGo}$) and no association with negative symptoms, suggesting that the explore parameter may be a larger contributor to patient-control differences in RL performance than alterations the "Go" system.

Given that patients showed a modest deficit in $\alpha^{Go}$ and DEV, but not $\alpha^{NoGo}$ or IEV, it would seem that patients with schizophrenia possess specific RL differences in speeding RTs for reward. Moreover, these deficits are at least partly more pronounced in patients with high number of negative symptoms (albeit not correlated with symptoms). This deficit is suggestive of reduced sensitivity to positive PEs potentially by impairments in the $D_1$-driven basal-ganglia pathway. Ultimately, this deficit would lead to abnormalities in using positive feedback to guide behavior with the clinical expression of this strategy manifesting as negative symptoms. By contrast the lack of differences between groups on the IEV condition is suggestive of intact $D_2$-driven NoGo-learning and processing of negative feedback. However, the lack of a noticeable IEV effect in controls dampens the robustness of this conclusion. The results also indicate that a reduced likelihood for exploration may be feature of schizophrenia associated with negative symptom expression. High avolition—anhedonia scores were associated with a failure to strategically adjust responses to gather more information about uncertain alternative behaviors; a function served by prefrontal cortex. It was therefore argued that patients with anhedonia may assign negative value to uncertainty, potentially limiting their capacity to explore actions that would lead to significantly better outcomes.

## 9.6 PAVLOVIAN BIAS—EXTENDED Q-LEARNING

### 9.6.1 Rationale

The TUIT and the probabilistic selection task presents evidence suggesting that negative symptoms might be due to an inability to use positive PEs, particularly for states associated with a "Go" action. However, despite seeming to be selective for active or "Go" responding, none of the above mentioned tasks has explicitly used the complete withholding of an action to indicate learning. There are significant innate "Pavlovian" biases that modulate the behavior observed during operant learning tasks, such that reward-predicting stimuli invigorate and loss-predicting stimuli inhibit active or "Go" responding (Guitart-Masip et al., 2012; Cavanagh et al., 2013). Just as in the TUIT task, action invigoration in response to positive PEs and action inhibition in response to negative PEs are linked with phasic dopamine firing and the propensity to take action. The Pavlovian bias discussed here describes a propensity for more efficient learning when requiring an active response to rewards and inhibiting a response to punishments. If we cross these associations, such that rewards are obtained by withholding action and losses are avoided by initiating an action, learning becomes less efficient requiring an external system to override the innate action bias. One region of override has been located in the prefrontal cortex, which is engaged more strongly when stimuli that possess an anti-Pavlovian bias are presented.

An investigation into the role of Pavlovian biases therefore has the potential to examine several possible competing accounts of poor RL performance in schizophrenia. For example, it is possible that reward-learning deficits could be explained by a reduction in Pavlovian biases modulating action, rather than due to degraded representations of action values. There is also the possibility that alterations in dopamine signaling in schizophrenia, inherent in the illness or related with dopamine D$_2$ receptor blockade, might dampen Pavlovian biases even in the presence of a prefrontal cortical dysfunction that fails to override the innate Pavlovian bias. Such an effect would potentially show in a situation where dampening action invigoration in response to rewarding stimuli is beneficial. This would be able to determine to some degree the generality of reward specific learning deficits in schizophrenia. Moreover, given the above findings of intact learning to negative outcomes, we may also be able to determine whether this is specific to anti-Pavlovian conditions or whether the behavior generalizes to a more explicit requirement to withhold responding.

The Pavlovian bias task reported on below presents the participant with four stimuli. Two stimuli yield rewards at a 0.8/0.2 ratio for correct responses, two stimuli yield punishments at the same ratio for incorrect responses. Two possible options are allowed by the participant, they are either required to press a button, or withhold pressing a button within the time frame allowed. Response options and stimulus-feedbacks are crossed, to give four conditions: "Go-to-Win," "Go-to-Avoid," "NoGo-to-Win," and "NoGo-to-Avoid." The Go-to-Win and the NoGo-to-Avoid conditions are the Pavlovian consistent conditions. The Go-to-Avoid and the NoGo-to-Win are the Pavlovian inconsistent conditions, with NoGo-to-Win being the most difficult condition for subjects to learn.

## 9.6.2 Pavlovian Bias Model

Like the above models, we begin with the backbone of a Q-learning module keeping track of state-action pairings, with PEs updating the Q-values as a function of the learning rate $\alpha$. Additionally, differential sensitivities to rewards and punishments are modeled in the PE calculation using separate reward and punishment PEs:

$$Q_{t-1}(s,a) + \alpha \left( \rho^{Rew|Pun} r_t + Q_{t-1}(s,a) \right)$$

$$\delta_t^{Rew} = \rho_t^{Rew} \times r_t - Q_t(s,a)$$

$$\delta_t^{Pun} = \rho_t^{Pun} \times r_t - Q_t(s,a)$$

where $\rho^{Rew|Pun}$ describe the scaling applied to rewards or punishments. There is now the possibility to withhold responding as a valid "action," we include a Go bias $b$ that modulates the state-action value for go stimuli, such that:

$$Q_t(s_{Go}) = Q_t(s_{Go}) + b$$

reflecting the degree to which participants favor active responding over withholding a response. A final addition to the model is the inclusion of a Pavlovian bias parameter $\pi$. To include this parameter, we calculate the overall experienced value of a stimulus (similar to the actor-critic model) by:

$$V_t(s) = V_{t-1}(s) + \alpha\left(\rho^{Rew|Pun}r_t + V_{t-1}(s)\right)$$

We then modify the state-action value of Go associated state-action pairs as a function of the stimulus value weighted by the Pavlovian bias $\pi$:

$$Q_t(s_{Go}) = Q_t(s^{Go}) + b + \pi V(s)$$

thus modeling the degree of Pavlovian bias by modifying state-action values as a function of their overall history of reward or punishment.

### 9.6.3 Findings

Fig. 9.4 presents the behavioral results comparing patients and controls across trials. Within each of the four conditions, patients performed worse than controls on the Pavlovian consistent conditions "Go-to-Win" and "NoGo-to-Avoid" and actually performed similarly or better on the Pavlovian inconsistent conditions "Go-to-Avoid" and "NoGo-to-Win" Fig. 9.5A (Albrecht et al., 2016). This is a complex mix of replication and nonreplication: we were able to replicate the poor reward-based learning in patients, but this only existed on the Go condition, with opposite results in the NoGo condition; similarly, we were able to replicate a relatively intact loss-avoidance learning in the Go condition, but there was a surprising performance impairment in the NoGo condition. Overall, this resulted in the result that patients had less Pavlovian bias compared with controls (Fig. 9.5B). However, the modeling indicated only weak differences between the two groups on the fitted Pavlovian bias parameter. Instead, the most robust differences between the two groups were reductions in the Go bias and Punishment sensitivity parameters in patients. It seems this combined reduction may be responsible for the observed Pavlovian bias reduction in the patient group as a whole, a reduction in Go bias gives a disadvantage to patients during Go-to-Win but an advantage during NoGo-to-Win conditions, with the reduced punishment sensitivity necessary to model the poor performance in the patients on the NoGo-to-Avoid condition.

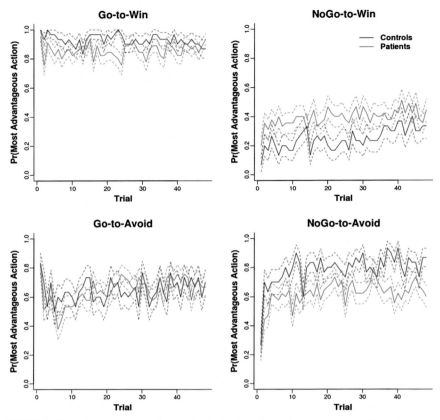

FIGURE 9.4    Behavioral results for the Pavlovian bias experiment on a trial-by-trial basis comparing patients and controls.

There were no correlations found between negative symptoms and reward sensitivity or punishment sensitivity. Only the learning rate parameter was significantly associated with negative symptoms, indicating a general impairment in updating EV, potentially similar to the $\alpha^{Go}$ findings from the TUIT task. Moreover, behavioral performance on the Go-to-Avoid condition was correlated negatively with negative symptoms, contrary to our previous results indicating that negative symptoms are associated with intact learning from punishments. There were, however, interesting effects in a subgroup of patients taking the antipsychotic clozapine. Notably, all of the above group differences were amplified in the patients on clozapine. Moreover, analysis of the modeling output managed to identify a significant reduction of the Pavlovian bias parameter in those taking clozapine, suggesting a significant impact of the type of antipsychotic in RL. With no definitive unique mechanism of

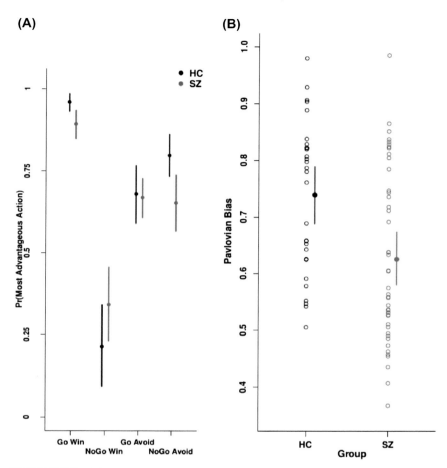

FIGURE 9.5 (A) Mean + 95% highest density interval for each condition and participant group for the summed accuracy over trials. (B) Pavlovian bias calculated from the behavioral results by averaging reward-based invigoration ((Go on Go-to-Win + NoGo-to-Win)/Total Go) and punishment-based suppression ((NoGo on Go-to-Avoid + NoGo-to-Avoid)/Total NoGo).

action for clozapine yet found, it is unknown exactly which target or targets may be responsible for the findings in this study. Overall, we suggested that the pattern of results seen in schizophrenia on this Pavlovian bias task is most likely due to disrupted communication between the striatum and frontal cortex and between striatal dopamine driven linkage of feedback with behavior that has been previously seen in RL tasks (Schlagenhauf et al., 2009; Quidé et al., 2013; Yoon et al., 2013). This is as opposed to enhanced functioning of the inferior frontal gyrus, one region that is likely responsible for overriding innate action biases in response to a history of feedback.

# 9.7 DIRECT ADDITION OF WORKING MEMORY TO REINFORCEMENT LEARNING MODELS

## 9.7.1 Rationale

The tasks presented above often use the repeated presentation of two to four stimuli interspersed amongst one other. This presents a moderate demand on WM systems that likely play a significant role during the kinds of RL tasks we have described above, particularly in patients with schizophrenia who frequently show deficits in WM compared with controls. This presents a potentially important confound when attempting to determine the relative contributions of different neural systems at play in schizophrenia and whether they play a role in negative symptom formation. To more explicitly assess the contribution of WM to RL we parametrically manipulated WM load. We have done this by using multiple sets of stimuli, each set containing between one and six stimuli (Collins and Frank, 2012). For example, low WM load Set A may possess two stimuli requiring an appropriate response (e.g., by pressing button 1 or button 2), while high WM load Set B possesses five interspersed stimuli. The response accuracy for Set B naturally shows significant learning delays compared to Set A, and takes many more stimuli to be presented before plateau performance is reached.

## 9.7.2 Reinforcement Learning and Working Memory Model

To model this data, we can again use a mixture of two models: a classic RL module and a fast learning but capacity-limited WM module. The RL module uses many of the standard Q-learning features described above, with Action-State pairings $Q(s,a)$ derived from PEs modulated by an estimated learning rate $\alpha$, and softmax temperature $\beta$. The fast-acting WM module again uses the same Q-learning model as a base, with some adjustments. The learning rate $\alpha$, instead of being estimated, is set to a value of 1 so that it can learn a state-action relationship based on a single trial. However, this ability to learn is limited by WM capacity. To model this limitation, the probability of a stimulus being in WM is given as:

$$p_{\text{WM}} = \rho \times \min\left(1, \frac{K}{n_s}\right)$$

where $K$ is an estimated parameter for WM capacity, $n_s$ is the set size or number of unique stimuli in the block, and $\rho$ is an estimated parameter (between 0 and 1) indicating reliance on the WM system versus the RL system. Two additional parameters are also estimated in the final model, a forgetting/decay parameter $\varphi$ for both the RL and WM modules and a perseveration parameter to account for tendencies to repeat actions despite negative feedback.

Similar to the Q-learning/Actor-Critic hybrid described above, the RL and WM modules are integrated into one model using a mixture function. Each module generates a policy for action, one for the straight RL module and one for the WM module. The final decision policy for the combined model is a mixture function given as:

$$\pi_{RLWM} = p_{WM}\pi_{WM} + (1 - p_{WM})\pi_{RL}$$

where $\pi_{RL}$ and $\pi_{WM}$ are the policies from the softmax function for the RL module and WM modules, respectively. The combined model therefore captures: (1) rapid and accurate encoding of information when a low amount of information is to be stored, and (2) the decrease in the likelihood of storing or maintaining items when more information is presented or when distractors are presented during the maintenance period.

### 9.7.3 Findings

Learning was much faster for low set sizes compared with high set sizes, as could be expected given an influence of WM on RL performance (Collins et al., 2014). Differences between the two groups were particularly notable for the time taken to reach plateau/speed of learning across set sizes. Controls reached plateau performance very early at set sizes two and three, generally within the first three exposures of each stimulus. Controls also managed to reach an almost asymptotic performance for all high set size blocks after seven exposures. By contrast, patients learned much slower, taking between four and six exposures at the low set sizes to reach plateau. The impaired performance also continued during the higher set size blocks, failing to reach plateau accuracy after seven exposures. Notably, patients appeared to learn slower at all set sizes and were more impaired relative to controls as set size was increased suggesting a substantial influence of WM deficits in patient performance.

The modeling results confirmed the role of WM deficits on patient RL performance. The fitted WM parameters of capacity, forgetting, and reliance were all found to be lower in patients compared with controls, indicating both general impairments in WM and specific impairments in using a deficient WM for RL performance. This was especially notable given the presence of a seemingly intact RL module and intact RL-associated parameters in patients. Intact RL processes suggest relatively normal functioning of the basal ganglia system in medicated patients, where slow, incremental is learning thought to be driven by dopaminergic PE signaling.

There were no correlations found between modeling parameters and negative symptoms, despite the number of computational parameters implicating WM and feedback processing deficits during RL in patients. The failure to find statistically significant correlations was suggested to be because the task is more related to rule-based WM process involving the lateral prefrontal cortex rather than the mPFC/OFC. Previous associations with negative symptoms are hypothesized to relate with tasks or constructs that are involved in representing specific reward values for each stimulus/action, functions attributed to more limbic portions of PFC, e.g., OFC (Gold et al., 2012). While this study did not show a specific negative symptom association, it does highlight alternative mechanisms for RL performance deficits in schizophrenia. Notably, these results suggest that, again, patients are less able to use feedback to guide decision-making due to a core deficit in WM appearing as an RL deficit on RL tasks. Interestingly, in a replication and extension of this study (Collins et al., submitted), patients again showed significant WM deficits but showed intact performance during a transfer phase that followed the above procedure. This transfer phase potentially reflects a more pure RL aspect of learning to probabilistic RL values.

## 9.8 SUMMARY

The above studies highlight the utility of computational modeling in the understanding of decision-making deficits in schizophrenia generally and the computational processes carried out by the brain that may be more specifically associated with negative symptoms. One common theme throughout these studies was a behavioral deficit in learning from rewards relative to punishments in patients with a greater negative symptom burden. This was shown by reduced ability to track value in the TUIT during the DEV condition, impaired learning and transfer performance on the probabilistic stimulus-selection task for high probability rewarding stimuli, and impaired performance on the Pavlovian bias experiment for the Go-to-Win condition. Moreover, there appeared to be intact processing of punishment stimuli for each of these tasks, as long as there was a response requirement necessitating an active response.

However, the computational modeling of the behavior identified an alternative explanation for this pattern of results. Namely, that higher order processes were probably more likely to be responsible for the associations between behavior and negative symptoms. This can be seen in the mixing parameter from the probabilistic stimulus-selection task and the explore parameter in the TUIT both being associated with negative symptom expression. These two parameters can be contrasted with the other parameters that would be more likely to reflect a differential sensitivity to rewards and punishments but were not associated with negative symptoms, including the $\alpha^{Go}$ parameter from the TUIT and the punishment sensitivity parameter $\rho^{Pun}$ from the Pavlovian bias experiment. Therefore, from the modeling evidence presented here, we would assume that negative symptoms are a failure in two processes: the ability to adequately engage the OFC over the basal ganglia during RL (mixing parameter) and the failure to adjust strategic responding to gather information during uncertainty (explore parameter). Both these parameters are putatively served by frontal cortical regions (OFC, dorsal anterior cingulate cortex) and represent higher order cognitive functions, rather than more basic responding to rewards and punishments. Indeed, a higher order explanation of negative symptoms appears to match the updated perceptions of negative symptoms mentioned in the introduction, i.e., in the moment hedonic processes seem to be intact, but assessment of future expected enjoyment/utility are dampened. Furthermore, such deficits in higher order processes might also be expected to alter cost-benefit decision-making in people with schizophrenia. The ability to represent prospective reward value is critical to the decision to expend effort to obtain a reward. There is replicated evidence that cost-benefit decision-making is altered in schizophrenia (Fervaha et al., 2013; Gold et al., 2013; Barch et al., 2014), often as a function of negative symptom severity (Gold et al., 2013; Barch et al., 2014). Such a model of symptoms provides a meeting of clinical experience with computational psychiatry.

Separate to the issue of negative symptoms, the reinforcement learning and working memory study highlighted a generic deficit in WM that seems to play a dominant role in RL performance in schizophrenia. This study provided a more direct merge between behavior and computational modeling: performance in patients was worse during conditions that placed increasing load on WM and computational modeling identified a consistent deficit in WM parameters. Furthermore, the RL module seemed to be intact in patients, again highlighting that patient-control differences are likely reflected by differences in higher order processes. Neurophysiological evidence for the notion of intact RL with deficit higher order functioning is emerging. A number of event-related potential (ERP) studies have shown intact feedback-related negativity (FRN) in patients (Morris et al., 2008, 2011; Koch et al., 2010), an evoked potential that is elicited in response to feedback and is proportional to the size of the PE (Holroyd and Coles, 2002). Indeed, during the Pavlovian bias experiment (Albrecht et al., 2016), we showed that the FRN tracked with PE in patients during the early stages of the ERP (i.e., during the FRN), but later tracking of the ERP with PE (around the P300 time region) associated with more complex processing of feedback (Fischer and Ullsperger, 2013; Ullsperger et al., 2014) showed marked differences from controls.

It is uncertain yet whether any of these parameters/behaviors are specific to the generation of negative symptoms in schizophrenia or whether they may also cross diagnostic boundaries. Indeed, even within the scope of schizophrenia and negative symptoms, the idea that there might be one common underlying mechanism for the emergence of negative symptoms seems implausible. This seems especially so given the highly heterogenous nature of schizophrenia (Tandon et al., 2013; Arnedo et al., 2014; Schizophrenia Working Group of the Psychiatric Genomics Consortium, 2014; Jablensky, 2015). Nevertheless, it is plausible that the mechanistic understandings of negative symptoms in schizophrenia obtained from computational modeling could apply to syndromes such as depression. In a meta-analysis summarizing the results of computational modeling during RL in depression and anhedonia, Huys et al. (2013) find evidence for a reduction in reward sensitivity over other modeled parameters. This suggests a substantially different origin of symptom formation between the two clinical groups. In depression, deficits may occur at lower order levels related more proximally with reward processing, while in schizophrenia, deficits seem to occur at higher order levels implicating more strategic deficits.

## 9.9 CONCLUSION

Our computational modeling work in schizophrenia has identified novel possible mechanisms for the differences seen in RL performance that would not otherwise be explicitly identifiable given a simple analysis of behavior. Although there has been a start at linking RL modeling with neurophysiology in schizophrenia, further investigations integrating multiple imaging modalities with behavior will likely refine our understandings of the affected processes responsible for the formation of negative symptoms. This offers a powerful approach for identifying mechanistic pathways, where our current evidence suggests that frontal cortical deficits are the most likely originator of negative symptom development. Notably, this appears to be to the exclusion of deficits in more basic reward processing function.

### References

Albrecht, M.A., Waltz, J.A., Cavanagh, J.F., Frank, M.J., Gold, J.M., 2016. Reduction of Pavlovian bias in schizophrenia: enhanced effects in clozapine-administered patients. PLoS One 11, e0152781.

Arnedo, J., Svrakic, D.M., del Val, C., Romero-Zaliz, R., Hernández-Cuervo, H., Fanous, A.H., et al., 2014. Uncovering the hidden risk architecture of the schizophrenias: confirmation in three independent genome-wide association studies. Am. J. Psychiatry 172, 139—153.

Barch, D.M., Treadway, M.T., Schoen, N., 2014. Effort, anhedonia, and function in schizophrenia: reduced effort allocation predicts amotivation and functional impairment. J. Abnorm. Psychol. 123, 387.

Bayer, H.M., Lau, B., Glimcher, P.W., 2007. Statistics of midbrain dopamine neuron spike trains in the awake primate. J. Neurophysiol. 98, 1428—1439.

Cavanagh, J.F., Eisenberg, I., Guitart-Masip, M., Huys, Q., Frank, M.J., 2013. Frontal theta overrides Pavlovian learning biases. J. Neurosci. 33, 8541—8548.

Cohen, A.S., Minor, K.S., 2010. Emotional experience in patients with schizophrenia revisited: meta-analysis of laboratory studies. Schizophr. Bull. 36, 143—150.

Collins, A.G.E., Brown, J.K., Gold, J.M., Waltz, J.A., Frank, M.J., 2014. Working memory contributions to reinforcement learning impairments in schizophrenia. J. Neurosci. 34, 13747—13756.

Collins, A.G.E., Frank, M.J., 2012. How much of reinforcement learning is working memory, not reinforcement learning? A behavioral, computational, and neurogenetic analysis. Eur. J. Neurosci. 35, 1024—1035.

Collins, A.G.E., Frank, M.J., 2014. Opponent actor learning (OpAL): modeling interactive effects of striatal dopamine on reinforcement learning and choice incentive. Psychol. Rev. 121, 337—366.

Evans, J.D., Bond, G.R., Meyer, P.S., Kim, H.W., Lysaker, P.H., Gibson, P.J., et al., 2004. Cognitive and clinical predictors of success in vocational rehabilitation in schizophrenia. Schizophr. Res. 70, 331—342.

Fellows, L.K., Farah, M.J., 2007. The role of ventromedial prefrontal cortex in decision making: judgment under uncertainty or judgment per se? Cereb. Cortex 17, 2669—2674.

Fervaha, G., Graff-Guerrero, A., Zakzanis, K.K., Foussias, G., Agid, O., Remington, G., 2013. Incentive motivation deficits in schizophrenia reflect effort computation impairments during cost-benefit decision-making. J. Psychiatr. Res. 47, 1590—1596.

Fischer, A.G., Ullsperger, M., 2013. Real and fictive outcomes are processed differently but converge on a common adaptive mechanism. Neuron 79, 1243—1255.

Frank, M.J., Claus, E.D., 2006. Anatomy of a decision: striato-orbitofrontal interactions in reinforcement learning, decision making, and reversal. Psychol. Rev. 113, 300.

Gold, J.M., Strauss, G.P., Waltz, J.A., Robinson, B.M., Brown, J.K., Frank, M.J., 2013. Negative symptoms of schizophrenia are associated with abnormal effort-cost computations. Biol. Psychiatry 74, 130—136.

Gold, J.M., Waltz, J.A., Matveeva, T.M., Kasanova, Z., Strauss, G.P., Herbener, E.S., et al., 2012. Negative symptoms and the failure to represent the expected reward value of actions: behavioral and computational modeling evidence. Arch. Gen. Psychiatry 69, 129—138.

Guitart-Masip, M., Huys, Q.J.M., Fuentemilla, L., Dayan, P., Duzel, E., Dolan, R.J., 2012. Go and no-go learning in reward and punishment: interactions between affect and effect. Neuroimage 62, 154—166.

Hart, A.S., Rutledge, R.B., Glimcher, P.W., Phillips, P.E.M., 2014. Phasic dopamine release in the rat nucleus accumbens symmetrically encodes a reward prediction error term. J. Neurosci. 34, 698—704.

Holroyd, C.B., Coles, M.G., 2002. The neural basis of human error processing: reinforcement learning, dopamine, and the error-related negativity. Psychol. Rev. 109, 679.

Huys, Q.J., Pizzagalli, D.A., Bogdan, R., Dayan, P., 2013. Mapping anhedonia onto reinforcement learning: a behavioural meta-analysis. Biol. Mood Anxiety Disord. 3, 12.

Huys, Q.J.M., Maia, T.V., Frank, M.J., 2016. Computational psychiatry as a bridge from neuroscience to clinical applications. Nat. Neurosci. 19, 404—413.

Jablensky, A., 2015. Schizophrenia or schizophrenias? The challenge of genetic parsing of a complex disorder. Am. J. Psychiatry 172, 105—107.

Katschnig, H., 2000. Schizophrenia and quality of life. Acta Psychiatr. Scand. 102, 33—37.

Koch, K., Schachtzabel, C., Wagner, G., Schikora, J., Schultz, C., Reichenbach, J.R., et al., 2010. Altered activation in association with reward-related trial-and-error learning in patients with schizophrenia. Neuroimage 50, 223—232.

Kring, A.M., Moran, E.K., 2008. Emotional response deficits in schizophrenia: insights from affective science. Schizophr. Bull. 34, 819—834.

Maia, T.V., Frank, M.J., 2017. An integrative perspective on the role of dopamine in schizophrenia. Biol. Psychiatry 81, 52–66.

Montague, P.R., Dayan, P., Sejnowski, T.J., 1996. A framework for mesencephalic dopamine systems based on predictive Hebbian learning. J. Neurosci. 16, 1936–1947.

Morris, S.E., Heerey, E.A., Gold, J.M., Holroyd, C.B., 2008. Learning-related changes in brain activity following errors and performance feedback in schizophrenia. Schizophr. Res. 99, 274–285.

Morris, S.E., Holroyd, C.B., Mann-Wrobel, M.C., Gold, J.M., 2011. Dissociation of response and feedback negativity in schizophrenia: electrophysiological and computational evidence for a deficit in the representation of value. Front. Hum. Neurosci. 5, 123.

Moustafa, A.A., Cohen, M.X., Sherman, S.J., Frank, M.J., 2008. A role for dopamine in temporal decision making and reward maximization in parkinsonism. J. Neurosci. 28, 12294–12304.

Norman, R.M.G., Malla, A.K., McLean, T., Voruganti, L.P.N., Cortese, L., McIntosh, E., et al., 2000. The relationship of symptoms and level of functioning in schizophrenia to general wellbeing and the quality of life scale. Acta Psychiatr. Scand. 102, 303–309.

O'Doherty, J., Dayan, P., Schultz, J., Deichmann, R., Friston, K., Dolan, R.J., 2004. Dissociable roles of ventral and dorsal striatum in instrumental conditioning. Science 304, 452–454.

Papp, M., Willner, P., Muscat, R., 1991. An animal model of anhedonia: attenuation of sucrose consumption and place preference conditioning by chronic unpredictable mild stress. Psychopharmacology (Berl.) 104, 255–259.

Quidé, Y., Morris, R.W., Shepherd, A.M., Rowland, J.E., Green, M.J., 2013. Task-related fronto-striatal functional connectivity during working memory performance in schizophrenia. Schizophr. Res. 150, 468–475.

Schizophrenia Working Group of the Psychiatric Genomics Consortium, 2014. Biological insights from 108 schizophrenia-associated genetic loci. Nature 511, 421–427.

Schlagenhauf, F., Sterzer, P., Schmack, K., Ballmaier, M., Rapp, M., Wrase, J., et al., 2009. Reward feedback alterations in unmedicated schizophrenia patients: relevance for delusions. Biol. Psychiatry 65, 1032–1039.

Schoenbaum, G., Roesch, M., 2005. Orbitofrontal cortex, associative learning, and expectancies. Neuron 47, 633–636.

Strauss, G.P., Frank, M.J., Waltz, J.A., Kasanova, Z., Herbener, E.S., Gold, J.M., 2011. Deficits in positive reinforcement learning and uncertainty-driven exploration are associated with distinct aspects of negative symptoms in schizophrenia. Biol. Psychiatry 69, 424–431.

Strauss, G.P., Gold, J.M., 2012. A new perspective on anhedonia in schizophrenia [internet]. Am. J. Psychiatry 169, 364–373. Available from: http://journals.psychiatryonline.org/article.aspx?volume=169&page=364.

Strauss, G.P., Harrow, M., Grossman, L.S., Rosen, C., 2010. Periods of recovery in deficit syndrome schizophrenia: a 20-year multi–follow-up longitudinal study. Schizophr. Bull. 36, 788–799.

Tandon, R., Gaebel, W., Barch, D.M., Bustillo, J., Gur, R.E., Heckers, S., et al., 2013. Definition and description of schizophrenia in the DSM-5. Schizophr. Res. 150, 3–10.

Ullsperger, M., Fischer, A.G., Nigbur, R., Endrass, T., 2014. Neural mechanisms and temporal dynamics of performance monitoring. Trends Cogn. Sci. 18, 259–267.

Waltz, J.A., Frank, M.J., Robinson, B.M., Gold, J.M., 2007. Selective reinforcement learning deficits in schizophrenia support predictions from computational models of striatal-cortical dysfunction. Biol. Psychiatry 62, 756–764.

Yoon, J.H., Minzenberg, M.J., Raouf, S., D'Esposito, M., Carter, C.S., 2013. Impaired prefrontal-basal ganglia functional connectivity and substantia nigra hyperactivity in schizophrenia. Biol. Psychiatry 74, 122–129.

# 10

# Bayesian Approaches to Learning and Decision-Making

## Quentin J.M. Huys[1,2]

[1] University Hospital of Psychiatry, Zürich, Switzerland; [2] University of Zürich and Swiss Federal Institute of Technology (ETH), Zürich, Switzerland

## O U T L I N E

## 10.1 INTRODUCTION

Learning and decision-making are highly intertwined processes. While learning influences what decisions are taken, the decisions taken determine what will be learned. Jointly, they serve the purpose of optimizing behavior and a breakdown in either will upset the functioning of the other. This vicious circle is often seen in mental illness, where poor decisions in mental illness lead to the self-selection of individuals into high-risk situations (Kendler et al., 1999) and thereby likely to more mental illness.

In this chapter, we will consider a series of approaches to the guidance of behavior. Some, mostly from Reinforcement Learning (RL; Sutton and Barto, 1998) involve "learning," while others, from the related field of Dynamic Programming, are more akin to inference (Bertsekas and Tsitsiklis, 1996). The key aspect to consider is that actions taken now do not just have rewarding or punishing consequences now, but also in the future. For instance, theft may lead to a short-term gain, but in the longer term may well lead to very significant losses that far outweigh the short-term gains. Identifying optimal behaviors at any one point in time, therefore, requires thinking ahead and considering the various possible consequences of any current behavior. This, however, is extremely difficult: first, the list of possible things that may happen in the future is vast, and second, the future is uncertain. RL is a field with a host of techniques for taking long-term outcomes into account when making decisions.

This chapter will first introduce so-called Markov Decision Problems (MDPs) and their solutions formally. In a second part, it will give the reader tools to use these models to examine choice behavior. In a third part, we will examine a few specific models as examples of decision-making in health and illness. In the following, we focus on the key concepts and omit a number of important details for the sake of simplicity. The interested reader is referred to Bertsekas and Tsitsiklis (1996) and Sutton and Barto (1998) for accessible but more in-depth treatments.

# 10.2 MARKOV DECISION PROBLEMS

Fig. 10.1A shows the general MDP setup that underlies RL and Dynamic Programming methods. An MDP is defined by five components that we will briefly introduce below:

- set of states $s \in \mathcal{S}$
- set of actions $a \in \mathcal{A}$ and an associated set of action transition matrices $\mathcal{T}^a$
- reward function $\mathcal{R}$
- policy $\pi$

The intuition is that an agent is in some particular state $s$. In this state, the agent can perform certain actions $a$. Depending on the environment, this leads to a new state $s'$ and a reinforcement $r$, which can be positive or

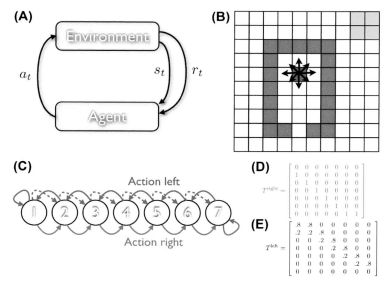

FIGURE 10.1   (A) The setting. An agent interacts with an environment by choosing actions which result in rewards and in turn influence its current state. (B) Grid world example. Each square in the grid is a different state $s$. The state of the agent is indicated by a *green square*, i.e., it is roughly in the middle of the grid. Actions correspond to moving around on this grid. In this example, the agent can move to all adjacent squares, i.e., has 8 actions available in each state (exemplified by the *black arrows* emerging from the *green square* in the middle). Some states lead to losses, here indicated by the color red, and some to gains, here indicated by yellow. A policy assigns each state preferences for particular actions. The aim is to find an optimal policy, i.e., one that maximizes long-term rather than just immediate reward. (C) Simple linear state space with two actions. While the red action "right" is deterministic and thus has only 0s and 1s in the transition matrix (D), the green action left is probabilistic, corresponding to a transition matrix with off-diagonal terms (E).

negative. Fig. 10.1B shows a more specific example: a so-called grid world, where the state is simply the position on the grid.

The techniques described below will typically focus on simple definitions of states within particular experiments, where the relevant states can simply be the stimuli presented during the experiment. However, the notion of state $s$ in RL is potentially very broad. In neuroscience terms, it could include internal states such as arousal or hunger, and as such is clearly a very complex construct.

The actions $a$ are defined in terms of their impact on states. In Fig. 10.1C, the action "going left" is defined in terms of moving from any one state to its left neighbor. More generally, actions are defined in terms of probability distributions over successor states (Fig. 10.1D and E). Putting all state succession probabilities for one action, next to each other, into one matrix results in the transition matrix $\mathcal{T}^a$ for that action (Fig. 10.1D and E). This describes the consequences of emitting that action in each of the existing states; it is generally assumed that the transition matrices are fixed and determined by the world, though they may not be known to the agent.

This definition of actions has an important consequence for how states are defined: The consequences of actions must depend only on the current state, and not on past states. Consider braking when driving a car. The impact of braking depends not only on the position of the care, but also on its speed. Hence, the impact of braking on transitions to other states cannot be described purely in terms of the current position. For the techniques below to apply, the problem must be a so-called MDP. For this to be true, speed should be part of how states are defined in the car example, such that the consequence of braking is clearly defined for each state independent of what the previous states were.

The reward $r$ is a scalar, i.e., a unidimensional number that takes on positive or negative values for rewards and losses, respectively. The richness of real rewards is captured by the dependence on actions and state transitions: Rewards $r$ are generated by a reward function $\mathcal{R}(s, a, s')$ that depends both on the action taken, and the current and next states. Just like ingesting food is rewarding when hungry but not when sated, taking a step to the right can lead to a loss in states left of the red punishing barrier in Fig. 10.1, and to reward when left of the yellow reward area. Just like the transition matrices $\mathcal{T}$, the reward function $\mathcal{R}$ is assumed to be a fixed part of the environment, though again it may not be known to the agent. The agent's estimates of the transition matrices and the reward function are referred to as the agent's *model* $\mathcal{M}$ of the world.

The aim is to find an optimal policy $\pi^*(a; s)$. A policy $\pi(a; s)$ describes the probability of taking an action $a$ in state $s$. A policy is optimal if it always chooses one of the optimal actions in each state, where the optimal action is the one that maximizes the total sum of rewards that can be earned in the long term. Conceptually the simplest approach to infer the

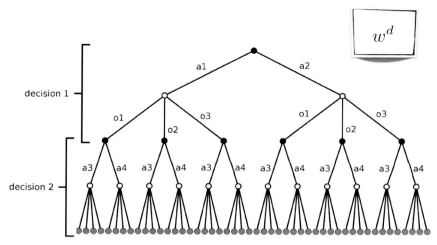

FIGURE 10.2    Decision tree. At the root of the tree, there are two available actions a1 and a2, each of which probabilistically leads to one of three outcomes (o1–o3). For each of these, there are new options a3 and a4. Overall, the size of the tree increases rapidly with the depth $d$ and width $w$ of the tree as $w^d$.

optimal policy is to consider all possible actions from a state; all the resulting state transitions and rewards; then all possible next actions for the successor states, etc. This results in a decision tree, with the root at the current state (Fig. 10.2). Unfortunately, these decision trees grow rapidly in size. For the simple grid world example, the number of actions and successor state to each state is 9 (disregarding the boundaries), and hence the decision tree corresponding to looking $d$ steps ahead has $9^d$ branches. Such an explicit tree search is hence prohibitive for all but the very simplest of problems.

## 10.2.1 Bellman Equation

Optimal, in RL, is defined in terms of achieving the maximal expected sum over rewards $r_{t'}$ in the future, i.e., for times $t' \geq t$. The expected total future reward from state $s$ at time $t$ when following a particular policy $\pi$ is called the value $\mathcal{V}^\pi(s)$ of the state and defined as:

$$\mathcal{V}^\pi(s_t) = \mathbb{E}\left[\sum_{t'=0}^{\infty} r_{t+t'}\gamma^{t'} \Big| s_t; \pi\right] \tag{10.1}$$

where the discounting factor $0 \leq \gamma < 1$ is not only necessary to ensure that the sum is finite, but also gives rewards in the near future more weight than rewards in the distant future. It is set to 1 if only finite problems are considered. The key insight is that Eq. (10.1) is a sum and due to the linearity of expectations (because the average of two means is the same as the mean of two averages), it can be rewritten into two terms as:

$$\mathcal{V}^{\pi}(s_t) = \underbrace{\mathbb{E}[r_t|s_t;\pi]}_{\text{immediate reward}} + \underbrace{\mathbb{E}\left[\sum_{t'=1}^{\infty} r_{t+t'}\gamma^{t'}\Big|s_t;\pi\right]}_{\gamma\cdot\text{reward from next timestep onwards}}$$

The total future reward from the next timestep onward, the second term in the equation above, is simply the value of the next state–action pair $\mathcal{V}^{\pi}(s_{t+1})$, and hence we can write as:

$$\mathcal{V}^{\pi}(s_t) = \mathbb{E}[r_t|s_t;\pi] + \mathbb{E}\left[\gamma\mathcal{V}^{\pi}(s_{t+1})\Big|s_t;\pi\right]$$

The rewards $r_t$ are drawn from the reward process $\mathcal{R}(s_t, a_t, s_{t+1})$. The expectations $\mathbb{E}[\cdot]$ are over two processes: first, the likely actions taken, and second, the likely consequences of those actions. Expanding these expectations and substituting the policy $\pi$ for the first, and the transition matrices $\mathcal{T}$ for the second, results in the so-called Bellman equation (Bellman, 1957; Sutton and Barto, 1998):

$$\mathcal{V}^{\pi}(s_t) = \sum_{a_t}\pi(a_t;s_t)\sum_{s_{t+1}}p(s_{t+1}|a_t,s_t)(\mathcal{R}(s_t,a_t,s_{t+1})+\gamma\mathcal{V}^{\pi}(s_{t+1})) \quad (10.2)$$

or, using a more compact notation:

$$\mathcal{V}^{\pi}(s) = \sum_{a}\pi_s(a)\sum_{s'}\mathcal{T}^a_{ss'}\left(\mathcal{R}^a_{ss'}+\gamma\mathcal{V}^{\pi}(s')\right)$$

## 10.2.2 Solving the Bellman Equation

Eq. (10.2) describes a consistency between values of states $s$ and its successor states $s'$ for a given policy $\pi$. If the reward function $\mathcal{R}$ and transition matrices $\mathcal{T}$ are known, then this consistency can be used to solve the equation and infer the values $\mathcal{V}^{\pi}(s)$ for all states $s$. The first and conceptually most straightforward way is to recognize that Eq. (10.2) is linear and can be rewritten in vector form. Dropping the subscript $t$ and letting the successor state be $s'$, we have:

$$[\mathbf{v}^{\pi}]_s = \mathcal{V}^{\pi}(s)$$
$$[\mathbf{r}^{\pi}]_s = \sum_{a}\pi(a;s)\sum_{s'}p\left(s'|a,s\right)\mathcal{R}(s,a,s')$$
$$[\mathbf{T}^{\pi}]_s = \sum_{a}\pi(a;s)\sum_{s'}p\left(s'|a,s\right)$$

We can now rewrite the Bellman equation as

$$\mathbf{v}^\pi = \mathbf{r}^\pi + \gamma \mathbf{T}^\pi \mathbf{v}^\pi \tag{10.3}$$

which is simply solved by:

$$\mathbf{v}^\pi = (\mathbf{I} - \gamma \mathbf{T}^\pi)^{-1} \mathbf{r}^\pi$$

Here, we note an important feature of the effective transition matrix $\mathbf{T}^\pi$ induced by the policy. It is a square stochastic matrix all columns of which are probability distributions. As such, its leading eigenvector is 1, and the steady-state distribution of state visits is the eigenvector corresponding to that leading eigenvalue. The values are hence only finite as long as $\gamma < 1$. An alternative is to have a matrix $\mathbf{T}^\pi$, the leading eigenvector of which $<1$. This is true if all states have a finite probability of leading to an absorbing state that cannot be left and which has zero reward. This latter setting effectively curtails the infinite sum of rewards in Eq. (10.1) to a finite sum of exponentially distributed length.

A different approach to solving the Bellman equation is to note that if the values assigned to states are incorrect, then there is a difference $\Delta$ between the left and the right side of Eq. (10.3):

$$\Delta = \mathbf{r}^\pi + \gamma \mathbf{T}^\pi \mathbf{v} - \mathbf{v}$$

This can be used to turn the Bellman equation into an update equation:

$$\begin{aligned} \mathbf{v}_{i+1} &= \mathbf{v}_i + \Delta_i \\ &= \mathbf{r}^\pi + \gamma \mathbf{T}^\pi \mathbf{v}_i \end{aligned} \tag{10.4}$$

which can be shown to converge to the true value $\mathbf{v}^\pi$ for the same reason as above (Bertsekas and Tsitsiklis, 1996).

### 10.2.2.1 Model-Free Temporal Difference Prediction Error Learning

These previous approaches to evaluating the value function require the model $\mathcal{M}$ of the world consisting of the transition matrices $\mathcal{T}$ and the reward function $\mathcal{R}$ to be known, and are hence instances of "model-based" value estimation. So-called model-free techniques do not require this. Instead, they only require that samples can be drawn from the transition matrix and the reward function. Drawing samples corresponds to observing the reward and state consequences of taking an action, i.e., drawing an action $a_t \sim \pi(a; s_t)$ given the current state $s_t$; and then observing a successor state $s_{t+1} \sim p(s_{t+1}|a_t, s_t)$, and a reward $r_t \sim \mathcal{R}(s_t, a_t, s_{t+1})$ (see Fig. 10.1A). The Bellman equation (Eq. 10.2) contains two expectations, one over the transition probabilities and one over the action probabilities, which can be approximated with samples drawn from the two distributions. Temporal difference learning effectively performs the iterative update of Eq. (10.4) after every sample, but includes a learning rate $0 \leq \alpha \leq 1$:

$$
\begin{aligned}
\mathcal{V}_{t+1}(s_t) &= \mathcal{V}_t(s_t) + \alpha \delta_t \\
&= \mathcal{V}_t(s_t) + \alpha(r_t + \mathcal{V}_t(s_{t+1}) - \mathcal{V}_t(s_t))
\end{aligned}
\tag{10.5}
$$

This fixed learning rate $\alpha$ effectively induces an exponentially decaying average over past samples. If it is chosen to decay with the number of times a particular state has been sampled, this procedure can be shown to converge to the true value function of the policy over time under some conditions (see toy example below).

### 10.2.2.2 Phasic Dopaminergic Signals

Notably, the long-term expected future reward can be learned over time by comparing the expected reward $\mathcal{V}_t(s_t)$ with the sum of the received reward and the expected reward of the successor state $\mathcal{V}_t(s_{t+1})$. The difference between the two, $\delta_t$, is the temporal difference prediction error thought to be reported by phasic dopaminergic firing (Schultz et al., 1997). We note here that this can be positive for a transition from a state of low-reward expectation to a state of high-reward expectation even if the immediate reward is zero. This is thought to explain the transfer of phasic firing observed during conditioning of a cue to predict reward. Early on in learning, dopaminergic neurons do not respond to the cue, but do respond to the (unexpected) reward. Over time, as the animal learns that the cue predicts the reward, the value $\mathcal{V}$ of the cue increases, and its unexpected presentation elicits a prediction error, and hence firing in the dopaminergic neurons. However, as the reward is predicted, the value $\mathcal{V}$ is equal to the reward $r$, and hence a prediction error no longer occurs at the time of reward, resulting in no dopaminergic firing.

## 10.2.3 Policy Updates

Given the value $\mathcal{V}^\pi$ of each state under a given behavioral policy $\pi$, the policy can now be improved in a very simple manner by choosing that action, which has the highest expected value in each state, i.e.,

$$
\pi^{\text{new}}(a; s) = \begin{cases} 1 & \text{if} \quad a = \text{argmax}_{a'} \mathcal{Q}^\pi\left(a', s\right) \\ 0 & \text{else} \end{cases}
$$

where

$$
\mathcal{Q}^\pi(a_t, s_t) = \sum_{s_{t+1}} p(s_{t+1} | a_t, s_t)(\mathcal{R}(s_t, a_t, s_{t+1}) + \gamma \mathcal{V}^\pi(s_{t+1}))
$$

is the state–action $\mathcal{Q}$ value of taking action $a_t$ in $s_t$ under the old policy $\pi$. Again, this can be shown to converge to the optimal policy under some conditions (Bertsekas and Tsitsiklis, 1996; Sutton and Barto, 1998). What is notable here, is that optimal policies are always deterministic—there is no reason ever to choose a suboptimal action.

Though conceptually simple, such policy updates are biologically unreasonable, as they would require completely evaluating the value function for a policy before any behavioral adaptation. Updating the policy before having performed a full evaluation of the value function has the potential of breaking many of the guarantees. In contrast, Q-learning (Watkins and Dayan, 1992) is an "off-policy" method. This means that the estimated values are not affected by the sampling process (the policy). It proceeds as follows:

$$\mathcal{Q}_{t+1}(a_t, s_t) = \mathcal{Q}_t(a_t, s_t) + \alpha\left(r_t + \gamma \max_a \mathcal{Q}_t(a, s_{t+1}) - \mathcal{Q}_t(a_t, s_t)\right)$$

The key differences are the maximum operation over the next actions to be taken, which requires some foresight and can be computationally challenging if the potential behavioral repertoire is large. As long as all state–action pairs continue to be sampled, this converges to the true state–action value for any policy, and hence the policy can be updated and learning occur online.

## 10.3 MODELING DATA

### 10.3.1 General Considerations

Having provided a brief overview over the key features of RL and dynamic programming, we now turn to a tutorial overview of how these techniques can be used to probe human (and animal) decision-making. The framework suggested here is distinct from the standard approach in a number of ways. First, it is a generative framework. This means that the model can be run on the experiment under scrutiny and used to simulate data akin to that obtained in the experiment. Rather than modeling only specific aspects of the data, such as the averages in different conditions, the approach is to model the process by which the data came about, and the data itself, in their "holistic" entirety. For this, the internal inference processes that give rise to the data have to be captured in sufficient detail. The result is that learning or inference processes can be tested on the data in their entirety. The test statistics are replaced by parameters determining the internal processes. Unlike traditional test statistics, their meaning is made explicit by their function in the model.

Model building The first step is to build a series of models. Each contains an internal process by which different choice options are valued, and a link function which describes how preferences turn into observed decisions. At least two models should be built: a model M0 of "no interest" that performs the task, but without involving the process of interest, and a model M1 that does contain the process of interest.

Validation on surrogate data

1. Data generation: Run each model on the experiment from which data will be examined. Do the generated data look reasonable?
2. Surrogate model fitting: Fit each model to the data generated from it. Are the true parameters readily recovered? Are some parameters not identifiable?
3. Surrogate model comparison: Does the model comparison procedure correctly identify the data generated by each model?

Real data analysis

1. Real model fitting: Fit each model to the real data.
2. Real model validation: Run each model with the fitted parameters on the exact experimental instance presented to that particular subject. Are the key features of the real data captured reasonably?
3. Real model comparison: choose the least complex model that best accounts for the data.
4. Parameter examination : only at this point should the parameters of the model be examined, and only the parameters of the most parsimonious model should be ascribed meaning.

FIGURE 10.3    Overview over modeling approach.

The freedom to build different models is huge and vastly extends the kinds of processes that can be inferred and tested. However, as each model has to be built separately, there is also ample scope for a variety of mishaps. As a result, the modeling should contain three general steps. In a first step, the model needs to be built; in a second step, this model should be validated with surrogate data; and in a third step, the model is applied to the real data. A general suggested framework is shown in Fig. 10.3 (Daw, 2009).

A few comments are worthwhile. The key first step clearly is the model building. Here, the valuation processes by which choice preferences arise in the models are the hypotheses to be tested. A reasonable approach is to build a series of models starting from a very simple "null" valuation process, and then adding in the various features of interest to examine to what extent they parsimoniously contribute toward explaining the data. The second component is the link function, which needs to be probabilistic to allow noisy experimental data to be fitted. We noted above that optimal policies are always deterministic. Making this assumption when fitting models makes them very brittle as errors due to other, unforeseen and maybe unrecorded events are interpreted as strong evidence. Hence, one role of the link function is to assimilate noise from a variety of sources, and inferring its parameters allows for individual variation in this. Nevertheless, its functional form should be checked, and we will return to this below.

Validation on surrogate data serves a number of purposes. First, it is important to check that the data the model generates are actually comparable to the data obtained in the experiment. Second, by fitting data from the surrogate model, the ability to identify and recover parameters is established. This is an important step before interpreting any parameters.

Third, the ability to reliably distinguish between different models can be established on surrogate data comparable to the one available in the experiment under scrutiny. Indeed, it is prudent to attempt to perform these steps before running the experiment in real as they may suggest changes in experimental parameters, such as the length of the tasks or the number of subjects to run.

Finally, the models need to also be validated on the actual data under scrutiny. One possibility is to compare data generated from the model (with fitted parameters) to the real data. For learning experiments, it is, for instance, often useful to plot learning curves and ask whether the model captures the shape of these curves well. Once the models have been thus validated, it is meaningful to ask which of the models provide the most parsimonious account of the data. This is the domain of model comparison. Note that a model comparison is always relative and does not provide any absolute information and even the best amongst a set of models may still be too poor to provide any meaningful information. The interpretation of parameters in the models should only follow at the end, once one model has been chosen as a good characterization of the data.

## 10.3.2 A Toy Example

As a first example, we consider very simple learning experiment in Fig. 10.4A. In this experiment, each action $a_t$ on trial $t$ yields an immediate reinforcement $r_t$, but does not have any influence on future options. Hence, the total summed future reward in this case is simply the average immediate reward offered by each of the stimuli.

The first model assumes that individuals perform temporal difference learning, adapted to this extremely simple scenario. Taking Eq. (10.5) and observing that there is no next state, but only immediate rewards, the temporal difference prediction error learning becomes simple prediction error learning $\mathcal{V}_{t+1}^{TD}(s_t) = \mathcal{V}_t^{TD}(s_t) + \alpha(r_t - \mathcal{V}_t^{TD}(s_t))$, as in Rescorla–Wagner learning (Rescorla and Wagner, 1972). The second model assumes that individuals simply perform averages over the reinforcements earned for each of the two stimuli, which is the correct inference to perform given how the outcomes are generated. The expected values $\mathcal{V}^{av}$ are hence,

$$
\mathcal{V}_{t+1}^{av}(s) = \frac{1}{t}\sum_{t'=1}^{t} r_{t'} = \frac{1}{t}\left(\sum_{t'=1}^{t-1} r_{t'} + r_t\right) = \frac{t-1}{t}\mathcal{V}_t^{av}(s) + \frac{1}{t}r_t
$$

$$
= \frac{1}{t}\left((t-1)\mathcal{V}_t^{av}(s) + r_t\right) + \mathcal{V}_t^{av}(s) - \mathcal{V}_t^{av}(s)
$$

$$
= \mathcal{V}_t^{av}(s) + \frac{1}{t}\left(r_t - \mathcal{V}_t^{av}(s)\right)
$$

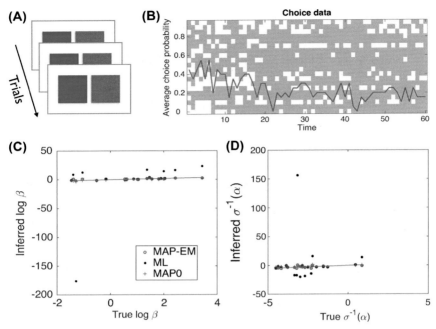

FIGURE 10.4   (A) Simple toy learning experiment. On each trial, individuals have to choose one of two squares. The *blue square* yields small rewards on 80% of trials, and the *red square* on 20% of trials. (B) Surrogate data generated from a simple learning model. Each of the horizontal rows shows the choice data for one subject, with gray indicating choice of the blue and white choice of the *red button*. The *red superimposed line* is the average probability of choosing the *red button* across subjects on that trial. (C) Plots of true parameters $\beta$ against the parameters inferred from data in panel (B). The *red line* indicates correct equality. (D) Plots of true learning rates $\alpha$ against those inferred from data in panel (B). Note that both parameters were transformed to deal with natural limits on their values: to ensure $\beta \geq 0$ all models are written in terms of $\beta = \exp(\beta')$, and to ensure $0 \leq \alpha \leq 1$ they are written in terms of $\alpha = 1/(1 + \exp(\alpha'))$. *MAP-EM*, maximum a posteriori using expectation maximization to infer the priors; *ML*, maximum likelihood.

The first line rewrites the sum over all past rewards as an iterative update. The second line then rewrites this into a form similar to that of the temporal difference (TD) learning rule. Comparing these, we see that the fixed learning rate $\alpha$ in the TD learning rule has been replaced by a decaying term $1/t$ in the average. While the averaging rule gives each of the $t$ samples the same weight, the TD rule always gives the most recent sample a weight $\alpha$, and the samples before that an exponentially smaller weight. While the TD rule has one free parameter $\alpha$, the averaging rule has no free parameters.

## 10.3.3 Generating Data

Given a model of the choice process, it is straightforward to generate data by using a link function that maps the values $\mathcal{V}$ onto probabilities of taking particular actions. A frequent choice is the use of a softmax link function whereby the probability of choosing stimulus $s$ on trial $t$ is:

$$p(a_t = s | \mathcal{V}_t) = \frac{e^{\beta \mathcal{V}_t(s)}}{e^{\beta \mathcal{V}_t(s)} + e^{\beta \mathcal{V}_t(\bar{s})}} \tag{10.6}$$

The data in Fig. 10.4B were generated from the TD model with this softmax.

## 10.3.4 Fitting Models

Having built a model and generated data from it, the next step is to fit the model to the generated data. Fitting a model means finding the set of parameters that are most compatible with the data. The maximum likelihood (ML) parameters are those under which the data are most likely. To find them, we must maximize the likelihood of all the $T$ actions $a_1, \ldots a_T$ by one subject given that subject's parameters:

$$\widehat{\theta}^{ML} = \underset{\theta}{\text{argmax}} \log p(a_1, \ldots a_T | \theta) \tag{10.7}$$

The question is how to compute the total likelihood of all choices. On first sight, this appears difficult because choices depend on previous choices and so cannot be treated independently. However, if every choice only depends on the value $\mathcal{V}_t$ at the time of the choice $t$, as assumed in Eq. (10.6), then the probability of observing a sequence of stimulus choices $a_1, \ldots a_T$ is simply:

$$\log p(a_1, \ldots a_T | \theta) = \log \prod_{t=1}^{T} p(a_t | \mathcal{V}_t) = \sum_{t=1}^{T} \log p(a_t | \mathcal{V}_t) \tag{10.8}$$

which is notable: Even though choices at any time $t$ clearly depend on the previous ones; once we condition on the values the choices become independent of the previous choices. The values can be updated iteratively before computing the likelihood of each choice, leading to an algorithm that takes the general and very simple form shown in Algorithm 10.1.

initialize values $\mathcal{V}$
**foreach** *trial t* **do**
    |   compute log likelihood of choice $a_t$ on trial $t$ given parameters : $l_t = \log p(a_t | \mathcal{V}_t, \theta)$
    |   update value $\mathcal{V}_{t+1}$ given outcomes on trial $t$
**end**
compute total log likelihood $l = \sum_t l_t$

ALGORITHM 10.1   Likelihood computation.

The total likelihood can now be passed to any of a number of optimization tools to solve Eq. (10.7). Fig. 10.4C and D shows the result of an ML fit in black for the TD model with the two parameters $\alpha$ and $\beta$. As can be seen, the black dots are sometimes very far off the diagonal, which unfortunately is relatively typical for these kinds of models. Although ML estimators are asymptotically unbiased, they do have high variance. This is often a prominent problem because parameters often do have overlapping effects and therefore can trade-off each other. In these examples, whenever $\beta$ was set to a very small value, the learning rate $\alpha$ was set to a very high value.

The blue circles show a very simple and often very powerful solution to this, which is to impose a soft prior on the parameters and performing maximum a posteriori (MAP) inference rather than ML. This is very simply achieved by replacing Eq. (10.7) with

$$\widehat{\theta}^{MAP} = \underset{\theta}{\operatorname{argmax}} \log p(a_1, \ldots a_T | \theta) p(\theta)$$

The computation of the posterior likelihood is thus just the same as that of Algorithm 10.1, but with the log likelihood of the prior added to the total log likelihood of the choices.

At times, however, the choice of the prior $p(\theta)$ can be difficult. In these situations, it can make sense to infer the prior from the data in an empirical Bayesian setting (Huys et al., 2012). There are a number of techniques available for this, and this is becoming a more common approach. Fig. 10.4C and D shows this in blue. For this simple example, little is gained over the basic MAP approach, but this changes for larger models.

## 10.3.5 Model Comparison

Having fitted the model to the data, we can ask how good an account it provides. When doing so, however, it is not sufficient to simply look at the model fit. Fig. 10.5A shows data generated from a straight line with some noise added. The top panel shows a linear fit, while the bottom panel shows a sixth order polynomial. Clearly the latter is a better fit despite the fact that the top is closer to the truth. To understand why the model with the better fit is nevertheless poorer, consider Fig. 10.5B and C. As the data (orange dots) bunch up toward the right, they are better fit by one of the triangular probability distributions in panel B than by the two uniform distributions in panel C. The model in panel B, is very powerful. Different parameter settings lead to wildly different distributions that often miss the data entirely and predict data which is never observed. Hence, the powerful model is likely to predict novel data less well. We can think of this as a trade-off between the different settings a model allows, and the fit it provides to the data. Fig. 10.5D illustrates that this problem exists for learning models, too.

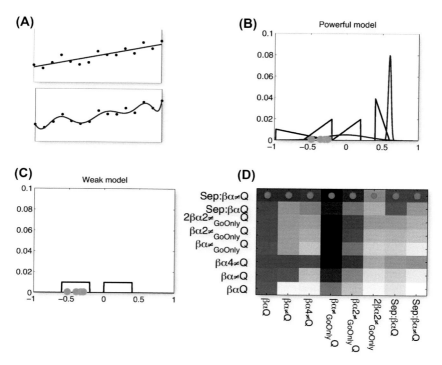

FIGURE 10.5 Model comparison. (A) Data (*black dots*) generated from a straight line with added noise is fit better by a complex sixth order polynomial (bottom) than by a *straight line* (top). This is overfitting. (B and C) Intuition for the need to average over all possible parameter settings to infer a model's parsimony. An overly complex model will contain many parameter settings that provide poor accounts of the data (orange), and only very few that provide a good fit. When averaging these, the many poor fits outweigh the few very good fits (B). Conversely, a simple model may not fit the data so well but is never far from the data and does not predict data that are never observed. (D) Learning data generated from models of increasing complexity (left to right), and fitted with models of increasing complexity. The best fitting model with best likelihood is always the most complex one at the top.

Bayesian model comparison takes this into account by using as a measure of fit not the best possible likelihood, but the average likelihood over all possible parameter settings:

$$p(\mathcal{A}|\mathcal{M}) = \int d\theta p(\mathcal{A}|\theta, \mathcal{M})p(\theta) \tag{10.9}$$

The Bayes factor between two models is then defined as

$$BF = \log_e \frac{p(\mathcal{A}|\mathcal{M}_1)}{p(\mathcal{A}|\mathcal{M}_2)} \tag{10.10}$$

and is considered substantial if greater than 3, and conclusive if greater than 5 (Kass and Raftery, 1995). Unfortunately, the integral in Eq. (10.9) is not always straightforward to evaluate, and there exist a number of approximations to it. The simplest ones are the Akaike Information Criterion $\text{AIC} = -2 \log p\left(\mathcal{A} \middle| \widehat{\theta}^{ML}\right) + 2d$, and the Bayesian Information Criterion $\text{BIC} = -2 \log p\left(\mathcal{A} \middle| \widehat{\theta}^{ML}\right) + d \log(n)$, where $d$ is the number of parameters in the model and $n$ is the number of data points. These penalizes models by counting their parameters. AIC tends to be less conservative, while BIC can be too conservative. Another possibility is to perform a Laplace approximation around the MAP parameters (Daw, 2009).

### 10.3.6 Group Studies

The methods so far have considered individual subjects. However, most studies, particularly in clinical settings, deal with group data. Fig. 10.6 shows different approaches to group data. Two simple approaches are to treat all individuals as using the same parameters, i.e., a fixed-effects treatment (panel A) or treating them entirely separately (B). While the former conflates different types of noise and is therefore not recommended, the latter can inflate noise depending on how the parameters are estimated. A more natural approach is to respect the fact that individuals in a group tend to be similar, and hence should have similar parameters (Fig. 10.6C; Huys et al., 2012). However, even this still assumes that all individuals use the same model. Two relaxations of this approach exist. First, one can employ a random-effects treatment over models (Fig. 10.6D; Stephan et al., 2009), or one can nest multiple models in a more complex model (Fig. 10.6E; Daw et al., 2011; Guitart-Masip et al., 2012). While the former assumes that individuals in a group may differ in terms of their internal processes, it assumes that these internal processes are homogeneous. The latter conversely assumes that individuals employ a mixture of strategies, but that this is true across the entire group. We note that nesting models are problematic in that there can be an overfitting by the more powerful component *within* individuals.

## 10.4 DISSECTING COMPONENTS OF DECISION-MAKING

Having described the theoretical core of decision-making and how to fit these valuation models to data, we turn to four examples. These are chosen to illustrate some of the insights gained from detailed modeling of behavioral data with a combination of RL and Bayesian techniques.

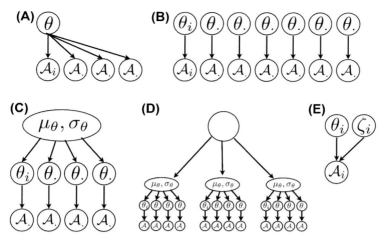

FIGURE 10.6    Group data. (A) A fixed-effects analysis would assume that all subjects share the same parameters. This is not recommended. (B) The extreme opposite is to perform separate maximum likelihood fits for each subject. This in effect assumes that all subjects are independent and have parameters that are not a priori related. (C) In a group design, it is natural to assume that individual subjects are drawn from a group that describes their similarity. For instance, parameters of individuals in a group could cluster around a particular value. However, although this model is a random-effects model in terms of the individual parameters, it is nevertheless still a fixed-effects treatment of the model itself; all individuals are assumed to be examples of the same model. (D) Next, it is possible to consider random-effects treatments of the models, i.e., that some individuals in a group will behave according to model 1, others according to model 2, and yet others according to model 3. (E) Finally, it is possible to examine whether individuals behave according to two different models. As this is simply a more complex model, it can be combined with the approaches in panel (A–D).

## 10.4.1 Reward Learning

Alterations to how rewards are processed are important in a number of psychiatric conditions. For instance, anhedonia is one of the core elements of depression and refers to an inability to experience pleasure. Pizzagalli et al. (2005) asked whether anhedonia might specifically influence the ability of people with depression to learn from rewards. They used a perceptual decision-making task where subjects had to report the length of a briefly presented mouth (Fig. 10.7A) as either short or long. Unbeknownst to the subjects, one option was rewarded more frequently than the other. Over time, subjects came to express a bias toward identifying the more rewarded stimulus, but this bias was abolished by anhedonia. This task raises two possibilities: either anhedonia blunts the sensitivity to rewards; or it blunts the ability to learn from the rewards. In principle, this might be testable by using a very simple prediction error learning to value the different choices:

$$\mathcal{Q}_{t+1}(a_t, s_t) = \mathcal{Q}_t(a_t, s_t) + \alpha(\rho r_t - \mathcal{Q}_t(a_t, s_t)) \tag{10.11}$$

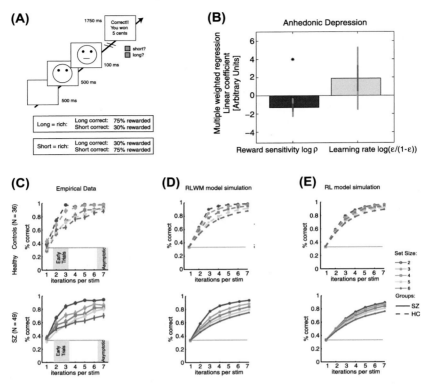

FIGURE 10.7     Reward learning. (A) Pizzagalli et al. (2005) perceptual decision-making task. Subjects have to indicate whether a briefly flashed mouth is long or short. Unbeknownst to them, one option is more frequently rewarded than the other, leading to a bias in reporting that option amongst healthy subjects. However, this bias could arise from either changes in the sensitivity to rewards, or changes in the ability to learn from rewarding events. (B) Across multiple studies using this task, anhedonia was related to reward sensitivity, but not to learning rate. (C) Requiring subjects to learn about multiple stimuli at the same time slows down learning both in controls (top) and patients with schizophrenia (bottom). (D) Including a working memory component in the model accounts for the pattern of data in controls (top); and its impairment for the pattern in patients (bottom). (E) A model without a working memory component is not able to account for the observed patterns. *HC*, hippocampus; *RL*, reinforcement learning; *RLWM*, RL model with working memory; *SZ*, patients with schizophrenia. *Panels (A and B) reproduced from Huys, Q.J.M., Pizzagalli, D.A., Bogdan, R., Dayan, P., 2013. Mapping anhedonia onto reinforcement learning: a behavioural meta-analysis. Biol. Mood Anxiety Disord. 3 (1), 12; Panels (C−E) from Collins, A.G.E., Brown, J.K., Gold, J.M., Waltz, J.A., Frank, M.J., 2014. Working memory contributions to reinforcement learning impairments in schizophrenia. J. Neurosci. 34 (41), 13747−13756.*

where $\rho$ scales the size of the received reward, while $\alpha$ is the learning rate. However, as alluded to above, this can be rewritten as:

$$\mathcal{Q}_{t+1}(a_t, s_t) = (1 - \alpha)^t \mathcal{Q}_0(a_t, s_t) + \alpha\rho \sum_{t'=0}^{t} (1 - \alpha)^{t'} r_{t-t'}.$$

Due to the product $\alpha\rho$, the two parameters are partially negatively correlated and specific statements about them require substantial data. Nevertheless, when pooling across multiple experiments, it appears that anhedonia is in fact related to a significant reduction in reward sensitivity but does not impact learning rate (Fig. 10.7B; Huys et al., 2013). Additional credence to this finding was given by the fact that a dopaminergic manipulation mostly affected the learning rate. This is consistent with a multiplicative change in the prediction error putatively reported by dopamine (Schultz et al., 1997). However, while an impact of anhedonia on the learning rate might have implied dopaminergic mechanisms, the origins of changes to reward sensitivity in depression remains uncertain (Treadway and Zald, 2011; Huys et al., 2015a).

The ability to learn from rewards is also thought to be affected in schizophrenia. The prominent involvement of dopamine suggested that this impairment may either arise through an impairment of striatal reward learning mechanisms, or alternatively also through impairment of prefrontal working memory mechanisms where dopamine also plays a key role (Durstewitz and Seamans, 2008). Collins et al. (2014) exploited a standard operant conditioning task which is, nevertheless, sensitive to both working memory and striatal prediction error learning mechanisms: when subjects are presented with increasing numbers of stimuli to learn about concurrently, a slowing of learning is observed (Fig. 10.7C). This pattern is not well accounted for by a simple change in learning rate and instead requires a working memory component to be postulated (Fig. 10.7D and E). Specifically, they consider a combination of two learners. The first is the reward learning module and is as in Eq. (10.11). The second, the working memory module, has a learning rate $\alpha$ set to 1. This means that the resulting $\mathcal{Q}_{wm}$ values store the previous event, and discard anything before that. After the choice, the $\mathcal{Q}_{wm}$ values are decayed to mimic forgetting. Strikingly, the impairment seen in schizophrenia was due mostly to the working memory component, rather than to the reward learning component.

## 10.4.2 Pavlovian Influences

We next turn to the distinction between two types of values: Pavlovian values of state $\mathcal{V}(s)$ and instrumental or operant values of state–action pairs $\mathcal{Q}(s,a)$. The former designate desirable states, but imply a policy or behavioral preference only via additional mechanisms, for instance, evolutionarily preprogrammed approach responses to appetitive states (Dayan et al., 2006). In contrast, the $\mathcal{Q}$ values measure the goodness of actions and hence can theoretically be used directly to motivate arbitrarily specific behaviors. There is a rich literature distinguishing these (see Dayan and Berridge, 2014; Huys et al., 2014 for reviews).

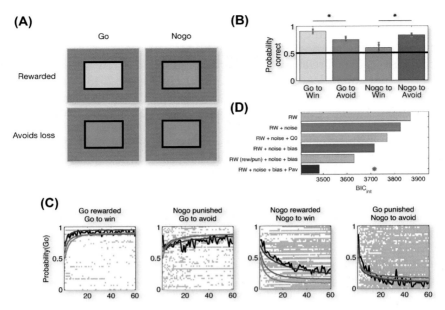

**FIGURE 10.8** Pavlovian and instrumental components of choice. (A) Subjects were presented with one of four stimuli on each trial. For the yellow stimulus, go responses were rewarded and nogo not rewarded. For the orange stimulus, nogo responses were rewarded and go not rewarded. Similarly, for the blue stimulus go responses led to avoidance of a loss, while nogo responses led to avoidance of the loss for purple stimuli. (B) Overall pattern of results: performance is impaired when go and loss are paired, and when nogo and rewards are paired. (C) Learning curves. The background shows individual choices (go white, nogo gray) for each participant; *black lines* show averages over subjects; and *colored lines* are data generated from different models. (D) Model comparison, with the most parsimonious model having the lowest score (indicated with a *red star*). BIC, Bayesian Information Criterion; *Pav*: Pavlovian influence; *RW*: Rescorla—Wagner. *From Guitart-Masip, M., Huys, Q.J.M., Fuentemilla, L., Dayan, P., Duzel, E., Dolan, R.J., 2012. Go and no-go learning in reward and punishment: interactions between affect and effect. Neuroimage 62 (1), 154—166.*

Fig. 10.8A shows a very simple task that shows these components concurrently at work during learning in humans; when subjects have to go and are rewarded, or when they have to withhold going and are in a punishment context, they perform well, whereas performing go responses to avoid losses or nogo responses for reward is far more difficult (Fig. 10.8B). Looking at the learning curves (Fig. 10.8C), it appears clear that learning is slower in the two difficult scenarios. A simple model (blue) that only incorporates instrumental learning of stimulus—action values cannot account for this pattern. Incorporating a bias toward or away from performing go responses also fails to capture the data (green lines). It is only when a second, Pavlovian, learning mechanism is added to the instrumental learner that the performance across the four contexts can be matched, and then does so in sufficiently great detail to merit the increase in complexity (Fig. 10.8D). This Pavlovian influence simply promotes the active go choice in proportion to the average reward experienced for each stimulus,

$$\mathcal{V}_{t+1}(s) = \mathcal{V}_t(s) + \alpha(\rho r_t - \mathcal{V}_t(s))$$

$$w(a,s) = \begin{cases} \mathcal{Q}(a,s) + \epsilon \mathcal{V}(s) & \text{if } a \text{ is go action} \\ \mathcal{Q}(a,s) & \text{else} \end{cases}$$

$$p(a_t|s_t) = \frac{\exp(w(a_t, s_t))}{\sum_{a'} \exp\left(w\left(a', s_t\right)\right)}$$

that is, when the stimulus leads to rewards, go is promoted, and when the stimulus tends to lead to losses, go is inhibited proportionally to the value of the stimulus. This is another instance where each individual appears to be influenced by multiple learning systems akin to Fig. 10.6E.

Though not examined with this particular task, the influence of Pavlovian stimulus-bound values on instrumental choices has been found to be aberrant in a variety of conditions ranging from alcoholism to depression. In alcoholism, for instance, Pavlovian influences are stronger, and the extent to which this involves the ventral striatum appears to predict relapse after detoxification (Garbusow et al., 2016).

### 10.4.3 Model-Based and Model-Free Decision-Making

A third example concerns the distinction between model-based and model-free decision-making. In model-based decision-making, the agent is assumed to know the consequences of actions and knows where rewards are located. This implies knowledge of transition matrices $\mathcal{T}$ and reward functions $\mathcal{R}$. At choice time, evaluations of different behavioral options are performed by searching the tree defined by $\mathcal{T}, \mathcal{R}$ (Daw et al., 2005; though see Daw and Dayan, 2014). In model-free decision-making the values $\mathcal{V}$ are accumulated over time through experience. At choice time, no further computation is required. The two types of decision-making thus trade computational costs for experiential costs. Daw et al. (2011) designed a task to measure the trade-off between the two types of learning within an individual.

Motivated by the suggestion that addictive and compulsive disorders might involve a shift from model-based toward model-free decision-making (Robbins et al., 2012), this task has since been examined extensively, with some supporting (Voon et al., 2015; Gillan et al., 2016), but also complicating evidence (Nebe et al., 2016). The difficulties stem particularly from the fact that the model-free component appears both poorly measured and unresponsive to any intervention (cf., Huys et al., 2016).

## 10.4.4 Complex Planning

We finally turn to a fourth example that uses RL techniques to examine how more complex planning tasks are solved (Huys et al., 2012, 2015c). The motivation for doing so is that many daily tasks involve planning problems that are extremely complex and easily overwhelm even powerful computers. They therefore cannot be solved fully, but mostly be approximated and simplified. Fig. 10.9A and B shows an example task that has to be solved by planning, but which is difficult. Fig. 10.9C and E show two possible strategies to approximate the task. The first, pruning, involves reflexively stopping the consideration of a plan if the plan requires transitioning through a salient loss (here, −70 points; cf. panel B). This means that large gains hiding behind the large losses are also missed. Indeed, subjects nearly never chose to transition through the path involving a large loss when there was another equally good path (Fig. 10.9D). Strikingly, when comparing the inferred tendency to stop thoughts at salient loss points, this effect appeared nearly independent of the size of the salient loss (Fig. 10.9E). If pruning were an adaptive response to the large loss, then this should have varied with loss size. This instead suggests a very simple, reflexive reaction to stop thoughts when salient losses are encountered. Further models examined how subjects subdivided the task (Fig. 10.9F). Strikingly, they subdivided the task in a manner that nearly optimally reduced the computational load (Fig. 10.9G).

## 10.5 DISCUSSION

Learning and decision-making are closely related facets of human affect and cognition. RL and dynamic programming provide principled approaches, which have been briefly reviewed here. This was followed by a brief, tutorial-like overview over how to fit such models to actual data. A point worth emphasizing is the importance of validating the model and of combining formal model comparison with informal comparisons of data generated from the model with the real data. Finally, the chapter covered a few prominent applications of the theory to psychiatric or neuroscience questions.

Taking a step back, one can ask what paths decision—theoretic accounts provide for psychiatric dysfunctions. One categorization is into three such paths (Huys et al., 2015b):

- Solving the wrong problem. This features the use of the wrong model of the world: either maximizing the wrong reward function (for instance, judging a short-term drug reward more important than long-term financial stability), or utilizing the wrong predictions about action consequences (wrongly believing that one becomes more socially adept when high), or interpreting events wrongly due to errors in the likelihood.

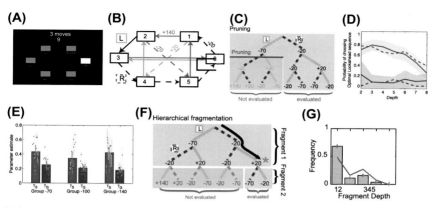

FIGURE 10.9 Task and approximations. (A) Subjects were shown six boxes. The randomly chosen starting location was indicated by the bright box and the number of moves to plan by the number at the top. Subjects were given time to plan, and then had to enter the entire planned sequence in terms of left/right button presses before seeing the chosen sequence and the rewards earned. (B) The task consisted of a maze, and subjects were placed in one of the six boxes at the beginning of each trial. They planned how to traverse the maze such as to maximize the sum of deterministic outcomes earned along the path. Each state had two successor states, which could be reached deterministically by right or left button presses. (C) Decision tree starting from state 3 and for a depth of 3 moves to plan. Pruning involves cutting off branches of the tree. A simple pruning strategy is to avoid transitions through large losses. In this particular setup with −70 as large losses, this would lead to the even larger gains being forfeited. (D) The lines show the fraction of optimal paths chosen for each depth of problem. In this version of the task, there were always two optimal paths: one through a salient loss (*blue line*), the other avoiding the salient loss (*green line*). When given the choice, subjects thus nearly deterministically avoided transitions through the large loss even when this had no impact on the outcome. (E) A computational measure of the probability of stopping the evaluation of a tree at a salient loss (blue) and at other points (red) for three groups with different salient losses of −70, −100, and −140. Strikingly, the stopping probabilities are barely different, suggesting that the inhibition of thoughts is reflexive rather than adaptively goal-directed itself. (F) Hierarchical decomposition. The complexity of the problem can be drastically reduced by approximating it with a subdivision of the task into smaller problems that are composed greedily. Here, for instance, first solving the depth-2 tree, and then solving whichever depth-1 tree this leads to. (G) The *blue line* shows the distribution of thought fragment lengths that would maximally reduce computational load without affecting performance. The *gray lines* are inferred from the data and show a close match, suggesting that individuals spontaneously near-optimally subdivided the task to minimize computational costs. *From Huys, Q.J.M., Eshel, N., O'Nions, E., Sheridan, L., Dayan, P., Roiser, J.P., 2012. Bonsai trees in your head: how the Pavlovian system sculpts goal-directed choices by pruning decision trees. PLoS Comput. Biol. 8 (3), e1002410; Huys, Q.J.M., Lally, N., Faulkner, P., Eshel, N., Seifritz, E., Gershman, S.J., Dayan, P., Roiser, J.P., 2015c. Interplay of approximate planning strategies. Proc. Natl. Acad. Sci. U.S.A. 112 (10), 3098−3103.*

- Solving the correct problem, but poorly or wrongly. As most decision problems are too hard to solve, some measure of approximation and error will naturally occur. The examples in the previous section show that these features are actively being investigated.
- Solving the correct problem, correctly, but based on poor experience. Trauma and stress are strongly associated with psychiatric ill-health. Behavior following traumatic exposure may well represent the "correct" solution even though it impairs well-being.

Finally, it should be mentioned that these techniques may well be useful in combination with other techniques. For instance, the extraction of meaningful parameters in a generative model may provide a very accurate and informationally efficient summary of complex, high-dimensional data. As such, these models can function preprocessing to reduce the dimensionality of data before applying other analyses (Wiecki et al., 2015, 2016; Huys et al., 2016).

## References

Bellman, R.E., 1957. Dynamic Programming. Princeton University Press.

Bertsekas, D.P., Tsitsiklis, J.N., 1996. Neuro-Dynamic Programming. Athena Scientific.

Collins, A.G.E., Brown, J.K., Gold, J.M., Waltz, J.A., Frank, M.J., 2014. Working memory contributions to reinforcement learning impairments in schizophrenia. J. Neurosci. 34 (41), 13747–13756.

Daw, N., 2009. Trial-by-trial data analysis using computational models. In: Delgado, M.R., Phelps, E.A., Robbins, T.W. (Eds.), Decision Making, Affect, and Learning: Attention and Performance XXIII. OUP.

Daw, N.D., Dayan, P., 2014. The algorithmic anatomy of model-based evaluation. Philos. Trans. R. Soc. Lond. B Biol. Sci. 369 (1655).

Daw, N.D., Gershman, S.J., Seymour, B., Dayan, P., Dolan, R.J., 2011. Model-based influences on humans' choices and striatal prediction errors. Neuron 69 (6), 1204–1215.

Daw, N.D., Niv, Y., Dayan, P., 2005. Uncertainty-based competition between prefrontal and dorsolateral striatal systems for behavioral control. Nat. Neurosci. 8 (12), 1704–1711.

Dayan, P., Berridge, K.C., 2014. Model-based and model-free pavlovian reward learning: revaluation, revision, and revelation. Cogn. Affect. Behav. Neurosci. 14 (2), 473–492.

Dayan, P., Niv, Y., Seymour, B., Daw, N.D., 2006. The misbehavior of value and the discipline of the will. Neural Netw. 19 (8), 1153–1160.

Durstewitz, D., Seamans, J.K., 2008. The dual-state theory of prefrontal cortex dopamine function with relevance to catechol-o-methyltransferase genotypes and schizophrenia. Biol. Psychiatry 64 (9), 739–749.

Garbusow, M., Schad, D.J., Sebold, M., Friedel, E., Bernhardt, N., Koch, S.P., Steinacher, B., Kathmann, N., Geurts, D.E.M., Sommer, C., Müller, D.K., Nebe, S., Paul, S., Wittchen, H.-U., Zimmermann, U.S., Walter, H., Smolka, M.N., Sterzer, P., Rapp, M.A., Huys, Q.J.M., Schlagenhauf, F., Heinz, A., 2016. Pavlovian-to-instrumental transfer effects in the nucleus accumbens relate to relapse in alcohol dependence. Addict. Biol. 21 (3), 719–731.

Gillan, C.M., Kosinski, M., Whelan, R., Phelps, E.A., Daw, N.D., 2016. Characterizing a psychiatric symptom dimension related to deficits in goal-directed control. Elife 5.

Guitart-Masip, M., Huys, Q.J.M., Fuentemilla, L., Dayan, P., Duzel, E., Dolan, R.J., 2012. Go and no-go learning in reward and punishment: interactions between affect and effect. Neuroimage 62 (1), 154–166.

Huys, Q.J.M., Dayan, P., Daw, 2015a. Depression: a decision-theoretic account. Ann. Rev. Neurosci. 38, 1–23.

Huys, Q.J.M., Eshel, N., O'Nions, E., Sheridan, L., Dayan, P., Roiser, J.P., 2012. Bonsai trees in your head: how the Pavlovian system sculpts goal-directed choices by pruning decision trees. PLoS Comput. Biol. 8 (3), e1002410.

Huys, Q.J.M., Guitart-Masip, M., Dolan, R.J., Dayan, P., 2015b. Decision-theoretic psychiatry. Clin. Psychol. Sci. 3 (3), 400–421.

Huys, Q.J.M., Lally, N., Faulkner, P., Eshel, N., Seifritz, E., Gershman, S.J., Dayan, P., Roiser, J.P., 2015c. Interplay of approximate planning strategies. Proc. Natl. Acad. Sci. U.S.A. 112 (10), 3098–3103.

Huys, Q.J.M., Maia, T.V., Frank, M.J., 2016. Computational psychiatry as a bridge from neuroscience to clinical applications. Nat. Neurosci. 19 (3), 404–413.

Huys, Q.J.M., Pizzagalli, D.A., Bogdan, R., Dayan, P., 2013. Mapping anhedonia onto reinforcement learning: a behavioural meta-analysis. Biol. Mood Anxiety Disord. 3 (1), 12.

Huys, Q.J.M., Tobler, P.N., Hasler, G., Flagel, S.B., 2014. The role of learning-related dopamine signals in addiction vulnerability. Prog. Brain Res. 211, 31–77.

Kass, R., Raftery, A., 1995. Bayes factors. J. Am. Stat. Assoc. 90 (430).

Kendler, K.S., Karkowski, L.M., Prescott, C.A., 1999. Causal relationship between stressful life events and the onset of major depression. Am. J. Psychiatry 156, 837–841.

Nebe, S., Kroemer, N.B., Schad, D.J., Bernhardt, N., Sebold, M., Müller, D.K., Scholl, L., Kuitunen-Paul, S., Heinz, A., Rapp, M., Huys, Q.J., Smolka, M.N., 2017. No association of goal-directed and habitual control with alcohol consumption in young adults, Addict. Biol. http://dx.doi.org/10.1111/adb.12490 (epub ahead of publication).

Pizzagalli, D.A., Jahn, A.L., O'Shea, J.P., 2005. Toward an objective characterization of an anhedonic phenotype: a signal-detection approach. Biol. Psychiatry 57 (4), 319–327.

Rescorla, R., Wagner, A., 1972. A theory of Pavlovian conditioning: variations in the effectiveness of reinforcement and nonreinforcement. Class. Cond. Curr. Res. Theory 64–99.

Robbins, T.W., Gillan, C.M., Smith, D.G., de Wit, S., Ersche, K.D., 2012. Neurocognitive endophenotypes of impulsivity and compulsivity: towards dimensional psychiatry. Trends Cogn. Sci. 16 (1), 81–91.

Schultz, W., Dayan, P., Montague, P.R., 1997. A neural substrate of prediction and reward. Science 275 (5306), 1593–1599.

Stephan, K.E., Penny, W.D., Daunizeau, J., Moran, R.J., Friston, K.J., 2009. Bayesian model selection for group studies. Neuroimage 46 (4), 1004–1017.

Sutton, R.S., Barto, A.G., 1998. Reinforcement Learning: An Introduction. MIT Press, Cambridge, MA.

Treadway, M.T., Zald, D.H., 2011. Reconsidering anhedonia in depression: lessons from translational neuroscience. Neurosci. Biobehav. Rev. 35 (3), 537–555.

Voon, V., Derbyshire, K., Rück, C., Irvine, M.A., Worbe, Y., Enander, J., Schreiber, L.R.N., Gillan, C., Fineberg, N.A., Sahakian, B.J., Robbins, T.W., Harrison, N.A., Wood, J., Daw, N.D., Dayan, P., Grant, J.E., Bullmore, E.T., 2015. Disorders of compulsivity: a common bias towards learning habits. Mol. Psychiatry 20 (3), 345–352.

Watkins, C., Dayan, P., 1992. Q-learning. Mach. Learn. 8 (3), 279–292.

Wiecki, T.V., Poland, J., Frank, M.J., 2015. Model-based cognitive neuroscience approaches to computational psychiatry clustering and classification. Clin. Psychol. Sci. 3 (3), 378–399.

Wiecki, T.V., Antoniades, C.A., Stevenson, A., Kennard, C., Borowsky, B., Owen, G., Leavitt, B., Roos, R., Durr, A., Tabrizi, S.J., Frank, M.J., 2016. A computational cognitive biomarker for early-stage huntington's disease. PLoS One 11 (2), e0148409.

III. CHARACTERIZING COMPLEX PSYCHIATRIC SYMPTOMS

# 11

# Computational Phenotypes Revealed by Interactive Economic Games

*P. Read Montague[1,2]*

[1] Virginia Tech, Roanoke, VA, United States; [2] University College London, London, United Kingdom

## OUTLINE

## 11.1 INTRODUCTION

What are the key questions for a computational psychiatry? What level of description provides the best route forward in computationalizing mental experience and its derangement by disease, injury, and developmental insult? Why will computational psychiatry provide something new not yet exploited by description-by-symptom clusters or even what many would call biological psychiatry? Do we hope or even imagine that a useful computational psychiatry will supplant these other approaches? In reverse order, "no," "I am not sure," "Who knows?," and "Too many to list." The translation of computational neuroscience to issues regarding ongoing mental function and dysfunction is a natural step at time when models can be built to address nervous system function at scales ranging from the synaptic to collections of interacting humans (Fig. 11.1).

Some meaning can get lost here if one is not careful. It's the computational process perspective that is new, not the mathematical modeling and its rendering in modern computers (Montague et al., 2004, 2012). The difference between what I will call mathematical phenomenology and computational modeling is sometimes subtle but always important to highlight. There is a difference between modeling an ion channel, neural membrane, neuron, or network of neurons using biophysical/biochemical parts and asking how the whole performs and proposing a computational process as being implemented in the dynamic interactions taking place in a piece of neural tissue. A hypothetical computational process, like the reward-dependent error signaling I will highlight below, provides a guide (here rendered as a differential equation) to organize the

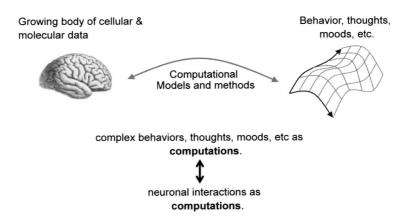

FIGURE 11.1 **The current ambition of computational neuroscience.** Computational process models "will" connect neurobiology to cognitive variables. It's an approach easy to state but hard to carry out.

underlying biophysical/biochemical dynamics rather than modeling them in the traditional sense. In many ways, a computational process approach is a more speculative approach to neural problems, but it's my and other investigator's instincts that this approach has a lot to contribute to the neurobiology of mental illness. How far it will go remains to be seen, it's currently in its infancy (Dayan et al., 2015; Maia and Frank, 2011; Montague et al., 2004, 2012).

## 11.2 REINFORCEMENT LEARNING SYSTEMS AND THE VALUATION OF STATES AND ACTIONS

In words of one investigator, "Reinforcement learning (RL) has become a dominant computational paradigm for modeling psychological and neural aspects of affectively charged decision-making tasks." (Dayan, 2012). To the uninitiated, the term reinforcement learning sounds profoundly behaviorist (think stimulus-response learning; Pavlov, 1927; Konorski, 1948; Hebb, 1949), but the modern use of reinforcement learning models shows them to be much more—including rich notions of internal reward, boundaries of an agent that interacts with the world, and how reinforcement system organize to integrate with cognitive control (reviewed in Dayan, 2012; but see Botvinick et al., 1999, 2001, 2009; Frank et al., 2001 for some of the seminal accounts surrounding issues of cognitive control not considered in detail here). Modern reinforcement learning models derive from parallel efforts in the mid-twentieth century: one from psychology and conditioning literature (Bush and Mosteller, 1951a,b, 1953, 1955) and the other from the world of optimizing control (Bellman, 1957).

The modern rendering of reinforcement learning as applied to neural systems began with the work of Robert Bush and Frederick Mosteller in the early 1950s. Their approach to animal learning was modern by emphasizing prediction learning, the animal as a multidimensional learning machine driven by statistical regularities in its world, and the history-independent assumption (the Markovian assumption) common to decision-making models today (Bush and Mosteller, 1953; also see Rescorla and Wagner, 1972; Dayan and Daw, 2008). Several sets of discoveries in the 1980s set the stage for the modern importance of reinforcement learning as a computational paradigm for understanding the neurocomputational basis of value-dependent choice. The first set of discoveries related to the clear rendering of how nervous tissue could carry out perceptual inference. This work involved the pioneering paper by John Hopfield (1982) on neural networks, augmented by the work of Hinton and Sejnowski (1983) showing how Hopfield networks could carry out **inference**, and followed by the paper by Hopfield and Tank in

1986 mapping many of the existing and emerging ideas in computation by neural networks onto potential components in real neural tissue. Collectively, this kind of work gave license to the idea that computational models could provide a new way to understand the extremely complex underlying neurobiology. Instead of simply working one's way out of the neurobiological detail toward more integrative function, one might make progress by seeing the system as being an evolved computational device where the details of the computation were the important feature on which to focus.

Many other investigators contributed to this climate of computation, but it was the work of Sutton and Barto that brought value-dependent decision-making to biology—almost unknowingly—with their work on a powerful approach to incremental learning called the method of temporal differences (Sutton and Barto, 1981, 1987, 1998). This work appealed to a deceptively simply learning algorithm that adjusted its learning in proportion to differences in successive predictions, rather than the Bush–Mosteller rule that learns based on a trial-based difference between a prediction and an outcome. The Sutton–Barto approach, such as similar methods developed in the area of optimizing control (Bellman, 1957), also explicitly posited a "goal of learning" and in doing so defined how an agent "should" value its states (Fig. 11.2). The value of a state at time $t$ should be

$$V(S_t) = E\big(r_t + \gamma r_{t+1} + \gamma^2 r_{t+2} + \cdots\big) \quad \text{for } 0 < \gamma \le 1 \qquad (11.1)$$

$E$ is expected value operation taken for each "tic" forward from present time $t$ and the $r$'s represent the distribution of rewards at each time into the distant future. $\gamma$ is a discount factor that builds in the notion that nearby times are more valuable than more distal times (and it helps immensely with convergence proofs! see Kushner and Clark, 1978; Dayan and Sejnowski, 1994). The first big take-home point is that the **value of a state depends on its future**. The second big take-home point is that once

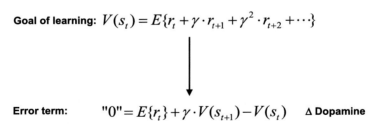

**Goal of learning:** $V(s_t) = E\{r_t + \gamma \cdot r_{t+1} + \gamma^2 \cdot r_{t+2} + \cdots\}$

**Error term:** $"\delta" = E\{r_t\} + \gamma \cdot V(s_{t+1}) - V(s_t)$ Δ **Dopamine**

FIGURE 11.2  **The goal of learning in reinforcement learning systems: the future is (almost) everything.** One virtue to reinforcement learning systems (however complex) is that they commit to a goal of learning. In the simplest settings the goal of learning is to adjust parameters to estimate the value $V$ of states where this value is defined by the future of that state: the average discounted reward expected from that state into the distal future. This assumption about the value of the state contains within it the "natural" definition of and error signal used to update the estimate. This latter quantity is a form of the Bellman equation and is recursive—connecting variables at time $t$ to variables at time $t + 1$ (Bellman, 1957).

one commits to this model for the value of a state, then there is a natural error signal latent in the definition. Take Eq. (11.1) and write it for $S_{t+1}$

$$V(S_{t+1}) = E\big(r_{t+1} + \gamma r_{t+2} + \gamma^2 r_{t+3} + \cdots\big) \qquad (11.2)$$

which means that

$$V(S_t) = E\{r_t\} + \gamma V(S_{t+1}) \qquad (11.3)$$

If a creature had such a future-looking valuation available to it, then it could use relationship Eq. (11.3) to define a natural error term for whether its evaluations at time $t$ were consistent with those at time $t + 1$.

$$0 = E\{r_t\} + \gamma V(S_{t+1}) - V(S_t) \qquad (11.4)$$

This is called the temporal difference error in the parlance of Sutton and Barto (1981, 1987) and Sutton (1988). The crucial difference with Bush—Mosteller was the successive prediction part. This is the second big discovery during the 1980s, a simple algorithm for valuing the world and learning how to value the world through prediction learning. The Bayesian rendering of these basic ideas retains their essentials but equips an agent with probability distributions over the states of the world and actions available from each state. One normative prescription in that context requires that the agent "should choose" the action that maximizes the average reward.

The third big realization emerged in the early 1990s with the proposal that diffuse ascending systems in the nervous systems—large systems of axons that deliver neuromodulators like dopamine, serotonin, norepinephrine, and so on—were implementing a form of temporal difference learning and that this was a general way that biological systems could learn to value states (Montague et al., 1993, 1995, 1996; Montague and Sejnowski, 1994; Schultz et al., 1997; also see Dayan et al., 2000; also see Montague et al., 2004 for early discussions). For this to be true, the dominant learning model, the idea of the Hebbian synapse (Konorski, 1948; Hebb, 1949), had to be modified. In 1993, Montague et al. proposed such a modification to traditional Hebbian learning: "We postulate a modification to Hebbian accounts of self-organization: Hebbian learning is conditional on an incorrect prediction of future delivered reinforcement from a diffuse neuromodulatory system." This modification allows the bidirectional synaptic change to store predictions rather than correlations. This group claimed that this same theoretical setting accounted for physiological recordings from dopamine neurons by Schultz et al.: "Recent data (Ljunberg et al., 1992) suggest that this latter influence is qualitatively similar to that predicted by Sutton and Barto's (1981, 1987) classical conditioning theory." (Montague et al., 1993). The detailed claim was finally published in 1996 (Montague et al., 1996), and the same theoretical proposal was applied successfully to account for important elements of bee learning also controlled by a

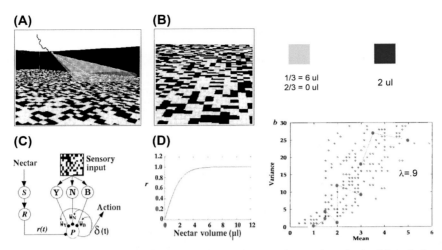

FIGURE 11.3    **Temporal difference learning accounts for bee learning and its relation to octopaminergic neuron(s) in bees.** (A, B) Simulated 'bee' agent takes in visual input in the form of blue and yellow squares (flowers) each color associated a specific set of statistics of nectar return from each color. (C) Activity in a neuron, VUMmx1, containing octopamine is necessary and sufficient for odorant conditioning in honeybees (Hammer, 1993). A temporal difference model of this arrangement connects this basic physiology to the statistics of flower sampling by bees (Montague et al., 1995; see Real, 1991 for artificial flower experiments). The same basic model also accounts for detailed electrophysiological recordings in primate dopamine neurons during conditioning experiments (Montague et al., 1996). (D) Subjective value function for predictor neuron P in panel C. A 'normal' saturating response to increasing nectar volume could be tuned to convey fitness value (given the bee's current state) of a volume of nectar. *Adapted from Montague, P.R., Dayan, P., Person, C., Sejnowski, T.J., 1995. Bee foraging in uncertain environments using predictive hebbian learning. Nature 377, 725–728.*

diffusely projecting biogenic amine system (Fig. 11.3; Montague et al., 1994, 1995. See Hammer, 1993). The theoretical framework was subsequently summarized in a review paper with the theoreticians (Montague and Dayan) breaking bread with the physiologist Schultz et al. (1997). This work connects a computational theory for how agents should value the world, generate errors, and update parameters based on this theory, and it links it directly to a neuromodulatory system (dopamine) involved in a number of psychiatric diseases. This connection forms a crucial piece in the theoretical approach to social exchange used below to probe psychopathology, and it's one starting point for translating this level of computational neuroscience model to human disease (Montague et al., 2004).

To summarize, the birth of applications of reinforcement learning models to dopamine systems had several crucial parts that converged in the early 1990s and up through the early 2000s that gave confidence that the models could be used to design and interpret experiments and that they should be stretched until they broke (with luck in fruitful ways). The remainder of this chapter will focus on how those models have inspired the use of economic games in humans to structure brain and behavioral responses in a way that gives computational insight into a number of traditional psychopathologies including major depression, autism spectrum disorder, borderline personality disorder, addiction, and attention-deficit hyperactivity disorder. I will focus primarily on the use of a (now) well-studied reciprocation game called the multiround trust game.

## 11.3 REACHING TOWARD HUMANS

The temporal difference framework was tested in humans using BOLD imaging and simple conditioning paradigms analogous to those used in nonhuman primates (e.g., Schultz et al., 1993). Fig. 11.4 shows strong activation in the ventral striatum when contrasting (nonequilibrated) predictable and unpredictable sequences of juice and water squirts during BOLD neuroimaging (Berns et al., 2001; Pagnoni et al., 2002; Montague and Berns, 2002). Fig. 11.5 shows a rather direct test of the model in human subjects during simple conditioning alongside a summary pictorial by Braver and Brown (2003) (McClure et al., 2003; O'Doherty et al., 2003). This is strong evidence but it asks for the model to extend—an area of active pursuit today. Altogether the early results in humans were found to be strongly consistent with a temporal difference-like signal in the striatum (see Glimcher, 2011 for review) (Figs. 11.4 and 11.5). Later work by Glimcher et al. using humans and nonhuman primates put the model and evidence on much firmer footing (Bayer and Glimcher, 2005; Rutledge

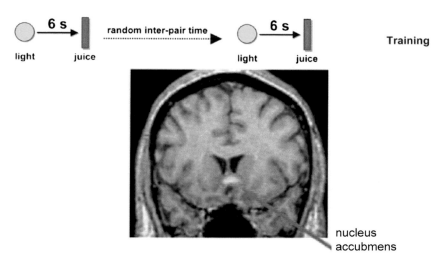

FIGURE 11.4 **Early predictability experiment in humans using BOLD imaging.** Comparing activity correlated with predictable sequences of juice (red) and water (black) shows strong responses in ventral striatum (Berns et al., 2001).

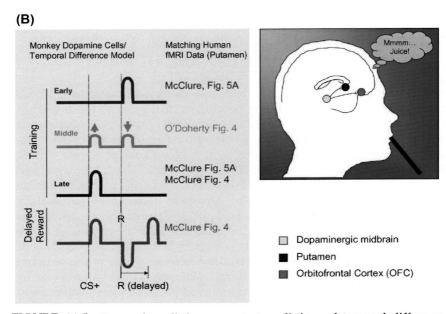

FIGURE 11.5    **Temporal prediction errors test predictions of temporal difference model of dopaminergic function during BOLD imaging in humans.** (A) Passive conditioning paradigm in humans overtrains on a specific time and (fixed) amount of juice delivery (6 s). Catch events allow experimenters to test three elements of the temporal difference model: (1) responses to the predictive cue (yellow light), (2) responses at the expected time of juice delivery but during those moments when it is not delivered, and (3) responses at the (unexpected) new time of juice delivery. (B) Pictorial summary of the tests of the temporal difference predictions for this experiment in humans. *(B) Adapted from McClure, S.M., Daw, N.D., Montague, P.R., 2003. A computational substrate for incentive salience. Trends Neurosci. 26 (8), 423–428; O'Doherty, J.P., Dayan, P., Friston, K., Critchley, H., Dolan, R.J., 2003. Temporal difference models and reward-related learning in the human brain. Neuron 38, 329–337. Pictorial summary from Braver, T.S., Brown, J.W., 2003. Principle of pleasure prediction: specifying the neural dynamics of human reward learning. Neuron 38 (2), 150–152.*

et al., 2010), but the consistent summary is that it is now a widely tested framework for one important computational process encoded by modulations in dopaminergic activity.

The framework has implications even for the complex act of interacting with other humans. In a series of BOLD imaging experiments using interactive economic games as a cognitive probe (Figs. 11.6 and 11.7),

Measuring reciprocity and model-building with a
**10-round** 'trust' game

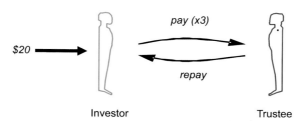

FIGURE 11.6   **Multiround reciprocation game (multiround trust game).** Two players exchange money under a transparent set of rules for 10 rounds. Each round the proposer (the investor) is given 20 money units and can send any fraction of this to the responder (the trustee). En route the amount is tripled (a return of 300%). The responder then sends back any fraction of the tripled amount. Round over. This cycle repeats for 10 rounds, and all the rules are transparently known to both players. This game has been studied in thousands of participants in fMRI devices (King-Casas et al., 2005, 2008; Montague et al., 2006; Koshelev et al., 2010; Xiang et al., 2012).

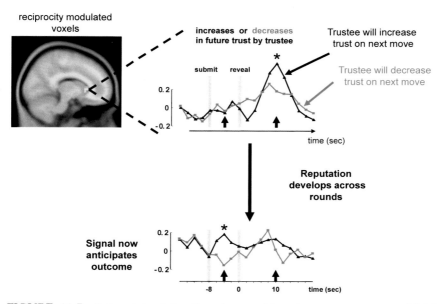

FIGURE 11.7   **Future intended actions in a social exchange game engage striatal responses consistent with a fashion consistent with a temporal difference model.** During the multiround trust game with another human, striatal responses (in regions modulated by reciprocity) in responders, when sorted on the responder's next action (which has not happened yet), shift from being reactive to the outcome to actually anticipating the offer of the proposer. This finding was the first to show the plans to act in a social context may also engage striatal prediction error responses that shift with learning in a fashion analogous to conditioning experiments (King-Casas et al., 2005).

Montague et al. showed that during a reciprocating exchange with another human, the plan to increase payments to one's partner correlates with striatal BOLD responses consistent with a dopamine-encoded temporal difference signal that shifts in time across trials in exactly the manner predicted from the nonhuman primate physiology experiments (King-Casas et al., 2005; also see King-Casas et al., 2008). This work showed that even relatively complex social settings and near-term plans for behavioral change (in the low-dimensional case of sending numbers to one another) can apparently engage reward prediction systems in a manner analogous to the basic conditioning experiments detailed above. However, this particular social exchange has now been used in the context of psychopathology groups to reveal new ways to classify subjects and perhaps to reveal ultimately some of the computational processes that are awry in traditionally defined psychopathology.

## 11.4 COMPUTATIONAL PROBES OF PSYCHOPATHOLOGY USING HUMAN SOCIAL EXCHANGE: HUMAN BIOSENSOR APPROACHES

The multiround reciprocation game shown in Fig. 11.6 is simple in execution—send some money to partner, it earns a return of 300%, partner sends back any fraction from 0% to 100% of the tripled amount. Despite this simplicity, the game requires an enormous amount of cognition to be intact including (1) responses to "fair" reciprocity, (2) sensitivity to the horizon (end of game), (3) sensitivity to history of play (intact working memory and valuation of histories), (4) ability to learn from partner's responses, and importantly (5) a capacity to model the partner and the partner's model of the subject. Without this last capacity intact, a subject can neither anticipate the impact of their monetary gestures on their partner nor can they react appropriately to the partner's response, which contains signals for the acceptability of the monetary gesture. So while the game is simple, it probes subtle and difficult-to-model features of human social exchange (Ray et al., 2008; Koshelev et al., 2010; Hula et al., 2015).

The basic idea behind our group's use of this reciprocating exchange is that it situates humans in an interactive setting that acts as a computational process primitive for the more complex way that humans sense model and update their models of other minds. Moreover, as shown in Fig. 11.7, the game also appears to engage midbrain prediction systems (putatively dopaminergic) in the same way that simple conditioning paradigms do. We saw this confluence of results as a way to probe a range of psychopathology groups. The hypothesis is that humans act as sensitive biosensors of exchange patterns during the game and that different traditional psychopathology groups might engender different behavioral

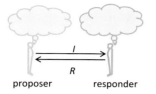

**psychopathology groups**

MDD, ADHD, ASD
BPD medicated
BPD unmedicated
healthy controls

proposer    responder

The idea: Neurotypical humans are sensitive detectors of **interpersonal exchange patterns** – exploit this capacity as a kind of device.

FIGURE 11.8    **Biosensor approach to dyadic exchange in humans.** Humans may act as sensitive biosensors when interacting with other humans that display psychopathology. *From Ray, D., King-Casas, B., Montague, P.R., Dayan, P., 2008. Bayesian model of behaviour in economic games. Adv. Neural Inf. Process. Syst. 21, 1345–1353; King-Casas, B., Sharp, C., Lomax-Bream, L., Lohrenz, T., Fonagy, P., Montague, P.R., 2008. The rupture and repair of cooperation in borderline personality disorder. Science 321, 806–810; Koshelev, M., Lohrenz, T., Vannucci, M., Montague, P.R., 2010. Biosensor approach to psychopathology classification. PLoS Comput. Biol. 6 (10), e1000966. http://dx.doi.org/10.1371/journal.pcbi.1000966; Xiang, T., Ray, D., Lohrenz, T., Dayan, P., Montague, P.R., 2012. Computational phenotyping of two-person interactions reveals differential neural response to depth-of-thought. PLoS Comput. Biol. 8 (12), e1002841. http://dx.doi.org/10.1371/journal.pcbi.1002841; Hula, A., Montague, P.R., Dayan, P., 2015. Monte Carlo planning method estimates planning horizons during interactive social exchange. PLoS Comput. Biol. 11 (6), e1004254. http://dx.doi.org/10.1371/journal.pcbi.1004254.*

trajectories through the exchange space. Perhaps useful biomarkers could emerge from such an effort. In testing this idea, one might find different behavior patterns and different BOLD imaging correlates of the patterns (Fig. 11.8).

Fig. 11.9 illustrates the basic idea in the context of a near model-free approach to the measured patterns of exchange (Koshelev et al., 2010; Xiang et al., 2012). As detailed in Fig. 11.6, each game consists of 10 rounds of exchange between the investor and the trustee making a complete game a collection of 10 investments and 10 repayments. As depicted in Fig. 11.9, these 20 numbers yield a vector {i1, r1, i2, r2,…i10, r10}; however, this is a reciprocating exchange, subjects respond to immediate partner responses and they compile in the previous history of responses, and so on. Thus there are less than 20 independent dimensions latent in the pattern of exchange. Without committing to a model of how humans in this setting plan forward their next move or model their partner in detail, Koshelev et al. used the game and a range of Diagnostic and Statistics Manual IV (DSM IV) classified partners, applied a Bayesian clustering scheme to classify the dyads (heathy control playing DSM IV partner), and generated a quantitative depiction of how a DSM-diagnosed subjects induces a pattern of play in healthy controls. The technical details of the clustering are beyond the purposes of this chapter, but the approach yielded a posterior distribution over the four clusters that emerged so that

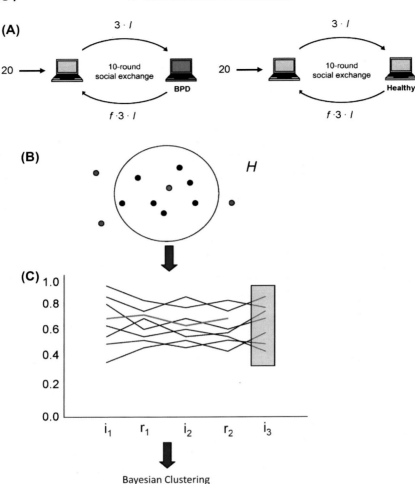

FIGURE 11.9 **Bayesian clustering on trajectories through the game space.** (A) Two sampling models: borderline personality disorder (BPD) subject and healthy, use a sampling procedure over real human data to play healthy human subjects. (B) The 'simulated agents' use a K-nearnest neighbor approach to choose the next move in the simulated agent play conditioned on the pattern of investments and repayments up to the current play. (C) Using only the pattern of investments and repayments in a game (20 numbers), Koshelev et al. (2010) developed a classification approach to the trajectories that revealed clusters related to traditional Diagnostic and Statistics Manual IV classification of subjects that played healthy subjects in the multiround game. This approach assumed no model for theory-of-mind and only examined the natural structure that emerged in the game trajectories.

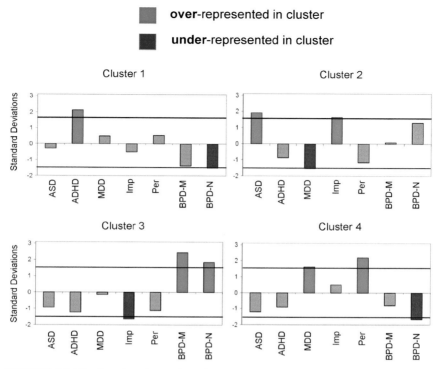

FIGURE 11.10 **Results of Bayesian cluster analysis on multiround trust trajectories using Diagnostic and Statistics Manual IV-defined subjects as partners to healthy controls.** A posterior distribution over the clusters was estimated such that a measure of the degree of over- or underrepresentation in each cluster could be computed. *ADHD*, attention-deficit hyperactivity disorder; *ASD*, autism spectrum disorder; *BPD*, borderline personality disorder; *Imp*, impersonal (subjects never meet); *MDD*, major depressive disorder; *Per*, personal version (subjects meet before and after). *From Koshelev, M., Lohrenz, T., Vannucci, M., Montague, P.R., 2010. Biosensor approach to psychopathology classification. PLoS Comput. Biol. 6 (10), e1000966. http://dx.doi.org/10.1371/journal.pcbi.1000966.*

one could estimate degree of over- or underrepresentation in each cluster. The results are backed by measured behavior (and brain BOLD responses) from n = 574 subjects, which of course took a while to gather. The large dataset is important in these kinds of new efforts because we do not yet know what or how big the "signals" will be. One important feature of the Koshelev work is that it provided classification insights into traditional psychopathology groups but using a probe not designed around any specific notion of how such groups would execute the game. As indicated in Fig. 11.10 the groups included autism spectrum disorder, borderline personality disorder (medicated and unmedicated), attention-deficit hyperactivity disorder, major depressive disorder, and healthy controls.

The biosensor hypothesis, which motivated the Koshelev et al. approach, was separately tested by the development of a computer agent that was substituted in the proposer role (the so-called healthy biosensor role). This agent had the structure of what we have called a "sampling bot" in that it used the history of investment and repayment exchange to condition a sampling of the next move from the recorded data on the multiround trust game. In summary, the bot was designed to play like the average human response conditioned the actual history of play up to that point. Koshelev et al. showed that the same basic clusters emerge containing the same over- or underrepresenting.

Using an overlapping dataset and the same social exchange game, Xiang et al. (2012) took a more in-depth approach by committing to a computational theory-of-mind model inspired by the work of Harsanyi (1967); model first described in Ray et al. (2008) on Bayesian players executing an exchange game with incomplete information. One goal of the Xiang et al. work was to track the impact of depth-of-thought on both the behavioral classifications and the associated BOLD responses. The idea of humans-thinking-about-humans generates lots of discussion about how lush and complex the ability to model others' minds could be. The approach by Xiang et al. is far less lush but commits to the use of a game of exchange to elicit quantifiable descriptions of depth-of-thought and other parameters that could classify the computations involved in modeling other minds. This is just one effort along these lines, but without committing to and testing a specific computational model of these capacities, the field will be left with narrative battles about what loosely described features may or may not be malfunctioning in a particular disease or injury state (for example, see the computational model of McClure et al., 2003 addressing the psychologically rendered idea of incentive salience as "learned wanting" but committing to equations that capture the effect). In one sense, this work is exceedingly narrow, but the results suggest that even a simple probe—like a simple two-party reciprocation where numbers fly back and forth—may carve off some of the computational primitives involved in the capacity to model other minds.

## 11.5 EPILOGUE: APPROACH AND AVOIDANCE IS NOT RICH ENOUGH

The preceding discussions have leaned heavily on my own group's work using reinforcement learning models (or perspectives) to capture basic features of social exchange between humans or human-like agents in pairs or even large groups. We have indicated above how this approach—guided by the use of structured games—may provide a new way to depict aspects of traditional symptom-cluster-defined psychopathology. However, it is our belief that looking at the valuation of drugs, the valuation of gestures or potential social gestures of other humans, and the valuation of mental states important for classifying the world around will have to reach well beyond simple ideas of approach and avoidance to provide models that undergird actual human mental disease. These cracks in the RL armor have long been noted, but we view them simply as expected shortcomings of early stage efforts to apply these models to real-world issues such as mental disease (Montague et al., 2012; Dayan et al., 2015) or cognitive control (reviewed succinctly in Dayan, 2012). The conceptual limitations of approach and avoidance are not a novel with this chapter, but one that has attracted the attention of leading RL investigators in the field for almost a decade (Daw and Doya, 2006; Dayan and Niv, 2008; Gershman et al., 2009; Dayan, 2012; Dayan et al., 2015). For our purposes, we use the issue to raise the question about the nature and structure of mental states and the way that they "couple" to lower level prediction and action choice systems.

In all the preceding discussion of reinforcement learning systems in the brain (both reward prediction systems and aversive prediction systems Montague et al., 2015) there was an implicit assumption that the prediction and error correcting systems were primary, low dimensional, and connected to a collection of devices (the cortex/striatum/hippocampus) that could flexibly represent the world possibly in useful hierarchical arrangements. In this sense, animals without a sophisticated cortex could still solve sophisticated tasks using their efficient prediction systems, but what is missing in such creatures is the capacity to represent a complex world in possibly flexibly complex ways. In a sense, this semantically "adds on" the representation piece as a kind of new feature that came along with an ever-increasing cortex. However, we would like to forward the case for the primacy of mental state representations.

In a strong sense such representations are devices for anticipating and responding to complex environmental challenges posed by the real-world including the challenges of dealing with other humans (probably the hardest problem). These representations surely need an intact cortex, corticostriatal loops, and hippocampus—entorhinal cortex; however, it also seems reasonable to suggest that the approach and avoidance rendering of RL models may be missing some key points about such representations and the way they are designed to interact with lower level rewarding and aversive events. Placebo effects, expectations, meditation-induced states (short and long term), and belief states conjured by instructions from other humans (or even internal voices) may need to be treated more "like primary sensory responses" than neural renderings that occur independent of but appended to lower level prediction systems. And should this be a fruitful direction then one could expect hierarchies or nosologies of such states to emerge quite naturally. I am not suggesting here a cognitive decomposition, which has long been underway, I am suggesting something like a dynamic neural decomposition that is stable, recallable, and maps naturally on what we might call belief states (in the vernacular human sense, not the Bayesian statistical sense). The levels of neural control available to such belief states would necessarily span many levels of the neuraxis and thus be responsible for cellular and subcellular signaling events at many levels—making such events difficult to comprehend in the absence of understanding their place in the structures supporting the state. To date, there is no systematic suggestion for how a computational psychiatry or any other human-focused effort should organize its ideas around the possible primacy of mental states. A few examples will help illustrate my point.

During simple instrumental reward task in Parkinson's disease patients Schmidt et al. (2014) found that the expectation that extra dopamine would be released enhanced behavioral measures of reward learning and provided strong modulation of BOLD learning–related signals. This was possible in these patients because they are routinely given dopamine precursor drugs as part of their treatment and these drugs enhance dopamine release in the striatum. Mere expectation of this effect appears—at the level of BOLD imaging and quantitative behavioral readouts—to enhance dopamine release as well. Now imagine that this is more like the normal operation of the state "I am getting dopamine" and that the whole point of egocentric reference in such states is to take control of powerful brainstem learning and attentional mechanisms. A similar, but not quite so biologically compelling finding, was reported by Gu et al. using a simple reinforcement learning task where subjects (who were smokers) were put in one of two expectation states "I am smoking a nicotinized cigarette" or "I am smoking a nonnicotinized cigarette." These investigators showed that in the presence of nicotine such beliefs differentially activated the striatum in a manner correlated with a value signal and a reward prediction error signal (Gu et al., 2015a,b; also see Volkow and Baler, 2015 for commentary and critique). Here the belief of the presence of nicotine was stronger than the actual presence of nicotine (a powerful neuroactive substance known to activated brainstem dopaminergic system among a number of its effects) in terms of the measured BOLD signals. In both these examples, the "semantic setup" is abstract and requires a subject to understand instructions from another human, and yet the impact of the mental states engendered by this maneuver has access to changing dopamine release and dopamine-modulated learning signals putatively generated in collaboration with the brainstem. This multilevel impact makes those mental states act like coherent devices fully equipped with sensory, effector, and reinforcer parts. How such assemblies are selected and remain stable is crucial, but so is understanding how such mental states are organized and fit into more comprehensive depictions of human cognition pertinent for disease and the sustenance of healthy mental function.

These two examples illustrate the sense in which approach and avoid is just one piece in the puzzle of how coherent behavior is organized and controlled by abstract mental states. This is a clear opportunity for cognitive and computational approaches to address and blend with what could be thought of as low-level neurobiological signaling approaches. There are many efforts reaching in this direction, but a useful and predictive computational psychiatry will require serious work in the area of mental states and their neurobiological support if progress, which feels like progress (i.e., the good kind of progress), is to be made.

# References

Bayer, H.M., Glimcher, P.W., 2005. Midbrain dopamine neurons encode a quantitative reward prediction error signal. Neuron 47 (1), 129−141.

Bellman, R., 1957. Dynamic Programming. Princeton University Press, Princeton.

Berns, G.S., McClure, S.M., Montague, P.R., 2001. Predictability modulates human brain response to reward. J. Neurosci. 21 (8), 2793−2798.

Braver, T.S., Brown, J.W., 2003. Principle of pleasure prediction: specifying the neural dynamics of human reward learning. Neuron 38 (2), 150−152.

Botvinick, M.M., Nystrom, L.E., Fissell, K., Carter, C.S., Cohen, J.D., 1999. Conflict monitoring versus selection-for-action in anterior cingulate cortex. Nature 402, 179−181.

Botvinick, M.M., Braver, T.S., Barch, D.M., Carter, C.S., Cohen, J.D., 2001. Conflict monitoring and cognitive control. Psychol. Rev. 108 (3), 624.

Botvinick, M.M., Niv, Y., Barto, A.C., 2009. Hierarchically organized behavior and its neural foundations: a reinforcement learning perspective. Cognition 113 (3), 262−280.

Bush, R.R., Mosteller, F., 1951a. A mathematical model for simple learning. Psychol. Rev. 58, 313−323.

Bush, R.R., Mosteller, F., 1951b. A model for stimulus generalization and discrimination. Psychol. Rev. 58, 413−423.

Bush, R.R., Mosteller, F., 1953. A stochastic model with applications to learning. Ann. Math. Stat. 24, 559−585.

Bush, R.R., Mosteller, F., 1955. Stochastic Models for Learning. Wiley, New York.

Daw, N.D., Doya, K., 2006. The computational neurobiology of learning and reward. Curr. Opin. Neurobiol. 16 (2), 199−204.

Dayan, P., Sejnowski, T.J., 1994. TD(l) converges with probability 1. Mach. Learn. 14, 295−301.

Dayan, P., Kakade, S., Montague, P.R., 2000. Learning and selective attention. Nat. Neurosci. 3, 1218−1223.

Dayan, P., Dolan, R.J., Friston, K.J., Montague, P.R., 2015. Taming the shrewdness of neural function: methodological challenges in computational psychiatry. Curr. Opin. Behav. Sci. 5, 128−132.

Dayan, P., Niv, Y., 2008. Reinforcement learning: the good, the bad, and the ugly. Curr. Opin. Neurobiol. 18 (2), 185−196.

Dayan, P., Daw, N.D., 2008. Decision theory, reinforcement learning, and the brain. Cogn. Affect. Behav. Neurosci. 8 (4), 429−453.

Dayan, P., 2012. Twenty-five lessons from computational neuromodulation. Neuron 76 (1), 240−256.

Frank, M.J., Loughry, B., O'Reilly, R.C., 2001. Interactions between frontal cortex and basal ganglia in working memory: a computational model. Cogn. Affect. Behav. Neurosci. 1, 137−160.

Gershman, S.J., Pesaran, B., Daw, N.D., 2009. Human reinforcement learning subdivides structured action spaces by learning effector-specific values. J. Neurosci. 29 (43), 13524—13531.

Glimcher, P.W., 2011. Understanding dopamine and reinforcement learning: the dopamine reward prediction error hypothesis. Proc. Natl. Acad. Sci. U.S.A. 108, 15647—15654.

Gu, X., Wang, X., Hula, A., Wang, S., Xu, S., Lohrenz, T., Knight, R., Gao, Z., Dayan, P., Montague, P.R., 2015a. Necessary, yet dissociable contributions of the insular and ventromedial prefrontal cortices to norm adaption: computational and lesion evidence in humans. J. Neurosci. 35 (2), 467—473.

Gu, X., Lohrenz, T., Salas, R., Baldwin, P.R., Soltani, A., Kirk, U., Cinciripini, P.M., Montague, P.R., 2015b. Belief about nicotine selectively modulates value and reward prediction error signals in smokers. Proc. Natl. Acad. Sci. U.S.A. 112 (8), 2529—2544.

Hammer, M., 1993. An identified neuron mediates the unconditioned stimulus in associative olfactory learning in honeybees. Nature 366, 59—63.

Harsanyi, J.C., 1967. Games with incomplete information played by "Bayesian" players. Manag. Sci. 14, 159—182.

Hebb, D.O., 1949. The Organization of Behavior. Wiley, New York.

Hinton, G.E., Sejnowski, T.J., 1983. Optimal perceptual inference. In: Proceedings of the IEEE Conference on Computer Vision and Pattern Recognition, Washington DC, pp. 448—453.

Hopfield, J.J., 1982. Neural networks and physical systems with emergent collective computational abilities. Proc. Natl. Acad. Sci. U.S.A. 79, 2554—2558.

Hopfield, J.J., Tank, D.W., 1986. Computing with neural circuits: a model. Science 233, 625—633.

Hula, A., Montague, P.R., Dayan, P., 2015. Monte Carlo planning method estimates planning horizons during interactive social exchange. PLoS Comput. Biol. 11 (6), e1004254. http://dx.doi.org/10.1371/journal.pcbi.1004254.

King-Casas, B., Tomlin, D., Anen, C., Camerer, C.F., Quartz, S.R., Montague, P.R., 2005. Getting to know you: reputation and trust in a two-person economic exchange. Science 308, 78—83.

King-Casas, B., Sharp, C., Lomax-Bream, L., Lohrenz, T., Fonagy, P., Montague, P.R., 2008. The rupture and repair of cooperation in borderline personality disorder. Science 321, 806—810.

Konorski, J., 1948. Conditioned reflexes and neuron organization. In: Tr. From the Polish Ms. under the Author's Supervision. Cambridge University Press, Cambridge.

Koshelev, M., Lohrenz, T., Vannucci, M., Montague, P.R., 2010. Biosensor approach to psychopathology classification. PLoS Comput. Biol. 6 (10), e1000966. http://dx.doi.org/10.1371/journal.pcbi.1000966.

Kushner, H.J., Clark, D., 1978. Stochastic Approximation Methods for Constrained and Unconstrained Systems. Springer-Verlag, Berlin.

Ljunberg, T., Apicella, P., Schultz, W., 1992. Responses of monkey dopamine neurons during learning of behavioral reactions. J. Neurophysiol. 67 (1), 145—163.

Maia, T.V., Frank, M.J., 2011. From reinforcement learning models to psychiatric and neurological disorders. Nature Neurosci. 14 (2), 154—162.

McClure, S.M., Daw, N.D., Montague, P.R., 2003. A computational substrate for incentive salience. Trends Neurosci. 26 (8), 423—428.

Montague, P.R., Dayan, P., Nowlan, S.J., Pouget, A., Sejnowski, T.J., 1993. Using aperiodic reinforcement for directed self-organization. Adv. Neural Inf. Process. Syst. 5, 969—976.

Montague, P.R., Dayan, P., Person, C., Sejnowski, T.J., 1994. Foraging in an uncertain environment using predictive hebbian learning. Adv. Neural Inf. Process. Syst. 6, 598—605.

Montague, P.R., Sejnowksi, T.J., 1994. The predictive brain: temporal coincidence and temporal order in synaptic learning mechanisms. Learn. Mem. 1 (1), 1−33.

Montague, P.R., Dayan, P., Person, C., Sejnowski, T.J., 1995. Bee foraging in uncertain environments using predictive hebbian learning. Nature 377, 725−728.

Montague, P.R., Dayan, P., Sejnowski, T.J., 1996. A framework for mesencephalic dopamine systems based on predictive hebbian learning. J. Neurosci. 16 (5), 1936−1947.

Montague, P.R., Berns, G.S., 2002. Neural economics and the biological substrates of valuation. Neuron 36, 265−284.

Montague, P.R., Hyman, S.E., Cohen, J.D., 2004. Computational roles for dopamine in behavioural control. Nature 431, 760−767.

Montague, P.R., King-Casas, B., Cohen, J.D., 2006. Imaging valuation models in human choice. Annu. Rev. Neurosci. 29, 417−448.

Montague, P.R., Dolan, R.J., Friston, K.J., Dayan, P., 2012. Computational psychiatry. Trends Cogn. Sci. 16 (1), 72−80.

Montague, P.R., Lohrenz, T., Dayan, P., 2015. The three R's of trust. Curr. Opin. Behav. Sci. 3, 102−106.

O'Doherty, J.P., Dayan, P., Friston, K., Critchley, H., Dolan, R.J., 2003. Temporal difference models and reward-related learning in the human brain. Neuron 38, 329−337.

Pagnoni, G., Zink, C.F., Montague, P.R., Berns, G.S., 2002. Activity in human ventral striatum locked to errors of reward prediction. Nat. Neurosci. 5, 97−98.

Pavlov, I.P., 1927. Conditioned Reflexes: An Investigation of the Physiological Activity of the Cerebral Cortex. New York, Dover.

Ray, D., King-Casas, B., Montague, P.R., Dayan, P., 2008. Bayesian model of behaviour in economic games. Adv. Neural Inf. Process. Syst. 21, 1345−1353.

Real, L., 1991. Animal choice behavior and the evolution of cognitive architecture. Science 253, 980−986.

Rescorla, R.A., Wagner, A.R., 1972. A theory of Pavlovian conditioning: variations in the effectiveness of reinforcement and nonreinforcement. In: Black, A.H., Prokasy, W.F. (Eds.), Classical Conditioning II. Appleton-Century-Crofts, pp. 64−99.

Rutledge, R.B., Dean, M., Caplin, A., Glimcher, P.W., 2010. Testing the reward prediction error hypothesis with an axiomatic model. J. Neurosci. 30 (40), 13525−13536.

Schmidt, L., Braun, E.K., Wager, T.D., Shohamy, D., 2014. Mind matters: placebo enhances reward learning in Parkinson's disease. Nat. Neurosci. 17 (12), 1793−1797.

Schultz, W., Apicella, P., Ljungberg, T., 1993. Responses of monkey dopamine neurons to reward and conditioned stimuli during successive steps of learning a delayed response task. J. Neurosci. 13 (3), 900−913.

Schultz, W., Dayan, P., Montague, P.R., 1997. A neural substrate of prediction and reward. Science 275, 1593−1599.

Sutton, R.S., Barto, A.G., 1981. Toward a modern theory of adaptive networks: expectation and prediction. Psychol. Rev. 88 (2), 135−170.

Sutton, R.S., Barto, A.G., 1987. A temporal-difference model of classical conditioning. In: Proceedings of the Ninth Annual Conference of the Cognitive Science Society. Seattle, WA.

Sutton, R.S., 1988. Learning to predict by the methods of temporal difference. Mach. Learn. 3, 9−44.

Sutton, R.S., Barto, A.G., 1998. Reinforcement Learning. MIT Press, Cambridge, MA.

Volkow, N.D., Baler, R., 2015. Beliefs modulate the effects of drugs on the human brain. Proc. Natl. Acad. Sci. U.S.A. 112 (8), 2301−2302.

Xiang, T., Ray, D., Lohrenz, T., Dayan, P., Montague, P.R., 2012. Computational phenotyping of two-person interactions reveals differential neural response to depth-of-thought. PLoS Comput. Biol. 8 (12), e1002841. http://dx.doi.org/10.1371/journal.pcbi.1002841.

# Further Reading

Ackley, D., Hinton, G., Sejnowski, T., 1985. A learning algorithm for Boltzmann machines. Cogn. Sci. 9 (1), 147—169.

Bhatt, M.A., Lohrenz, T., Camerer, C.F., Montague, P.R., 2010. Neural signatures of strategic types in a two-person bargaining game. Proc. Natl. Acad. Sci. U.S.A. 107 (46), 19720—19725. http://dx.doi.org/10.1073/pnas.1009625107.

Braver, T.S., 2012. The variable nature of cognitive control: a dual mechanisms framework. Trends Cogn. Sci. 16, 106—113.

Carter, C.S., Braver, T.S., Barch, D.M., Botvinick, M.M., Noll, D., Cohen, J.D., 1998. Anterior cingulate cortex, error detection, and the online monitoring of performance. Science 280 (5364), 747—749.

Chiu, P.H., Kayali, M.A., Kishida, K.T., Tomlin, D., Klinger, L.G., Klinger, M.R., Montague, P.R., 2008a. Self responses along cingulate cortex reveal quantitative neural phenotype for high-functioning autism. Neuron 57 (3), 463—473.

Chiu, P.H., Lohrenz, T.M., Montague, P.R., 2008b. Smokers' brains compute, but ignore, a fictive error signal in a sequential investment task. Nat. Neurosci. 11 (4), 514—520.

Daw, N.D., Niv, Y., Dayan, P., 2005. Uncertainty-based competition between prefrontal and dorsolateral striatal systems for behavioral control. Nat. Neurosci. 8 (12), 1704—1711.

Daw, N.D., O'Doherty, J.P., Dayan, P., Seymour, B., Dolan, R.J., 2006. Cortical substrates for exploratory decisions in humans. Nature 441, 876—879.

Gu, X., Kirk, U., Lohrenz, T., Montague, P.R., 2013. Cognitive strategies regulate fictive, but not reward prediction error signals in a sequential investment task. Hum. Brain Mapp. 35 (8), 3738—3740. http://dx.doi.org/10.1002/hbm.22433. Epub December 31, 2013.

Kirk, U., Gu, X., Harvey, A.H., Fonagy, P., Montague, P.R., 2014. Mindfulness training modulates value signals in ventromedial prefrontal cortex through input from insular cortex. NeuroImage 100, 254—262. http://dx.doi.org/10.1016/j.neuroimage.2014.06.035. Epub June 21, 2014.

Kirk, U., Gu, X., Sharp, C., Hula, A., Fonagy, P., Montague, P.R., 2016. Mindfulness training increases cooperative decision making in economic exchanges: evidence from fMRI. NeuroImage 138, 274—283.

Kishida, K.T., Montague, P.R., 2012. Imaging models of valuation during social interaction in humans. Biol. Psychiatry 72 (2), 93—100. http://dx.doi.org/10.1016/j.biopsych.2012.02.037.

Kishida, K.T., King-Casas, B., Montague, P.R., 2010. Neuroeconomic approaches to mental disorders. Neuron 67 (4), 543—554.

Montague, P.R., 2012. The Scylla and Charybdis of neuroeconomic approaches to psychopathology. Biol. Psychiatry 72 (2), 80—81.

Mosteller, F., 1974. Robert R. Bush, early career. J. Math. Psychol. 11 (3), 163—178.

Niv, Y., Montague, P.R., 2008. Theoretical and empirical studies of learning. In: Glimcher, P.W., et al. (Eds.), Neuroeconomics: Decision Making and the Brain. Academic Press, New York, pp. 329—349.

Samuel, A.L., 1959. Some studies in machine learning using the game of checkers. IBM J. Res. Dev. 3, 210—229.

# Index

Printed in the United States
By Bookmasters